A CULTURAL HISTORY OF INSECTS

VOLUME 5

A Cultural History of Insects
General Editor: Gene Kritsky

Volume 1
A Cultural History of Insects in Antiquity
Edited by Gene Kritsky

Volume 2
A Cultural History of Insects in the Medieval Age
Edited by Gene Kritsky

Volume 3
A Cultural History of Insects in the Renaissance
Edited by Sherilyn G.F. Smith

Volume 4
A Cultural History of Insects in the Age of Enlightenment
Edited by Matthew S. Lehnert

Volume 5
A Cultural History of Insects in the Age of Industry
Edited by Carol M. Anelli and Susan W. Fisher

Volume 6
A Cultural History of Insects in the Modern Age
Edited by Robert K.D. Peterson

A CULTURAL HISTORY OF INSECTS

IN THE AGE OF INDUSTRY

VOLUME 5

Edited by Carol M. Anelli and Susan W. Fisher

BLOOMSBURY ACADEMIC
LONDON • NEW YORK • OXFORD • NEW DELHI • SYDNEY

BLOOMSBURY ACADEMIC
Bloomsbury Publishing Plc
50 Bedford Square, London, WC1B 3DP, UK
1385 Broadway, New York, NY 10018, USA
29 Earlsfort Terrace, Dublin 2, Ireland

BLOOMSBURY, BLOOMSBURY ACADEMIC and the Diana logo are trademarks of Bloomsbury Publishing Plc

First published in Great Britain 2024

Series design by Raven Design
Cover image: Alice meets the Blue Caterpillar/British Library, London, UK © British Library Board. All Rights Reserved/Bridgeman

A catalogue record for this book is available from the British Library.

Library of Congress Cataloging-in-Publication Data

LCCN 2023029669

ISBN: HB: 978-13500-0314-9
Pack: 978-13500-0321-7
ePDF: 978-13502-6667-4
eBook: 978-13502-6666-7

Series: The Cultural Histories Series

Typeset by Integra Software Services Pvt. Ltd.
Printed and bound in Great Britain

CONTENTS

List of Illustrations vii

List of Contributors xiii

Series Preface xiv

Introduction: Origins, Transformations, Impacts 1
Carol M. Anelli and Susan W. Fisher

1 Insect Knowledge: The Discovery and Description of Insect Diversity 27
Norman F. Johnson and Andrew Polaszek

2 Insects and Disease: A Formidable War Alliance 47
Melissa K. Miller and Gary L. Miller

3 Insects and Food: A Global View of Edible Insects 79
Florence Vaccarello Dunkel

4 Insect Products: Apiculture, Sericulture, and Other Insect-Derived Commodities 101
Walter S. Sheppard and Carol M. Anelli

5 Insects in Mythology and Religion: Natural Theology 125
Megan E. Meuti and Susan W. Fisher

6 Insects as Symbols: Making Meaning with Insects in the Nineteenth Century 149
Adam Dodd

7 Insects in Literature and Language: "Crawling at your feet, you may observe a Bread-and-Butterfly" 173
Rosalind White

8 Insects in Art: The Age of Industry 199
Deirdre A. Prischmann-Voldseth

Notes 229

Bibliography 235

Acknowledgments 265

Index 266

ILLUSTRATIONS

0.1 Industrial melanism in the peppered moth. Warren Photographic Ltd 6

1.1 *Idea durvillei*, originally described by Boisduval (1832) from specimens collected in New Guinea (Irian Jaya) on the voyage of the *Astrolabe*. Wikimedia Commons 31

1.2 The Austrian frigate *Novara* in the teeth of a storm. From *Novara Expedition 1857–1859* (1861). Biodiversity Heritage Library 32

1.3 Beetles from the *Biologia Centrali-Americana* volume authored by the Rev. Henry Stephen Gorham in 1886. Biodiversity Heritage Library 36

2.1 Petersburg, VA. Water-filled ditch on west side of Fort Sedgwick. Insects abetted the panoply of horrors in warfare. The confined spaces of the trenches with poor sanitation and drainage were bordered by a no man's land that held the dead and dying. The decomposing waste created prime breeding grounds for filth flies along with rodents and their associated fleas. This trench with its bombproof shelter (seen in background) was photographed outside of Petersburg, VA in 1865. But, the scene was unfortunately replicated during other wars. Library of Congress, LC-DIG-cwpb-02573 48

2.2 Cholera "Tramples the victors & the vanquish'd both." The role of insects and disease was not fully recognized in the first half of the Age of Industry. Instead, mantles of miasmatic contagions were considered as the cause for many of the diseases. In 1831, Robert Seymour illustrated the specter of cholera inflicting death to all.

US National Library of Medicine. US National Library of Health, NLM Unique ID: 101393375 49

2.3 Camp sanitation. Anti-fly campaigns infiltrated both the public and military awareness by the turn of the nineteenth century. This postcard, printed prior to the U.S. involvement in the First World War, overzealously underscored the importance fly control. R.R. Donnelley. Authors' collection 51

2.4 Hospital scene. American soldiers receive the typhoid vaccination during the Mexican Expedition (1916–1917). The vaccine was a breakthrough but entomologists such as the UDSA's L.O. Howard were concerned people would become complacent to the dangers of the house fly and poor sanitation. UCR Arts. Authors' collection 55

2.5 Wounded in a shelter tent three miles from Russian batteries. Wounded Japanese soldiers during the Russo-Japanese War (1905) shelter near Russian artillery. The photograph records not only head netting to deter bothersome insects but also documents what are apparently flies on the dressed wound and arm of one soldier and the legs and soles of footgear from another. Exposed wounds could become fly-blown if untreated or infected when exposed to microbes carried by filth flies. *A Photographic Record of the Russo-Japanese War* by J. H. Hare (1905). P.F. Collier & Son, New York 57

2.6 The Rebel Flea. A Union political cartoonist likened Confederate General P.G.T. Beauregard to the flea after Southern forces withdrew from Corinth, Mississippi in May 1862. The sub-caption reflecting the jumping abilities of the flea was also observed in other wars. Insects were universal in their ability to span reactions and associations. *Harper's Weekly* 6, no. 289 (1862): 435. Authors' collection 61

2.7 *Le Poilu* (First World War French infantryman). This French postcard from the First World War whimsically illustrates how insects (and disease) can affect civilian populations too. The "petit souvenir" (little souvenir) of head lice from a soldier on furlough would make for a memorable visit. Since human lice are specific to humans, the scratching dog, cat, and bird would not likely be harboring human lice. However, the animals provide comic relief while emphasizing the degree of some infestations. Authors' collection 63

2.8 Läuse. This 1916 German postcard was postmarked from a "Fieldpoststation" probably on the Eastern front. The billeted soldier is surrounded by a family all busily scratching from their louse infestations. Military and civilians were all suffering from lice. The insect Illustrations in the lower corners served to familiarize folks on the home front. Authors' collection 64

2.9 *Our New Major-general.* In 1861, *Harper's Weekly* illustrated General Jack Frost clearing away malaria for the Union's march southward. Cooler weather meant mosquito-borne diseases would diminish but the connection between mosquitoes and their associated diseases were not fully understood at the time. *Harper's Weekly* 5, no. 249 (1861): 640. Authors' collection 70

2.10a, 2.10b Top: Washington, District of Columbia. Interior of ward in Harewood Hospital. Mosquito bars or nets shown in a soldiers' hospital ward in 1864 were used only as a relief from the annoyance flying insects. Library of Congress LC-B815-1006; Bottom: Stricken with fever. By the end of the Philippine-American War (1899–1902), mosquito bars were being used as an insect deterrent and a means to reduce transmission of insect-borne disease such as malaria and yellow fever. Here, some soldiers in a fever ward in Manila, Philippines are attended by both orderlies and netting in 1900. Authors' collection 73

3.1 A 170-year-old mulberry tree and the Hudgins family, descendants of the French Protestant Huguenots, sitting under it survived, but not the Virginia sericulture industry that immigrated with the trees and the weavers. Author's collection 81

3.2 The Great Basin defined topographically. Courtesy of Donald K. Grayson (2006) 83

3.3 The Great Basin defined by Native American nation and Traditional Ecological Knowledge base. These simplified ranges of nations are a generalization of the situation in the nineteenth century, with those of the Washoe, Northern Paiute, and Ute more for the first part of the century. D'Azevedo (1986). Redrawn by Gene Kritsky 84

3.4 Riley Locust Catcher developed by entomologist Charles V. Riley to facilitate large scale collection of the Rocky Mountain locust during outbreaks in the western US. United States Entomological Commission. 1878 85

4.1 Ancient Egyptian beekeeper in a tomb near Thebes. Gene Kritsky 102

4.2 Honey bee skeps being inspected in a market in Holland, 1920. Author's collection 103

4.3 Advertisement of honey bees for sale, 1866; Berlin Acclimatization Society. Author's collection 105

4.4 Beekeeper holding frame of newly capped queen cells. Walter S. Sheppard 108

4.5 "Gathering the silk ends," silk cocoon reeling. Syria, Mount Lebanon, 1914. Keystone View Company. Carol M. Anelli 113

4.6 "Insectes Utiles." Nineteenth-century French trade card depicting cochineal insects on cactus, produced by Aralette and Co. Courtesy of James P. Strange 119

4.7 Shellac thermoplastic union case, Berg #1-158, "Chasing Butterflies." Berg (2016) 121

4.8 Cynipid wasp, oak host, and resultant gall. *Medical Botany* by John Stephenson (1831), pl. 152. From the Collection of The Lloyd Library and Museum 122

5.1 What Newton labeled as a migratory locust in his *Little and Wise* book of sermons is actually a katydid. Migratory locusts have short antennae and females lack the long sword-like ovipositor, both of which are present in the illustration that Newton included. Newton (1878: 54) 131

5.2 Illustrations from Kirby and Spence's *Introduction to Entomology* (1826: pl. III) 134

5.3 Illustrations of trapdoor spiders and their burrows from Kirby's *Bridgewater Treatise, On the Power, Wisdom and Goodness of God, as Manifested in the Creation of Animals, and in their History, Habits and Instincts* (1836: pl. XI B) 138

5.4 Differences in appearance between male (top) and female (bottom) slender-bodied digger wasps (*Crabro cribrarius*) from Darwin's *Descent of Man* (1871, vol. I: 333). Illustration by Edward A. Smith 141

6.1 *The Adventures of Pinnochio*, C. Collodi, trans. M.A. Murrat, London, 1892. Author's collection 155

6.2 *Life and Her Children: Glimpses of Animal Life*, Arabella Buckley, London, 1917. Author's collection 160

6.3 *The Population of an Old Pear-Tree, or Stories of Insect Life*, E. van Bruyssel, London, 1888. Author's collection 162

6.4 *Insects and their Habitations: A Book for Children*, London, 1834. Author's collection 165

6.5 *Fairy Frisket; or, Peeps at Insect Life*, Charlotte Maria Tucker, London, 1893. Author's collection 168

6.6 *The House-Fly at the Bar*, New York, 1909. Author's collection 171

7.1 Sir John Lubbock, by Edward Linley Sambourne, August 19, 1882, *Punch*'s Fancy Portraits, no. 97. Wikimedia Commons 181

7.2 Alice Meets the Blue Caterpillar, by John Tenniel. The Granger Collection/Alamy Stock Photo 183

7.3 "Darwin and his Hobby," by Albert Way. Cambridge University Library. CUL DAR 204, f. 29. Reproduced by kind permission of the Syndics of Cambridge University Library 187

7.4 The Rocking-Horse Fly, by John Tenniel, *Through the Looking Glass*, 1871. Wikimedia Commons 191

7.5 Magnified studies of a ground beetle (*Carabus nemoralis*), from *Studies of a Ground Beetle* by Beatrix Potter, Penguin Random House, 1887. Reproduced with permission of Frederick Warne & Co. Ltd 194

7.6 "The Butterfly was a 'Lady'," by Edward Linley Sambourne, *Punch*, June 17, 1871 198

8.1 *Beach at Cape May*, Alfred Jacob Miller, 1840. Ink on paper. The Walters Art Museum, Baltimore, MD (37.2675). Gift of Decatur and Sally Miller, 1994. Wikimedia Commons 201

8.2 *Un Pouilleux* (Flea Seeker), Mariano Fortuny y Carbó (1838–1874). Etching. National Gallery of Art, Washington, DC (2003.118.18). Gift of Frank Anderson Trapp. Wikimedia Commons 203

8.3 *Die Pest*, Arnold Böcklin, 1898. Tempera on wood. Kunstmuseum Basel, Switzerland (Inv. 114). Deposit of the Gottfried Keller Foundation, Federal Office of Culture, Bern, 1902. Wikimedia Commons 205

8.4 *Cupid and a Butterfly*, Bertel Thorvaldsen, 1831. Marble. Thorvaldsens Museum, Copenhagen, Denmark (A397) 207

8.5 *The Fairy Queen Takes an Airy Drive in a Light Carriage, a Twelve-in-Hand, Drawn by Thoroughbred Butterflies*, Richard Doyle,

1870. Color wood engraving and color lithography. Engraved and printed in color by Edmund Evans. The Metropolitan Museum of Art, New York. Gift of Lincoln Kirstein, 1970 (1970.565.74-13) 209

8.6 *Butterflies: Design for the Gentle Art of Making Enemies*, James Abbott McNeill Whistler (1834–1903). Drawing, crayon. Los Angeles County Museum of Art, CA, www.lacma.org (M.2015.73.48). Gift of Anita and Julius L. Zelman in honor of the museum's fiftieth anniversary 211

8.7 Brooch, *c.* 1882–1892, Jaques & Marcus. Gold, pearl, demantoid garnets, and diamonds. Length 1 1/8 in. (2.9 cm). Brooch marked on hook: J & M. Box marked inside lid: JAQUES & MARCUS/ BROADWAY/CORNER 17th ST./NEW YORK. The Metropolitan Museum of Art, New York. (2001.246). Purchase, Susan and Jon Rotenstreich Gift, 2001 213

8.8 *Butterfly Hunting*, Thomas Rowlandson, 1806. Print, hand-colored etching. National Gallery of Art, Washington, DC (1945.5.890). Rosenwald Collection 217

8.9 *Poppies and Bees*, United States, 1906. Paul de Longpré (France, Lyons, active United States, 1855–1911). Watercolor on paper, mounted on cardboard. Sheet: 25 3/4 × 16 in. (65.41 × 40.64 cm), image: 23 × 13 3/8 in. (58.42 × 33.97 cm). Los Angeles County Museum of Art, CA, www.lacma.org. Charles H. Quinn Bequest (75.4.9) 220

8.10 *The Honey-Bee*, Edward Bevan, 1827. Illustration in *The Honey-Bee; Its Natural History, Physiology and Management*, frontispiece. CC BY 4.0 (.b12817946). Wellcome Collection 222

8.11 *Child Chasing Beetle with Butterfly Net*, Allen & Ginter Company, 1887. Color lithograph, tobacco card from the Girls and Children series (N58) promoting Our Little Beauties Cigarettes. The Metropolitan Museum of Art, New York (63.350.202.58.8). The Jefferson R. Burdick Collection, Gift of Jefferson R. Burdick 224

8.12 *Self-Portrait with a Straw Hat*, Vincent van Gogh, 1887. Oil on canvas. The Metropolitan Museum of Art, New York. (67.187.70a). Bequest of Miss Adelaide Milton de Groot (1876–1967), 1967 225

CONTRIBUTORS

Carol M. Anelli is Professor of Entomology at The Ohio State University, USA

Adam Dodd is Sessional Tutor at School of Communication and Arts, The University of Queensland, Australia

Florence Vaccarello Dunkel is Associate Professor Emerita at Montana State University, USA

Susan W. Fisher is Professor Emerita at The Ohio State University, USA

Norman F. Johnson is Professor, Moser Chair in Arthropod Biosystematics & Biological Diversity, The Ohio State University, USA

Megan E. Meuti is Associate Professor of Entomology at The Ohio State University, USA

Gary L. Miller is Research Entomologist in the Agricultural Research Service at the United States Dept of Agriculture, USA

Melissa K. Miller is a Civilian Entomologist (ret.), U.S. Army, USA

Andrew Polaszek is Researcher in the Department of Life Sciences at the Natural History Museum, UK

Deirdre A. Prischmann-Voldseth is Professor of Entomology at North Dakota State University, USA

Walter S. Sheppard is Professor of Entomology and Thurber Memorial Chair at Washington State University, USA

Rosalind White is Early Career Researcher in Victorian Studies & Corpus Linguistics at Royal Holloway, University of London and at the University of Birmingham, UK

SERIES PREFACE

This six-volume set, *The Cultural History of Insects*, is a unique examination of how insects have infested the human experience. The set is divided into six time periods: Antiquity, the Medieval Age, the Renaissance, the Age of Enlightenment, the Age of Industry, and the Modern Age. Within each time period, eight topics are addressed: the state of entomological knowledge during the period; insects and disease; insects and food; insect products; insects' role in mythology and religion; insects as symbols; the portrayal of insects in literature and language; and the representation of insects in art. This structure is meant to allow readers to tailor their reading to their specific interests: for example, if insects' influence in the history of art is of primary interest, the reader can select the art chapter from each volume; if a reader wants to understand the impact of insects on a particular period in human history, then reading the volume on that time period would be appropriate.

The coverage of the topics presented is not encyclopedic. Presented are histories and case studies describing how insects influenced human culture and survival. The geographic focus is generally the Western Hemisphere, but details are included of geographic intersections in which cross-cultural influence plays an important role in various topics and time periods.

Insects were here before humans. They have influenced our evolution, eaten our food, provided us with raw materials, changed our history, and been the muses for our creativity. These six volumes illustrate how insects have helped to shape our culture over the past five thousand years.

Gene Kritsky,
Mount St. Joseph University,
Cincinnati, OH

Introduction

Origins, Transformations, Impacts

CAROL M. ANELLI AND SUSAN W. FISHER

"Words are witnesses which often speak louder than documents," observed British historian Eric Hobsbawm (1962). During the period under consideration in this volume (*c.* 1820–1920), several words either entered the lexicon or acquired their modern meanings, including: scientist, railway, jerkwater, industrialist, pauper, deadline, capitalism, nation-state, jazz, Impressionism, and Cubism. If we ponder as novel the concepts and developments each word embodies, the Age of Industry materializes as a time of unprecedented global transformation. In the chapters that follow, readers will encounter many of these words and the extraordinary changes they represent, all within an entomological context. Over their long history on earth, insects have taken up residence in virtually every habitat and situation occupied by humans. Consequently, their impact on human affairs has been pervasive and complex, often to our benefit (though it may seem otherwise), sometimes to our astonishment or perhaps distress, and even to our demise. But the players involved are always intertwined. Our goal in this chapter is to set the stage for the timeframe and contents of this volume by reminding readers of major pertinent events in world affairs and their outcomes. Occasionally, we will cite a given chapter herein that bears on the particular topic at hand, to guide readers to additional information. The predominance of Great Britain spans our era, and we discuss it in some detail as well as the US, which also underwent radical transformations and rose significantly in prominence and influence.

The Age of Industry brought advances in manufacturing, transportation, agriculture, and other sectors that ushered in the modern economy, with its factory systems and intricate networks of international commerce. New technologies and inventions enabled rapid information transfer across previously inconceivable distances, shrinking geographic expanses. The remarkable human invention and discovery that characterize the era can stir feelings of awe. Yet any celebration of achievement is diminished by the reality that many whose labor produced the desired goods, services, and monetary gains never shared in them. This cruel asymmetry included international trade in human beings, with millions forced to toil under inconceivable conditions and suffer reprehensible acts of inhumanity.

GROWTH AND STATUS OF MAJOR WORLD POWERS, *c.* 1820

Long-established patterns of commerce shaped the wealth and influence of nations in the Age of Industry. During the late fifteenth century European states, notably Spain, Portugal, France, Britain, and the Netherlands, mounted seafaring voyages of discovery in search of exotic goods and lucrative trade (Barraclough 1982; Eltis and Richardson 2010). The endeavors proved profitable and intercontinental trade systems developed and evolved in the Atlantic along prevailing winds and ocean currents. Merchant vessels departed Europe with exports of manufactured and luxury goods, including guns and ammunition. They sailed to West Africa to pick up and transport captive slaves to the Caribbean via the Middle Passage, then voyaged back to Europe with goods produced by slave labor in the New World. South of the equator, slaves from southwest Africa were transported to Portuguese Brazil and the British West Indies, where sugar plantations absorbed a large portion of the transatlantic slave trade. As the demand for cotton textiles increased, more enslaved people were sent to the southern US to plant and pick cotton. New World commodities included sugar, rum, corn, molasses, tobacco, coffee, indigo, and cotton, while Africa supplied gold, ivory, spices, and approximately 12.5 million slaves, of whom about 10.7 million survived the transatlantic passage.

The transatlantic slave trade operated for 366 years, well into the era encompassed in this chapter, and all European maritime powers outfitted their colonies in the Americas with enslaved Africans. It bears noting that those who profited from the slave trade—nations, banks, investors, insurance companies, and others—were well positioned to parlay their gains into greater revenues and influence. Britain abolished its slave trade in 1807 but not slavery in its colonies, and a year later the United States followed suit, although illegal importation, a domestic slave trade, and slavery itself persisted until 1865.[1]

Prompted by abolitionist movements, other nations also outlawed their slave trades, Brazil being last in the Americas to do so in 1888.

In 1820, the beginning of our timeframe of interest, Great Britain dominated in naval power, international trade, and investments, and wielded the greatest economic influence. Explaining its primacy has been likened to a "search for the Holy Grail" (Crouzet 1990; Crafts 2011; Federico 2016), with scholars debating the relative importance of multiple factors, including: favorable geographic location on the Atlantic Ocean; fast-growing shipbuilding industry, coupled with navigable rivers and canal systems for transport; ready supply of natural resources (particularly coal and iron); rising population and pre-existence of a middle-class (merchants, skilled artisans, professionals, prosperous farmers) with ever-expanding demand for higher quality products; colonies with raw materials and the requisite labor force; and enclosure of the common fields,[2] impelling poorer peasants to seek jobs in the proliferating manufacturing towns. We must also note that Britain had maintained stability through political and social revolutions and costly wars, which had disrupted France and the Continent. Mokyr (2009), in his comprehensive examination of Britain's industrial era predominance, emphasizes what he terms the "Industrial Enlightenment" in the nation's rise to ascendancy. At its core is a realization of Baconian natural philosophy. Thus, careful investigation of the natural world and its phenomena expands knowledge (what we would term "science"), the application of which ("technology") leads to material progress and economic growth. Mokyr further argues that the marriage of scientific knowledge with technology reinforces both entities and produces a positive feedback loop, yielding ever more improvements and innovations. Humanity benefits and, with time, society becomes transformed. However, Mokyr stresses that such a transformation cannot occur without the dissemination of useful knowledge to those most likely to put it to practical ends (engineers, manufacturers, physicians, etc.). As he succinctly puts it, the Industrial Enlightenment "created a set of bridges between intellectuals and producers, between the *savants* and the *fabricants*," resulting in an unparalleled era of sustained economic growth (Mokyr 2009: 54).

Regardless of the significance one ascribes to some or all of the foregoing factors, the fact remains: Great Britain achieved the world's first industrialized economy and presided over a *Pax Britannica* from 1815 (end of the Napoleonic Wars) to 1914 (onset of the First World War). France had been a dominant power under Napoleon, but in 1820 its exiled emperor languished at St. Helena while western European nations were experiencing a period of general peace. The US, having gained its independence from Britain, issued its Monroe Doctrine in 1823, which stated that it would not interfere with European nations or their colonies, with the expectation that European nations would refrain from establishing additional New World colonies. In the Far

East, although China had been trading with Westerners since the sixteenth century (including insect-derived silk, see Chapter 4 in this volume), in 1820 its foreign trade was restricted to the walled port city of Canton (current-day Guangzhou). Japan was operating under complete diplomatic seclusion—no Japanese could leave the country and trade was limited to China and Korea. The Russian Empire lagged greatly behind the industrialization underway in parts of Western Europe. Countries in Central and South America were independent, operating under various political regimes, with little international commerce. The continent of Africa was largely independent of foreign influence except for some European coastal colonies of trade; by the end of the Age of Industry, hegemonic nations would dramatically transform its geopolitical landscape.

GREAT BRITAIN: INDUSTRIAL REVOLUTIONS IN HISTORICAL CONTEXT

Historians generally recognize the years 1760–1830 as comprising the first Industrial Revolution. This originated in England, whose nonpareil textile industry produced high-demand goods for export aboard merchant fleets. The powerful British East India Company dominated the global trade in cotton and silk goods during the eighteenth century (Dalrymple 2019). Incorporated in 1600 as a London trading company, it operated autonomously and ever more aggressively, gaining control over the great Mughal Empire in India. By 1820 the "Company" (as it became known) had amassed a private army of about 250,000—larger than that of many nations, including Great Britain,[3] with accrued territories more populous than Britain itself. The Company's monopoly on trade with China had enriched many London investors and ensured that well-heeled Brits could don silk attire and sip tea sweetened with sugar from British-owned plantations.[4] It was the Company's tea that American colonists famously dumped into Boston harbor in 1773, a year that marked a downward turning point in the Company's empire building and influence. Amidst rampant accusations of abuse in its territories, corruption, and bribe-taking, the Company was deep in debt and required a massive government bailout. This brought increased parliamentary oversight and regulatory control, culminating in the Company's liquidation in 1858.

During its transition from cottage industry to factory-based production, the British textile industry advanced with innovations like the flying shuttle, spinning jenny, and spinning mule. The steam engine and its improvements would propel the industry (discussed below), stimulating dual demands for coal as an energy source and greater efficiency in the transportation of raw materials and finished products. Roadways and inland canals proliferated. Parliament instituted a system of turnpike trusts to maintain road quality, and the innovations of three celebrated engineers—Metcalfe, Telford, and MacAdam (whence the term, "macadam")—yielded about twenty-two thousand miles of greatly improved roads by the mid-1830s (Mokyr 2009). With better-designed carriages, passenger service expanded, becoming faster and more reliable. Canals, while expensive to build and maintain, were the preferred mode of transport for heavy, bulky cargo such as ore, timber, brick, and limestone and for fragile goods like pottery. To meet increasing conveyance demands, rivers were made more navigable. Canals were constructed with needed locks, aqueducts, embankments, and other engineering innovations. The canal network spread further inland, promoting access to interior coalfields and stimulating nearby industrial growth. By 1830 British inland waterways amounted to nearly four thousand miles, an almost threefold increase above that in 1760 (Mokyr 2009). But innovations in roadways and canals were eclipsed by the invention of the steam engine train and establishment of rail systems.

Allen (1976) reckons the railways to be the "single greatest symbol of industrial progress" of the era. Once begun, railway construction proceeded rapidly, with ever-increasing traveler usage. The first railway line appeared in 1825, with 1,600 miles of track operational in Britain by 1841, a number that grew more than tenfold by 1886 (Mingay 1986). At mid-century over a hundred million passengers rode the trains yearly (Cannadine 2017). The liberating effects of this travel mode can be difficult to fathom but consider that, prior to the railway, no human had traveled at a rate greater than the speed of the fastest horse. Journey times were drastically reduced. For example, in 1750, travel by the fastest stagecoach from London to Bristol—114 miles—took forty hours, reduced to about a full day by 1821; in 1845, the trip took a mere four hours by rail (Daunton 1995). Escape from polluted towns and cities became practicable. One could take a vacation, visit relatives, or pursue educational and professional advancement in far-off places with relative ease. Easier methods of travel gave impetus to seaside recreation, collection of marine organisms, and the establishment of marine aquaria for both amusement and educational purposes (Allen 1976). Naturalists' interest in insects also intensified, and those with leisure time and the means traveled to more distant locations in search of unusual specimens. Such activities were seen as both enlightening and a way to honor God (discussed subsequently). Transatlantic steamships (the first regularly scheduled in 1838) brought even more distant lands and continents within reach. By 1910, the fastest ships could traverse the Atlantic within five days. In addition to transportation marvels, the era also brought thrilling achievements in communication across great geographic expanses. A list of notable "firsts" includes the electric telegraph (1835), underseas cable (1851), telephone (1876), and transatlantic radio communication (1901).

Steam technology was vital for the Industrial Revolutions. Its applications pumped water out of mines and ran blast furnaces for iron and steel fabrication.

During the second Industrial Revolution, starting in 1860 or thereabout (Mokyr 2009), advances in metallurgy enabled the manufacture of large quantities of high-grade, affordable steel. Railway systems expanded worldwide, expediting global trade and travel. Steam-powered foundries ran machines that produced parts for building ships; and steam-operated printing presses churned out newspapers and popular books, plus science-focused texts for an increasingly interested public (see Chapters 6 and 7). But steam was generated by burning coal, intensifying pollution and establishing conditions for what is termed "industrial melanism" in the peppered moth (Figure 0.1). This storied case lies at the intersection of industrialization and entomology and is considered one of the clearest examples of Darwinian evolution in action (Cook et al. 2012; Walton and Stevens 2018). In its pale form, which dominated during pre-industrialization, the peppered moth was well camouflaged visually from predatory birds when at rest on lichen-covered tree bark. However, pollution from burning coal killed off the lichens and deposited soot particulates, and the previously unknown dark "melanic" form of the moth gained a selective advantage and came to predominate.

The Age of Industry manifested an enhanced sense of civic pride in the erection of public buildings such as town halls, museums, libraries, and concert halls. A fondness for the natural world gave rise to stunningly beautiful botanical landscapes such as Kew Gardens, which became a national botanical garden in 1840 and opened to the public.[5] But British society remained

FIGURE 0.1 Industrial melanism in the peppered moth. Warren Photographic Ltd.

rigidly stratified and the era was marked by frequently antagonistic social and scientific forces pushing in opposite directions, reflecting the profound changes that transpired. The ruling class in England consisted of three hundred or so elite families of landed gentry, who increased their coffers by investments in banking, commerce, and shipbuilding. So-called additional, "invisible earnings" might accrue from overseas investments and the shipping of goods by British vessels (Mingay 1986). The elite believed that they were superior and, thus, entitled to write and enforce the rules. The uneven application of those rules and privileges was destined to fuel social unrest and challenge the status quo. As historian Janet Browne (2006: 35) observed, "It is often hard to remember just how unstable British society was in these first four decades of the nineteenth century."

Rapid population growth characterized the Victorian Era. In 1801 England had a population of nine million, which quadrupled to thirty-six million by 1911. Industrialized cities grew increasingly populous as farmers and country craftsmen sought employment opportunities, resulting in teeming cities bereft of urban planning and adequate housing. London, the world's largest city, had a population exceeding two million at mid-century yet lacked a central authority (Cannadine 2017). A string of bad harvests and a severe economic depression between 1837 and 1843 had worsened living conditions for many. In Liverpool and Manchester, life expectancy was twenty-eight years, and even lower for tradesmen and laborers, versus forty years nationwide (Cannadine 2017).[6] British historian David Cannadine paints a grim picture:

> All large industrial towns had brothels, gin shops, alehouses, thieves' dens, and filthy courts, rookeries, communal privies, cesspools, middens, dung heaps and ill-paved streets crawling with wild dogs, feral cats and rats. Most of them were noisy, smelly, smoky and polluted; they were dark at night-time, freezing cold in winter, and flea- and lice-ridden at all times.
>
> (Cannadine 2017: 204)

Since at least the seventeenth century the underemployed and those unable to work (e.g., the sick, elderly, orphans, and widows) could receive poverty relief financed by a local tax called the poor rate, administered through the local parish. But this system changed in 1834 with the Poor Law Amendment Act, which organized parishes into unions, each being required to operate a workhouse to feed and house paupers. Workhouse conditions were by intent severe, to discourage the needy from seeking help. Families were divided and housed in different workhouse sections, rules were strict, and labor strenuous for all. Children might be sent to work in the mines or factories. After comparing mortality rates for workhouse residents (nearly 21 percent) versus those for people receiving relief who lived at home (about 3 percent), the editor of the medical journal *The Lancet* railed against workhouses, calling them "ANTE-CHAMBERS OF THE GRAVE" [emphasis his] (Wakely 1841).

But a socioeconomic transformation was underway, fueled by new employment prospects in industry, mercantile exchanges, shipbuilding, wholesale, retail, and various professions that could boost one's living standard beyond subsistence levels. A diverse "middling" class had emerged and was growing, occupying the social stratum below the aristocrats and gentry and above the working classes. The wealthiest of this middle class included London's financiers and bankers; affluent professionals, such as lawyers; and industrialists and entrepreneurs. The next social class down included less well-off bankers, professionals, notable scientists, and academics of the best universities. Lower still were shopkeepers and other retailers, situated above the working classes, who constituted as much as 75 percent of the UK population (Cannadine 2017). Domestic help was cheap and within reach of middle-class families, and about three times as many women worked in domestic service as in the textile industry, their second largest employer. Hours were long and labor physically demanding for house servants who hauled water, chamber pots, coal, and other amenities up and down flights of stairs and laid fires, polished silver, and scrubbed floors and kitchenware. These women knew little of free time and comfortable living conditions. Most of the populace still worked in agriculture at mid-century, but by about 1890 an astonishing 85 percent of farm acreage was owned by wealthy landlords, who rented their land to tenant farmers (Mingay 1986). Joining farm laborers in this lowest social stratum were workers at dockyards, on fishing fleets, in coal mines, and in factories.

Steam-operated printing presses increased the written material in circulation, but in 1837 only about half the English population was literate. An intolerance for intermingling among social classes mitigated against any means for the poor to gain erudition, and the few schools in existence were run by the clergy, who emphasized biblical texts and the development of literacy, moral character, and discipline. This close relationship between secular education and religion persisted well into the nineteenth century.[7] In society's lower reaches, children's education was often home-based and practical rather than academic, and the very young might not receive any formal education, particularly in the first half of the nineteenth century. To help their parents eke out a living, children often worked in town or the village or perhaps for an estate owner. In 1833 Parliament allocated funds to construct schools for the poor and a few years later, a bill was advanced to establish a system of public education. The Grammar Schools Act (1840) broadened the classics curriculum to include literature and science; other bills provided for fee-paying schools for the well-off. Three decades later the Education Act (often called "Forster's Act") established a plan for teaching science to young students at all government-funded schools. The prodigious scientist T.H. Huxley, known as Darwin's "Bulldog," served on the London School Board (1870–1872), where his education committee proposed curricular changes that established the framework subsequently adopted (Huxley 1900; Desmond 2015). Huxley himself spearheaded summer classes for science teachers, which featured hands-on laboratory dissections of representative types of organisms, a revolutionary approach at a time when both teachers and students learned solely from books. The new curriculum advanced work with microscopes, anatomy, and physiology. But compulsory education was not mandated until 1880 and even then, only children aged five to ten years were required to attend, and employed children under age thirteen were regularly exempted from school by their parents. In contrast, upper-class children were schooled by tutors and governesses in the classics, history, and literature. For male children of privilege, study at university was common, buttressed by a "Grand Tour" of the Continent to absorb its art and cultural offerings. Eventually universities underwent major reform. Scholarship in the classics, moral philosophy, and history had been the staples of a college education until 1851, when the University of Cambridge offered the first natural science degree, evidencing the growing value placed on science as a discipline.

Changes to the British political system occurred in 1832 with the Great Reform Act, which removed the rotten boroughs,[8] extended the borough vote to men who paid a yearly rental fee of £10 or more (and some lodgers), and changed the county property qualifications to enable small landowners, tenant farmers, and shopkeepers to vote (UK Parliament 2022). But this excluded most working-class men (and all women), whose discontent with their lot in life was rising and who saw the Act as a betrayal, as it had no impact on their lives. Into this increasingly troubled milieu came a nascent protest group who took up the cause of "manhood suffrage," giving birth to the Chartist Movement (1838–1848).[9] Its name derived from a set of reforms published in a document called "The People's Charter," which articulated the goal of securing the vote for every male aged twenty-one, of sound mind, who had committed no crimes. The movement sponsored rallies around Great Britain and after a national convention in 1839, the Chartists collected and submitted 1.3 million petition signatures to the House of Commons, which voted by a large margin not to hear the petition. Angry Chartists took to the streets to demonstrate their pique. Another petition followed in 1842, with a remarkable three million signatures, but which failed to secure the proposed changes.[10] Yet the Chartists had put wheels in motion to ultimately enact many of their goals. Under the Reform Act of 1867, many more working men were enfranchised, although full manhood suffrage was not achieved until 1918, when women over age thirty who met the property requirement also gained the vote[11] (Johnston 2013). A lasting impact of the Chartist movement was to raise the "Condition of England" problem, a phrase coined by Thomas Carlyle (Bloom 1986) to describe the dismal condition of the English working class.

Whereas conservative members of Parliament focused on repressive measures to quell the violence incited by the Chartist movement, Carlyle recognized that social inequities were fueling the conflagration and published an influential tract outlining the spiritual, political, and societal changes needed to make the country whole. Carlyle's views would be widely circulated as Britain negotiated a second debacle caused by the growing disparity between rich and poor: the Corn Laws.

Passed in 1815 and later modified, the Corn Laws privileged the landed and agricultural sectors by imposing an import tariff on grain crops, thereby excluding foreign competition and maintaining high domestic prices. Opposition to protectionism was consistently an undercurrent in Parliament; however, no progress was made in changing the Corn Laws to an open trade system until political fallout of the Irish Potato Famine reached the halls of Parliament. Ireland had been the unwitting recipient of a plant pathogen, *Phytophora infestans*, likely a stowaway on clipper ships traveling from the US. Arriving in Ireland in 1845 it spread rapidly, resulting in the complete failure of the potato crop in the south and west of Ireland. At least one-third of Ireland depended on the potato as its sole food source, and the potato famine was devastatingly deadly: between 1845 and 1849, a million people died due to starvation and disease; another 1.5 million emigrated to escape the famine (Cannadine 2017). Parliament was filled with anti-Catholic bias, but Prime Minister Robert Peel broke from landowners in his conservative party, siding with the Whigs and the Radicals to support repeal of the Corn Laws.[12] Their abolition signaled an important change in British social mores, likely made more palatable to landowners as they could safeguard their assets by alternative investment in turnpikes, canals, textile mills, and rail systems (Mokyr 2009).

Additional pressure for societal and political change came from the popular literature. Social justice became a theme in the writings of Charles Dickens, Oscar Wilde, and Thomas Hardy, whose impact reached well beyond Britain's borders. For example, Dutch painter Vincent van Gogh was deeply enamored of Dickens's books, and greatly admired the depictions of common laborers and the poor found in the weekly British illustrated newspaper *The Graphic*. These works influenced Van Gogh's ideas regarding social reform and found expression in his empathetic paintings of peasant life. Insect metamorphosis also figured in his ruminations about human existence, and insects—both as artistic renderings and actual carcasses—appeared in some Van Gogh paintings (see Chapter 8). The serialization of novels and improvements in the printing press brought important social ideas to all levels of British society. Female writers began to demand a voice, albeit one often disguised by a male pseudonyme—e.g., George Eliot (née Mary Anne Evans) and the Brontë sisters, known to their publishers as Currer (Charlotte), Ellis (Emily), and Acton (Anne) Bell, respectively.[13] Complaints about unequal status from the distaff side were all filed under the risible category of "the woman's question" and were largely ignored, even though the voices were insistent. But the social unrest could not be suppressed forever and as is often the case, departure from accepted norms and standards profoundly troubled those favored by the status quo. At the same time, major scientific advances in Europe and the US added to growing apprehension that scientists were increasingly pursuing a materialist approach in their work and neglecting the First Cause, or the supremacy and necessity of God in all things. As Charles Darwin labored on his *magnum opus* at home in Down House, he and his closest confidants knew that more unrest was brewing.

BRITISH SCIENCE AND MEDICINE

The early decades of the nineteenth century witnessed a revival of interest in Natural Theology—the notion that the wisdom of God was manifest in His works; thus, one could study nature to reveal the mind and majesty of the Creator. This combination of science and religion appealed greatly to what Victorians viewed as "rational amusement," i.e., an activity with a high-minded purpose, not merely a pastime to titillate the senses (Barber 1980). Natural Theology was expansive, as exemplified by the *Bridgewater Treatises* (Topham 1992), which called on multiple scientific disciplines including biology, physics, geology, and cosmology to reveal the wonder of God.[14] Reverend William Paley's book *Natural Theology* (1802), which included a chapter on insects, enjoyed great popularity in scientific, secular, and religious circles (see Chapter 5). Darwin took the volume with him on the HMS *Beagle* voyage and claimed to be able to recite lengthy passages by heart (Darwin Correspondence Project, DCP-LETT-2532). But by the mid-nineteenth century Natural Theology was being sidelined by advances in natural science that increasingly found material bases for explaining nature. For example, across Europe the idea that species were fixed and unchanged over time was being replaced by an understanding that some species had gone extinct and contemporary species could change with time. The startling view that humans were not special creations of the Deity and were subject to the same forces that affected other organisms was gaining currency. Even the previously heretical idea that the tree of life was not simply a grand progressive march to the ultimate goal of creating man was being discussed. The age of the earth, once calculated in biblical generations, was yielding to a view that the earth was at least millions of years old and undergoing changes due to the constant application of geophysical forces. Collectively, these new findings significantly challenged the accepted status quo.

Also influential was the enthusiasm of Prince Albert (Queen Victoria's husband) for promoting learning of all kinds, from science to technology to the arts (Weintraub 1997). As elected Chancellor of the University of Cambridge he actively promoted modernity and stringent pedagogical standards. He further argued that social improvement was achievable through adoption of modern scientific principles and implementation of technologies that came from that knowledge. Albert's enthusiasm along these lines was the driving force behind the Great Exhibition of 1851, held in Hyde Park in an architectural marvel known as the "Crystal Palace," the largest structure then built, at four times the size of St. Peter's Cathedral in Rome and six times the size of St. Paul's Cathedral, London (Anonymous 1970). This elegant, cast-iron and plate glass structure enclosed about eighteen acres and housed more than a hundred thousand exhibits from nations around the world.[15] Conceived to celebrate and inspire human artistic and industrial achievement, the Great Exhibition likely also served as a rejoinder to Britain's traditional rival, France, which two years prior had held an *Exposition des produits de l'industrie française* in Paris. Initially, admission costs to the Great Exhibition were quite high, ensuring that only aristocratic and middle-class visitors would enter, but by the end the price was dropped to one shilling, making the event affordable for the working class (Auerbach 1999). All told, over six million admissions were taken at entry gates and the event yielded a surplus of about £200,000 (Cannadine 2017). Prince Albert used the profits to create a scientific and cultural campus by purchasing land around Exhibition Road in South Kensington and founding what would become the Victoria and Albert Museum[16] and the Science Museum.[17] The Natural History Museum, erected after Albert's passing, was located nearby and became the repository of the flea collection of Nathanial (Charles) Rothschild, conservationist and entomologist from the family of international financiers. This world-famous collection, estimated at 260,000 specimens collected globally, represents more than 70 percent of all described flea species.[18] Charles's daughter, Dame Miriam Rothschild, would herself become a world authority on fleas and ecological biochemistry.[19]

The establishment of scientific societies also attested to the Victorian zeal for science.[20] The Entomological Society of London was founded in 1833, whose luminaries included John Curtis, J.O. Westwood, and F.W. Hope. In addition to bringing minds together and promoting discussion and publication of scientific works, scientific societies played the important role of helping to fund scientific expeditions to foreign lands. Such expeditions typically collected animals and plants, both extant and fossilized forms, plus rocks and other inorganic specimens. European museums gained acquisitions from around the globe, affording analysis of how similar species change over a broad swath of geography, prefiguring the reality that some species that once inhabited the earth were no longer present. Expeditions undoubtedly accelerated the pace and broadened the scope of scientific discovery in the nineteenth and early twentieth centuries (see Chapter 1). Scott's final expedition to Antarctica (*Terra Nova*, 1910–1913) is illustrative of the kinds of radical leaps in scientific knowledge that could be made from such quests (Martin 2012).[21] The scientists aboard performed extensive work in terrestrial magnetism and made important contributions to botany, geology, chemistry, and physics that culminated in sixteen separate scientific volumes;[22] collected more than two thousand specimens of fossils, plants, and animals, curated in the Natural History Museum (London), of which 20 percent were new to science; established definitively that Antarctica is a continent, a fact previously under debate; generated data on glaciology, then a nascent science, that yielded a classic work by Wright and Priestly (1922) (Fogg 1992); used weather balloons to gather temperature, wind, and barometric pressure data that today serve as a baseline for climate change studies, and which lay the groundwork for understanding that high-altitude winds circling Antarctica affect weather globally (Niiler 2012); and unearthed pounds of ancient plant fossils, which evidenced that Antarctica had undergone extreme temperature fluctuations and that an abundant flora had once carpeted the region. (We now know that Antarctica was part of an immense supercontinent called Gondwana (Coyne 2010).) The *Terra Nova* mission was but one scientific expedition. If we consider the cumulative impact of other societies, institutions, and benefactors dispatching scientists globally to discover what was out there, the significance is astounding. Indeed, if one simply reflects upon two other expeditions: the five-year voyage of the *Beagle* to circumnavigate South America with the young Darwin aboard, and Alfred Russel Wallace's collecting excursion in the Malay Archipelago (where he independently arrived at an evolutionary theory uncannily similar to Darwin's), the value of scientific expeditions comes into sharp focus.[23]

Undergoing profound and welcome change during this time was the field of public health. The eponymous term "Dickensian" derives from the author's lurid description of the wretched conditions that beset Victorian society, especially in overcrowded cities. It is equally descriptive of the status of public health, which advanced markedly in the second half of the nineteenth century as sanitation reforms became codified into law and new scientific information came to light. For instance, the Public Health Act (1848) created local Health Boards to investigate unsanitary conditions with the goal of preventing communicable diseases such as cholera, typhoid, diphtheria, and smallpox (Paneth et al. 1998). That year fourteen thousand people perished in London's cholera epidemic and it is thought that typhoid, caused by the bacterium *Salmonella typhi*, may have claimed the life of Prince Albert at the premature age of forty-two (Cannadine 2017). When in 1854 England's third cholera epidemic broke out in London, the Act enabled appointment of an investigatory committee to track the epidemic's progress and quell its spread. While not a committee member, Dr. John Snow famously undertook pioneering epidemiological studies, correctly linking the source of the outbreak to a contaminated drinking water pump handle. This conflicted with the committee's view that cholera was spread by inhaling odors arising from decomposing animals and other organic matter ("miasmatic theory"). The ill-placed belief in this notion is understandable. On a particularly hot day in June 1858, the stench from raw sewage sluiced into the Thames and the resulting "Great Stink" forced members of Parliament to retreat from portions of the building overlooking the river (*Times*, 1858).

Major scientific breakthroughs also occurred in medicine and bacteriology, bringing to the fore approaches now considered standard operating procedure. In 1846, Dr. H.J. Bigelow published a paper entitled "Insensibility during surgical operations induced by inhalation," thereby reporting freedom from surgical pain through anesthesia, generally delivered as ether or chloroform.[24] Queen Victoria made its use acceptable for childbirth as it was administered during the birth of her eighth and ninth children. The great French scientist Louis Pasteur enunciated the Germ Theory of Disease, giving an identity to the source of many otherwise mysterious diseases (see Chapters 2 and 8). (Pasteur also saved his country's silk industry by elucidating the causes of infectious diseases of silkworms and recommending preventative husbandry methodologies; see Chapter 4). Sir Joseph Lister, surgeon to Queen Victoria, applied Germ Theory tenets to develop an antiseptic process to prevent the spread of disease and infection during medical procedures. The first person to demonstrate unequivocally that a specific pathogenic microbe was the cause of an infectious disease was Robert Koch. This he did in research with the anthrax bacterium, which generally afflicted sheep and cattle but could infect humans and be lethal. Koch elucidated the complex etiology of anthrax, and also discovered the bacterium that causes cholera. With the publication of Koch's postulates in 1884, four essential criteria for linking a specific microbe to a particular disease were identified. Thus, instead of fighting unidentified phantoms or ghostly miasma, sights could be trained on verifiable sources. For his discovery of the tuberculosis bacterium, a disease then claiming the lives of about one in seven people, Koch received a Nobel Prize (1905).

One area notably absent of advances concerns the medical treatment of women. Clearly, one of the major life events of a typical woman would be the bearing of children. The upper classes could summon a physician who might be able to apply forceps in a difficult birth or even offer anesthesia. But no working-class woman could afford the fee charged by a male medical professional—and they were all male. An indigent woman might deliver on her own or under the care of a self-taught midwife with no medical training and certainly no awareness of the Germ Theory of Disease. Undoubtedly many lives were lost due to filthy conditions as evidenced by a maternal mortality rate of 35–40 percent in the mid-nineteenth century and a shocking infant mortality rate of 60 percent. Even after Florence Nightingale demonstrated the skill and heroism of female nurses during the Crimean War (1854–1856), little appetite persisted for admitting female practitioners to any part of the medical profession.[25] After decades of persistent opposition, in 1902 Parliament passed the first Midwives Act, which professionalized midwifery and created the Royal College of Midwives to educate future generations.

UNITED STATES: TRANSFORMATION FROM AGRARIAN TO INDUSTRIALIZED SOCIETY

The American colonists who broke from England had established a new nation. Seeking religious liberty and freedom from a dictatorial king, the colonists endeavored to tame a new continent, in the process subjugating Indigenous peoples and nearly annihilating their cultures through war, disease, and forced relocation. The challenges presented by life in their new world, the expansive resources available, and the desire to be free of restrictions on personal liberty, begat a society that would be substantially different from the European forebears from which it sprang. The United States would have no monarchy nor aristocracy, nor an established state church. Deism, a religion built on rational thought, was adopted by many of the nation's most influential intellectuals, e.g., Thomas Jefferson, Thomas Paine, and Benjamin Franklin (Mapp 2003). But in the early nineteenth century, evangelical Protestantism underwent a resurgence in the Second Great Awakening, when church membership and attendance swelled, with lasting impact. As Alexis de Tocqueville (1838) aptly noted, "There is no country in the world where the Christian religion retains a greater hold over the souls of men than in America." Comparatively isolated in their new land and galvanized by the need to make their own way, the settlers saw merit in hard work, sobriety, individualism, and self-reliance (Howe 1975). The surfeit of land enabled ordinary folks to purchase large tracts of property where the entire family lived and engaged in subsistence farming (Hurt 2002). Agricultural practices were handed down as traditions, informed by publications such as farmers' almanacs, and guided by superstition. The first undertaking was to clear the land, a highly labor-intensive task. As practically nothing was known about fertilizing crops or crop rotation, the cultivated land became overworked to a point of non-productivity, spurring a westward exodus. As the settlers moved, the labor-intensive clearing of land and draining of swamps to render the new land arable rapidly exceeded their capacity to perform the necessary labor. New sources of human power were required. Indentured servitude, in which one's passage to America was paid by a third party who then claimed the labor of the recipient for a period of four to seven years, became common. It had the benefit of bringing individuals, most from Britain, who were culturally similar and spoke a common language, into service in America. However, some Americans ultimately preferred another form of labor: chattel slavery. The possession of chattel slaves became a contentious issue as additional states sought admission to the Union. The Missouri Compromise (1820) aimed to maintain the legislative balance of power by admitting pairs of states—one free, one slave—to the Union. It stated that no territory north of Missouri's southern border could enter as a slave state, thus assuring a north/south divide on the issue. And while it is reckoned to have delayed the US Civil War by a decade, with the underlying issues unresolved conflicts continued, resulting ultimately in an estimated 750,000 Civil War deaths (Hacker 2011).

The migration that sparked tension over slaveholding resulted from a confluence of factors. As noted, exhaustion of arable land and the massive availability of land in the West were critical. A second factor was that Napoleon, tired of trying to field armies in yellow-fever-ridden lands, offered to sell the vast Louisiana Territory for a price of $15 million.[26] The deal, known as the Louisiana Purchase, was finalized in 1812. Functionally, vast new territory became available for exploitation by enterprising Americans willing to move westward. Economic depressions provided additional incentive. But an overriding impetus behind the rapid expanse to the west and north was the American mindset of Manifest Destiny, a descriptor coined in 1845 to denote the extraordinary energy with which millions of Americans left their homes in the east and voluntarily set off to establish a new home in uncharted territory. The doctrine of Manifest Destiny argued that God had ordained the US to expand its dominion across the entire North American continent and forge a society based on American exceptionalism and democracy wherever it lay down roots. This philosophy ably accommodated the uniquely American psychology of wanting to escape tyranny and create an independent existence, to thereby achieve one's destiny through individual effort as granted through God's grace.

Nineteenth-century technology and innovations rapidly transformed the American landscape (and the nation's economy) by providing farmers with revolutionary options for crop production and distribution. The cast iron plow entered use. Rather than preparing the ground for sowing solely by human labor, a couple of oxen attached to the plow made the job considerably easier. However, the high clay content of much of the soil stuck to the plow, requiring constant cleaning. Then in 1837 the iconic John Deere steel plow appeared, changing agriculture permanently. Not only was it much lighter than its cast-iron progenitor, the Deere plow could also easily break through ground (particularly the sod of the prairie) and had the added benefit of being self-scouring, i.e., soil didn't stick to it. Other innovations followed. The Pennock drill (1850) replaced the hoe for seeding small grains such as wheat. By mid-century, weeds could be removed with horse-drawn cultivators. The harvest of crops with a hand-held scythe was replaced by the mechanized McCormick reaper, and by the latter third of the nineteenth century steam power could be used instead of horses for threshing machines at harvest. By the end of the nineteenth century, the internal combustion gas engine began to find use as an alternative to steam, horses, or windmills for stationary power (a.k.a. belt power) (Meyer 2013). Gas engines, which could be used to power a broad array of farm machines, reduced labor costs by requiring fewer people to operate equipment. Crop yields per acre increased and food was produced at levels well beyond that necessary to sustain one's family. Consequently, subsistence agriculture gave way to production agriculture, with the need to get one's farm products to market being met by transportation innovations. Completion of the Erie Canal (1825) enabled farmers to cheaply ship grain from points of origin in the Old Northwest (OH, IN, MN, WI) to markets in the eastern US; post-Civil War, the laying of railroad tracks from the East to the West, followed by improvements in refrigeration, trucks, and vessels, meant that highly perishable produce could reach far-off destinations in good shape. By 1914 the US had more railroad tracks than all of Europe combined. Fewer farmers were producing greater volumes of food and selling it on the world market.

The economic importance of production agriculture did not escape the attention of the US Congress, which did much to encourage westward expansion and advances in agricultural science, both seen as essential to the future of the country. As the population increased, various pieces of legislation were devised to spur settlement in remote places like the Great Plains, starting with the first Homestead Act (1862).[27] Passed during the Civil War, the Act set aside more than a billion acres of public land as an expression of the homestead principle, a Lockean concept whereby ownership of property is a natural right warranted by the application of labor. The Homestead Act provided 160 acres of land to any adult who had never taken up arms against the US (thereby eliminating southern Rebels), including immigrants and women; revision of the Act in 1866 included Black Americans. Great challenges awaited the homesteaders, including little water, no wood for fuel or construction, and the ravages of grasshoppers and other insect pests that devoured crops. All told, ownership of 270 million acres, 10 percent of the total US at the time, was given to 1.6 million homestead applicants at no cost other than a nominal filing fee and a promise to occupy the land for at least five years and improve it (Dick 1970). Between 1863 and 1880, nearly 60 percent of farmers on the frontier were homesteaders. Additional migration booms resulted in the settling of the central and northern Great Plains. The Homestead Act provided an opportunity for many US immigrants fleeing disasters in their ancestral homes. In the decades prior, waves of immigrants, especially from western Europe, had added significantly to the population. Between 1815 and 1820 approximately twenty thousand immigrants arrived yearly, swelling to eighty thousand per year after 1820. The California Gold Rush (1848–1855) induced many Chinese to immigrate to the western US: in 1849, only seventy-six people did so, compared with twenty thousand just three years later. Many more immigrants arrived during the last two decades of the Age of Industry (discussed subsequently).

US SCIENCE IN AGRICULTURE AND ENTOMOLOGY

Other federal actions advanced the cause of agriculture, sustained its growth, and augmented its importance to American society. Despite intense pressure owing to the US Civil War, in 1862 Lincoln prioritized the establishment of a Department of Agriculture, headed by a commissioner but lacking cabinet level status. Among its many important functions were those to conduct experiments of scientific and practical importance, collect data on agricultural practices and crop yields, and distribute new varieties of seeds and plants to growers. As time passed, additional foci in plant and animal nutrition, methods for insect control, and identification of sources of disease were added to the mission. President Cleveland elevated the US Department of Agriculture to cabinet level status (1889) and by the early twentieth century, the USDA was one of the largest and most influential arms of the federal government.

Westward expansion prompted the founding of post-secondary institutions beyond the elite colleges and universities in the East. The desire to secure a practical education in agriculture was debated in state legislatures and commanded attention at the federal level. The Morrill Land Grant Act (1862) was the federal answer to this call. Under the Act, each participating state received acreages of public land to support learning related to the mechanical arts. By the 1870s, agricultural colleges had been established in several states across the nation and courses were being offered in entomology, a relatively new academic discipline. The Morrill Act was amended (1890) to provide land for colleges for African Americans, establishing institutions now referred to as "historically black." Passage of two additional Acts ensured the nation's sustained commitment to agricultural science: the Hatch Act (1887) established state agricultural experiment stations for research, whose findings are disseminated to the public through cooperative extension established by the Smith/Lever Act (1914).

By the late 1800s the US had asserted its authority in agriculture convincingly, but it is generally reckoned that the country was roughly fifty years behind Europe in other disciplinary sciences in the first half of the nineteenth century (Barnes 1985). Several reasons can be cited. Outside of the eastern US, colleges and universities were just getting started in expanded western territories, and nearly all institutions lacked proper scientific apparatus and instruments as late as 1874 (Loewenberg 1941). Additionally, the focus of the new institutions was to generate practical results of direct use in producing food rather than deepening the philosophical/theoretical understanding of various disciplines. Further, outside of major cities, very few sites existed where interested people could meet, exchange ideas, and increase levels of interest in and understanding of the disciplines. Lyceum centers hosted occasional lecture tours with prominent international speakers like Charles Lyell and Louis Agassiz, broadening public exposure to science (Smallwood and Smallwood 1941). But libraries containing scientific papers and research reports were largely restricted to eastern institutions and could only be procured in the hinterlands by mail, assuming that the material could be loaned or copied for the recipient.[28] This hampered the transfer of scientific knowledge. The fledgling federal government did manage to fund and task exploratory parties with gathering data about the natural world, and several notable entomologists joined these expeditions and/or worked to identify their collections, including Asa Fitch, Baron Osten-Sacken, Hermann Hagen, Cyrus Thomas, W.H. Edwards, and A.S. Packard (Weiss 1930; Mallis 1971). The enthusiasm for collecting and cataloging specimens was driven in part by the desire to compare New World species to those in the Old World. US scientists were aware that they were in possession of a great resource—a virtually undescribed, vast land potentially filled with new fauna and flora. Cataloging nature in the early years, however, begat something of an American "inferiority complex" about the quality of its science. Thomas Jefferson spent years attempting to rebut Buffon's claim that the New World was inferior to the Old World by trying to prove that various animals in the US were larger than their European counterparts. Pioneer entomologist Thomas Say,[29] who described over 1,150 species of insects, exemplified the nationalistic spirit of post-Revolutionary War naturalists who aimed to place American science on par with that of Europe.[30] For his three-volume work, *American Entomology* (1824–1828), Say collected several thousand insects on Major Long's expedition to the Rocky Mountains (1819–1820) and Long's second expedition (1823) into British Canada to Lake Winnepeck (now Winnipeg). The collecting and cataloging of nature paralleled that underway in England, and two British texts helped promulgate a theological interpretation of nature in the new nation. Reverend William Paley's *Natural Theology* (1802) was widely employed at Princeton and at US academies (secondary schools), which numbered about four hundred in 1833. The classic, four-volume textbook by Kirby and Spence, *An Introduction to Entomology* (1815–1826),[31] which is littered with natural theological allusions (see Chapter 5), also found usage in the US. Thoreau (1854) cited the book in *Walden*.

Progress in US entomology from emergent to professional status is evidenced by several developments. Entomological articles began to appear in the agriculture press and in new periodicals devoted to insects.[32] Scientific societies, organized according to local need and resources, were founded. The path blazed by the venerable American Philosophical Society, established in colonial America (1743), was followed by the Academy of Natural Sciences of Philadelphia (1812), of which Say was both curator and a founding member. Insects formed an important part of the collection of the Boston Society of Natural History (1830), and societies soon arose that focused solely on insects, e.g., the Entomological Society of Pennsylvania (1842) and American

Entomological Society (1859). These and various state agricultural societies provided venues for meeting and debating scientific ideas; exchanging specimens, which supported growth of collections; and sharing information about insects locally and abroad. The transactions of societies were generally published, yielding archival information about insects that could be disseminated broadly. The professional growth of entomology can also be gauged by the perceived value and impact of its practitioners. Most were self-taught in the discipline and many were trained in medicine; several gained national and even international recognition (Sorensen 1995). Thaddeus W. Harris, MD, wrote papers on insect taxonomy and natural history and his *Report on the Insects Injurious to Vegetation* (1841), commissioned by the Massachusetts State Legislature, captivated future leading entomologists J.H. Comstock and L.O. Howard in their boyhoods (Mallis 1971). Asa Fitch, MD, described hundreds of insect species in a series of fourteen annual reports as America's first "state entomologist" in New York (appointed 1854). Soon other states passed legislation to appoint their own entomologists, whose annual reports focused largely on pests deemed most injurious to the citizenry. Charles V. Riley, Missouri state entomologist and later the most prominent nineteenth-century federal entomologist, engaged with renowned foreign scientists and received a gold medal from the French minister of agriculture and commerce for research on the grape phylloxera, a root-feeding pest of European grapevines that threatened to destroy the French wine industry. Riley recommended grafting susceptible French grapevines onto naturally resistant American root stock. His advice was followed and proved valid; by 1890, nearly 720,000 acres had been replanted (Sorensen et al. 2018).

Unquestionably a highly valuable service provided by entomologists was to identify important species, describe basic life histories, and begin to formulate sound strategies for controlling insect pests. With the westward migration of people and planting of large monocultural crop stands, effective control measures grew increasingly urgent. Plagues of grasshoppers appeared in 1873, prompting the US Department of the Interior to appoint an Entomological Commission headed by Riley[33] (see Chapter 3), and the very destructive boll weevil invaded the South, resulting in great economic hardship for cotton growers. The cultural control methods recommended by the USDA were not a panacea, however, the boll weevil invasion had the salutary benefit of encouraging diversification of crops being grown. Moreover, Lincoln's foresight in establishing the USDA and land grant universities put in place mechanisms to promote higher education and disseminate needed information. Federal entomologist L.O. Howard contrasted the great change in the discipline from 1875 to 1925, when the US went from having "almost no teachers of entomology" to having more teachers of entomology than "all the rest of the world put together" (Howard 1930). Further, research on insects had broadened to yield ground-breaking advances in genetics. Largely through efforts spearheaded in the early twentieth century by T.H. Morgan at Columbia University, *Drosophila melanogaster* became a model experimental organism that today serves as a workhorse for genomics, developmental biology, behavior, and neuroscience, among other disciplines. The list of Morgan's students and their students and associates reads like a "who's who" of eminent geneticists and includes many Nobelists.[34] Our understanding of human diseases and disorders owes much to research on the tiny fruit fly.

ENTOMOLOGY IN OTHER NATIONS; WOMEN'S CONTRIBUTIONS

Distinguished entomologists elsewhere also made contributions to insect natural history and taxonomy, and many of their insect collections reside in museums around the globe and continue to be studied by scientists. Here we highlight but a few individuals (Essig 1931 provides many notable biographies). The eminent French entomologist Dejean, *aide-de-camp* to Napoleon, remarkably collected specimens on the battlefield at Waterloo. Upon his death, his nearly 25,000 beetle (Coleoptera) species and more than 118,000 specimens constituted the largest personal beetle collection in the world (Bousquet and Bouchard 2013a, 2013b). His compatriots, Boisduval and Latreille, co-founded the *Société entomologique de France* (1832). Boisduval described many beetle and lepidopteran species (Essig 1931) and Latreille established several insect orders (Thysanura, Siphonaptera, Heteroptera, Homoptera) (Dupuis 1974). In Germany the distinguished fly (Diptera) taxonomist Hermann Loew also examined specimens in Baltic amber, and Burmeister published on insect taxonomy, anatomy, and physiology (Lindroth 1973). Sélys Longchamps of Belgium was a prominent student of dragonflies and damselflies (Odonata) and lacewings, dobsonflies, and their relatives (Neuroptera) (Essig 1931).

Women's contributions to entomological science were severely limited by social norms and erroneous beliefs about their intellectual and physical capacities, which limited their options for formal education, independent interactions, and self-determination. Consequently, male authors dominated the entomological literature. Historian Margaret W. Rossiter (1982) explains that the pursuit of science had stereotypically been viewed as "tough, rigorous, rational, impersonal, masculine, competitive, and unemotional"—the antithesis to the idealized embodiment of womanhood: "soft, delicate, emotional, noncompetitive, nurturing." Women's efforts were restricted to nature writing and artistic renderings, often for their husband's publications—Lucy Say comes to mind.[35] The great majority of entomological publications authored by women between 1800 and 1900 appeared in the last two decades of the century; about 90 percent derived from the US and Britain, with the remainder coming from Germany, Canada, and Ireland (Creese 1998). The increasing level of female professionalism is reflected in the careers of several notable women. Anna Botsford Comstock, spouse and close collaborator of Cornell University entomologist John Henry Comstock, began as an illustrator of his work and gained renown in her own right as a science author, college professor, and leader of the Nature Study Movement (Smith 1976). Mary Treat, an internationally known American expert in "applied" (economic) entomology, corresponded with Darwin, Riley, and Harvard botanist Asa Gray and struck an unusual stance by supporting herself financially after separating from her husband in 1874 (Evans 2017). Eleanor Ormerod, a member of the British landed gentry, never needed the financial security of a husband and remained unwed, unlike 88 percent of her contemporary females, who either were or had been married (Clark 2009). Ormerod pressed for the creation of professional employment opportunities in agricultural science, including money to support them, at a time when such positions would have been unavailable to her as a female. For her pioneering efforts in economic entomology (including fifty-nine journal articles and scientific reports published between 1873 and 1900 (Creese 1998)), Ormerod garnered silver and gold medals from scientific societies in Britain, France, and Russia. Her autobiography chronicles her lifelong interest in insects, which began as a child on her father's 800-acre estate in Gloucestershire, England (Ormerod 1904). Her steadfast commitment to scientific pursuits flouted Victorian norms, as when she published a pamphlet advising farmers that house sparrows should be destroyed, not protected. Ormerod based her reasoning on research by Canadian and USDA scientists and the Association of American Ornithologists, which showed that house sparrows consumed few pest insects (or any insects, being seed feeders) and aggressively ran off more serviceable birds. Ormerod's pamphlet appalled the Reverend J.E. Walker, who advised her to instead turn her great talents to philanthropic work and fulfill her duty as a woman. But Ormerod remained adamant, writing to W.B. Tegetmeier,[36] "I quite reckon on being violently attacked, but it did me no harm before to be threatened to be shot at, also hanged in effigy, and other little attentions." Like Ormerod, Edith Marion Patch's fascination with insects began in childhood. In the early decades of the twentieth century, Patch earned distinction as the first department head of entomology at the University of Maine, first female head of the Maine Agricultural Experiment Station, and first woman president of the Entomological Society of America (Gibbs 2014). In addition to achieving prominence as a world authority on aphids, Patch gained recognition as an author of children's nature books and adult popular science articles. She remained unmarried and independently supported four family members for twenty years—atypical for demographics of her contemporary females.

CLOSING DECADES OF THE AGE OF INDUSTRY: CAPITALISM, IMPERIALISM, THE FIRST WORLD WAR

Great Britain's predominance was such that by 1880, an astounding 44 percent of the world's exported manufactured goods originated there (Cannadine 2017), earning the nation the moniker "workshop of the world." Britain manufactured the rails that ran across India and Australia and fabricated many of the traditional garments worn by Indians, Arabs, and Africans. At its peak, the British shipbuilding industry manufactured more than 80 percent of the world's tonnage (Mingay 1986). Reflecting the nation's formidable global footprint, in 1905 about one-third of all its exports went to the empire and about one-fourth of its imports came from the empire. No other nation rivaled Britain's overseas assets, but competition from the US and Germany was increasing.

In stark contrast to its fledgling nation status in 1820, by the end of the Age of Industry the US had become a world power with its own imperialistic ambitions. Its range had expanded from the Atlantic Ocean to the Pacific, and it had seized colonial territories during the Spanish-American War. The nation's ascendancy was swift, its rising prominence in world trade bolstered by a robust manufacturing base and floods of immigrant labor owing to the "greatest migration of peoples in history" (Hobsbawm 1975). The vast majority of these individuals, poor and from rural European areas, resettled in the US, where no federal immigration laws existed prior to 1882.[37] Between 1820 and 1910 an estimated 33.6 million people emigrated from Europe and the Far East (Barraclough 1982); astonishingly, in the two decades from 1900–1920, nearly fifteen million immigrated to the US.[38] These various diasporas were fueled by metaphoric visions of the US as a "great melting pot" and the "land of opportunity," a place where anyone could find safe harbor, gain citizenship, and make a better life through dint of hard work. In the heyday of the US silk industry, textile mills dotted urban areas of the northeastern US and many immigrants found work in Paterson, New Jersey, then known as "Silk City" (see Chapter 4). A thriving US steel industry employed many more immigrants and brought tremendous wealth to industrialist Andrew Carnegie, who at age thirteen had emigrated from Scotland and began work as a bobbin boy in a Pennsylvania textile mill. His philanthropic legacy in public works, the arts, and science endures.[39] Entomology, too, benefitted from the scientific contributions of many US immigrants.[40]

During the last two decades of the nineteenth century, the European colonial powers began to shift the focus of their international ventures toward the search for raw materials of industry such as tin, rubber, copper, and nickel. These efforts, coupled with imperialistic rivalries, spurred the Scramble for Africa, whereby Great Britain, France, Germany, Portugal, Belgium, Spain, and Italy partitioned and gained full control of Africa except for two countries: Abyssinia

(Ethiopia and Eritrea) and Liberia. The partitioning of Africa—20 percent of the global land mass—was complete by 1914. An interdependent capitalist world economy was in place, with Britain holding by far the most foreign investments of any nation. International commerce was booming, spurred by completion of the Suez (1869) and Panama (1914) canals. The latter provided a seafaring trade connection between the Atlantic and Pacific Oceans, thereby obviating the need for ships coming from Europe to circumnavigate around the southern tip of South America. A worldwide standard time system was established.

The Age of Industry drew to a close soon after the end of the First World War (1914–1918), a human tragedy of immense scope in which more than ten million men perished, with twice as many wounded. Multinational empires had toppled in the "war to end all wars" and the old international order ceased to exist. The roundly maligned Paris Peace Treaties failed to restore global stability or a lasting peace, and the influenza pandemic (1918–1919) claimed approximately fifty million lives worldwide.[41] The postwar zeitgeist was captured by W.B. Yeats (1919) in his poem "The Second Coming," in which "Things fall apart; the centre cannot hold; mere anarchy is loosed upon the world." Despite remarkable technological and scientific breakthroughs, and advances in human rights, the evils of modernity stood out in sharp relief. Progress and prosperity would return, marbled predictably with the hostilities of nations and the enduring achievements of creative, irrepressible individuals.

INSECTS MANIFESTED IN CULTURAL CONTEXTS

In the chapters that follow, authors recount the fascinating, diverse, and at times unexpected role of insects in human culture during the Age of Industry. Johnson and Polaszek (Chapter 1) show how the museum collections that today serve as valuable repositories for entomologists globally are rooted in the expeditions of intrepid nineteenth-century naturalists. Among these naturalists were Charles Darwin and Alfred Russel Wallace, whose theoretical work proved foundational for our understanding of species evolution and diversity. Miller and Miller (Chapter 2) discuss insects in military operations war, elucidating how insects could affect campaigns and torment, sicken, and even take the lives of soldiers and civilians by vectoring human diseases. Dunkel (Chapter 3) examines the practice of consuming edible insects (termed "entomophagy") in various cuisines, tracing its origins to ancient traditions, while also recounting approaches to gathering, processing, and utilizing insects as valuable sources of nutrition. Sheppard and Anelli (Chapter 4) review apiculture (beekeeping) and insect-derived products. They describe advances in hive design and queen production that proved critical for current day honey bee pollination services, and discuss the significance of silk, cochineal, shellac, and gall-based inks/dyes as globally traded commodities. Meuti and Fisher (Chapter 5) explore insects in mythology and religion, with particular attention to Natural Theology and the paradigm shift effected by the theory of organic evolution by means of natural selection. Dodd (Chapter 6) focuses on the use of insects as symbols, reflections of the human response to the profound changes of the industrial era, when printed matter about the natural world (including insects) grew in availability and popularity. White (Chapter 7) considers insects in literature and language, focusing on several authors who used elements of time, scale, and domesticity to illustrate the very different spheres occupied by insects versus humans. Prischmann-Voldseth (Chapter 8) surveys insect representation in art, including insects of medical importance, mythological significance, and aesthetic appeal, drawing on an array of artists and their works to illustrate the complicated relationship we humans share with these six-legged creatures. With these authors as expert guides, we invite readers to discover a world of connections between insects and people.

CHAPTER ONE

Insect Knowledge

The Discovery and Description of Insect Diversity

NORMAN F. JOHNSON AND ANDREW POLASZEK

The simplest questions can sometimes be the hardest: how many different kinds of insects are there? The surprising answer is that we do not know, probably not to even the nearest order of magnitude. Hundreds of thousands of species, the basic unit of evolution, have been described and named by scientists over the last 260 years. Thousands of new species are formally recognized each year, and there is no sign that this pace is slackening.

The hundred-year period that is the subject of this chapter is characterized, as far as knowledge of insect diversity is concerned, as a period of accelerating discovery and description. The early history of insect taxonomy has been reviewed by Tuxen (1973) and Lindroth (1973), covering the period roughly from 1700–1859. Following that, growth in knowledge of the true extent of insect diversity grew rapidly. Those earlier reviews were written with an emphasis on the personalities of the scientists involved. Our focus, in contrast, is on identifying some of the common trends in the field at this time, roughly 1820–1920, and in examining how the process of taxonomic discovery was conducted and how it changed.

The start of formal naming of species of insects, and implicitly the documentation of their discovery, is January 1, 1758, the accepted date of publication of the tenth edition of *Systema Naturae* by the Swedish biologist Carl von Linné (born Carl Linnaeus, 1707–1778). Although it would seem that "in the beginning" we would be starting from a position of total ignorance, a good deal was already known at that time about insect diversity. In 1758

Linnaeus described and named 1,929 species of insects from all continents save Australia and Antarctica. Over the remainder of his career he described and named about three thousand insects from around the world.

By the first decades of the nineteenth century the number of recognized species had multiplied many times over, most notably through the work of Johann Christian Fabricius (1745–1808). Fabricius was a student of Linnaeus for two years, 1762–1764, at the University in Uppsala, and a Professor of Natural History in his own right first in Copenhagen and later in Kiel. He alone described some ten thousand species of insects, setting the bar, at least in terms of numbers of taxa described, for the century following his death.

Although devoted to Linnaeus, Fabricius is not counted among the Swedish professor's so-called "Apostles." These students, numbering seventeen, were encouraged and sometimes financially supported by Linnaeus to travel the world collecting specimens, with the goal of allowing Linnaeus the opportunity to study them and select material to add to his personal collection. Expeditions were made to the Far East, North America, North Africa and the Mediterranean, South America, Russia, and Australia. Several of the apostles died in the course of these expeditions, and not all were particularly successful in acquiring specimens. One of them who is particularly worthy of mention in the context of our present subject is Daniel Carl Solander (1733–1782).

Solander was initially sent by Linnaeus to England to bring the gardens of correspondents in line with his latest classification of plants. But while there, Solander acquired a position at the British Museum, and later invited himself to accompany his friend Joseph Banks on Captain James Cook's *Endeavour* voyage to Australia and Oceania (O'Brian 1987). Banks himself sought to take part on the voyage, initially and officially chartered to witness the transit of Venus near Tahiti, so as "... to enrich science and become famous." Thus a precedent was set to have a naturalist present on epic sea voyages, a precedent probably most famously followed by Charles Darwin aboard the *Beagle* in 1831.

The work of Pierre André Latreille (1762–1833) was not focused on species, but rather on the use of an eclectic methodology to develop a natural classification of invertebrates, particularly insects. While Linnaeus initially classified all insects into sixty-five genera, Latreille (1807–1809) recognized an order of magnitude more, 675. The quality of this work was recognized even in his own time. Fabricius himself was reported to have called Latreille "the foremost entomologist of his time" (Geoffroy Saint-Hilaire 1833; Dupuis 1974): "Ce premier rang parmi les entomologistes de notre âge, Fabricius, comme un autre Élie, en avait de son vivant investi l'héritier de son talent; j'ai entendu de la bouche même du professeur de Keil [sic] cette solennelle designation." The caliber of the work is also reflected in the fact that these genera remain the foundation for the classification in use now more than two hundred years later.

Thus, the beginning of the period of time that is the focus of this chapter was not marked by a quantum leap in knowledge of the extent of insect biodiversity around the world. The broad outlines, the shorelines if you will, had been sketched out by the generations of natural philosophers in the sixty-plus years following Linnaeus. But the depths of the seas surrounding those continents had yet to be plumbed. The extent of the richness of the insect world, not only the number of different kinds, but also their diversity in size, shapes, colors, life history, and behavior, could not yet be appreciated fully. Today, the estimates of the total number of species of insects are a matter of some dispute (Stork et al. 2015). However, all agree that more than one million have already been described, and that may be only 10–20 percent of the true number. The century following Fabricius and Latreille was a time in which the true scope of insect diversity began to be sensed, and the scale of the richness encountered, in turn, prompted changes in the approach with which it is studied.

SPECIMENS AND COLLECTIONS: THE BASIS OF BIODIVERSITY DISCOVERY

Wandersee and Schussler (1999) introduced the term *plant blindness*, defining it as

> (a) the inability to see or notice the plants in one's environment; (b) the inability to recognize the importance of plants in the biosphere and in human affairs; (c) the inability to appreciate the aesthetic and unique biological features of life forms that belong to the Plant Kingdom; and (d) the misguided anthropocentric ranking of plants as inferior to animals and thus, as unworthy of consideration.

This same kind of cognitive filter applies to insects, both on the level of an individual person, and that of society as a whole. The underlying reasons are likely complex, but it seems reasonable to include on the list the perception that insects—with a few notorious exceptions—are just not that important. That perception, in turn, is likely driven by the fact that most insects are small in size.

The diminutive body size of insects means that in order to study them in any detail and to appreciate their diversity it is necessary to have specimens "in hand." Microscopes had been available since at least the mid-seventeenth century, but such a tool was next to useless in the field. The build up of collections of new material from around the world was a prerequisite to learning how to distinguish one minute creature from another and, so, to document insect biodiversity. The specimens then had to be stored somewhere, in a collection.

The nature of both collecting and collections changed significantly in the nineteenth century.

The great Linnaeus famously did relatively little personal collecting. In the early part of his career he traveled and collected widely in Sweden, from Lapland in the north to Skåne in the south. He did live in the Netherlands and England for some time, but after his return to Sweden in 1738 he never again left the country. Nevertheless, specimens were funneled to him from all over the world, most of which were incorporated into his private collection. Similarly, Fabricius based much of his work on the contents of private collections. These "natural history cabinets" were the predominant repositories of insect specimens in 1820.

Although it is intuitive that the number of specimens available for study increases over time, it is difficult to document this in a granular fashion. Unlike many other types of collections, it is fairly unusual for an insect collection to catalog individual specimens in the process of accession. As a result, exact numbers are rare, and the process of digital capture of specimen data—primarily the place, time, and means of the collecting—is hindered. The Global Biodiversity Information Facility in Copenhagen (www.gbif.org) acts as a central clearinghouse for the primary occurrence data associated with specimens in collections. With an abundance of caveats in mind, 319 specimens are recorded in GBIF that were collected by 1820. By 1920 the total increases to 26,240. Both numbers are certainly gross underestimates. Most insect collections are not digitized and their data are not available. Standards for recording specimen metadata have changed over the years, and the earliest collections often lacked any data whatsoever. But the data that are available suggest a fairly consistent increase in numbers of specimens in collections, particularly after 1870. These are the raw materials of biodiversity studies.

In the tradition of the *Endeavour*, government-sponsored expeditions often included personnel whose responsibilities including collecting natural history specimens. In the US Thomas Say (1787–1834) traveled as zoologist on the 1823 expedition of Maj. Stephen H. Long to the Lake of the Woods in the, then, little explored Northwest of the country at that time (modern-day Minnesota). The first expedition of the *Astrolabe* under the command of Dumont d'Urville, with Joseph Paul Gaimard (1796–1858) and Jean René Constant Quoy (1790–1869) as physician-naturalists, returned to France in 1829 with a wealth of material from the Pacific region. Jean Baptiste Boisduval (1799–1879) published nearly a thousand pages in two volumes (1832, 1835) describing the many new species of insects collected (Figure 1.1). Darwin's own voyage around the world aboard the *Beagle* is probably the best known example of such an expedition, and it has been written about extensively, most prominently by Darwin himself (1839). In the same vein, the army of France supported expeditions, such as that to Morée, the Peloponnese, between 1828

FIGURE 1.1 *Idea durvillei*, originally described by Boisduval (1832) from specimens collected in New Guinea (Irian Jaya) on the voyage of the *Astrolabe*. Wikimedia Commons.

and 1833. Auguste Brullé (1809–1873) documented 880 species of insects among the collections, over 40 percent of them (364) new to science (1832).

Other noteworthy state-sponsored expeditions include that of Spix and Martius to Brazil from 1817–1820, accompanying the Hapsburg princess Maria Leopoldina as she traveled there to become the bride of the future Emperor Pedro I. The entomological results of the trip were published in 1830–1834 by Josef Anton Maximilian Perty (1804–1884). As a last example, the *Novara*, a frigate of the Austrian Imperial Navy, sailed around the world in 1857–1859 (Figure 1.2). Its primary goal was to develop maritime expertise of the officers and men, but it also had the mission to make observations and collections of natural history. These materials, including many insect specimens, became a significant component of the collections of the *Naturhistorisches Museum* in Vienna.

The expeditions cited are but a few of the high-profile examples of voyages around the world and to unexplored regions. Natural history was not their primary purpose, but one of several goals that could include mapping, astronomy, geology, and ethnography. There were also notable privately financed ventures. Those with the greatest notoriety are the travels of Alfred Russel Wallace (1823–1913), first to the Amazon Basin (1848–1852) and then to the Malay Archipelago (1854–1862). The latter trip was jump-started by the Royal Geographic Society to which he had applied seeking free passage.

FIGURE 1.2 The Austrian frigate *Novara* in the teeth of a storm. From *Novara Expedition 1857–1859* (1861). Biodiversity Heritage Library.

Eventually he secured transport to Singapore through their auspices. Otherwise, his travels were self-funded. According to Henry Walter Bates (1825–1892), with whom the trip to South America was undertaken:

> In the autumn of 1847 Mr A.R. Wallace, who has since acquired wide fame in connection with the Darwinian theory of Natural Selection, proposed to me a joint expedition to the river Amazons, for the purpose of exploring the Natural History of its banks; the plan being to make for ourselves a collection of objects, dispose of the duplicates in London to pay expenses, and gather facts ....
>
> (Bates 1863)

Much of Wallace's South American materials were lost when the ship transporting him back to England caught fire and sank. Bates, however, was more successful. Every few months Bates would forward specimens to London, attesting to a rather lively trade network, given the remoteness of the areas in which he lived. He estimated that during his eleven years in the region he collected fourteen thousand species of insects. Wallace's specimens from the

Malay Archipelago, many of which are now still housed within the Natural History Museum, London, were the basis for a series of "catalogs" written by Frederick Smith (1805–1879) and Francis Walker (1809–1874) from 1857–1873. Walker published ten papers on Diptera (true flies), four on Lepidoptera (butterflies and moths), and three on Homoptera (leafhoppers, planthoppers, treehoppers, cicadas, aphids, whiteflies, scale insects, psyllids). Smith focused on the Hymenoptera (sawflies, wasps, bees, ants): in ten articles he described 471 new species, including 267 species of ants alone. Although he was mildly criticized in his obituary (Dunning 1879), "... it may perhaps be doubted whether he kept himself sufficiently acquainted with all that was being done by his contemporaries abroad ...," it was Walker who received the greater criticism and opprobrium. Evenhuis (2018) has published a recent biography of Walker in which the basis for Walker's sad modern reputation is discussed. His impact, for good or bad, on the understanding of insect diversity by the 1870s cannot be dismissed. In the series of catalogs of the insects of the British Museum Walker treated over forty-six thousand species and described nearly sixteen thousand as new to science.

Another Smith, Herbert H. Smith, was a prominent collector in the New World. He made five expeditions to Brazil between 1870 and 1886, collecting not only insects, but vertebrates, and documenting the geography and geology encountered. He also collected extensively in Colombia, Venezuela, and the West Indies. Some of his Brazilian material, now in the Carnegie Museum of Natural History, formed the basis of William H. Ashmead's monograph on chalcidoid wasps, and Smith's collections on the Caribbean islands of St. Vincent and Grenada are housed in the Smithsonian and the Natural History Museum.

By the end of the nineteenth century the flow of specimens into the hands of museums and taxonomists had turned from a trickle into a gusher. This is reflected in a perusal of a listing of Frederick Smith's publications in the last few years of his life. There we find articles based on a trip to Japan by Mr George Lewis, species collected by Mr. G.R. James Rothney in India, species collected "in the Island of Rodriguez by the naturalists accompanying the Transit-of-Venus Expedition," insects from New Zealand collected by C.M. Wakefield, Esq., the zoological collection of the HMS *Petrel* to the Galápagos, more New Zealand material collected by Prof. Hutton, the Yarkand Mission of Ferdinand Stoliczka, and species from the Sandwich Islands collected by the Rev. Thos. Blackburn.

The rate at which new species were being discovered and described had entered a period of rapid acceleration. This has not abated to this day, save for downturns during periods of war. While the quality of some of the science may be debated, the general message was clear: the diversity of insects surpasses, by orders of magnitude, that which anyone at the time dared to suspect.

Natural history museums, as we know them today, were not unknown, but were relatively unusual. The *Muséum National d'Histoire Naturelle* in Paris was founded already in 1635, the Zoological Museum in St. Petersburg in 1714, and the *Museo di Storia Naturale* in Florence dates from 1775. The beginnings of the modern Natural History Museum in London date from the collections of Sir Hans Sloane and the founding of the British Museum in 1753. But the great imperial and national museums were largely established in the nineteenth century. As examples of European institutions with large entomological collections, there is the *Naturhistorisches Museum* in Vienna founded in 1876, the *Muséum des Sciences Naturelles de Belgique* in Brussels in 1846, the *Národní Muzeum* in Prague in 1818, the *Zoologisk Museum* in Copenhagen in 1862, the *Museum für Naturkunde* in Berlin in 1886, *Magyar Természettudoményi Múzeum* in Budapest in 1802, the Natural History Museum in Dublin in 1856, the *Rijksmuseum van Natuurlijke Historie* in Leiden in 1820, and the *Naturhistoriska Riksmuseet* in Stockholm. Similar institutions appeared in the Americas, South Africa, Indonesia, and Australia in the same time period.

The holdings of the national museums often began as donations, bequests, or sales of privately held collections. The millions of specimens in the entomological collection of the *Naturhistoriska Riksmuseet* were seeded by the private collections of Carl Clerck (1709–1765), Charles De Geer (1720–1778), and Gustaf von Paykull (1757–1826). The massive collection of the Smithsonian Institution in Washington was jump-started with specimens provided by Charles Valentine Riley (1843–1895). Private cabinets continued to exist, and still do to this day, but the emphasis had clearly changed.

Concentrations of specimens in large, national institutions also led to an increase in the staff devoted to their care, curation, cataloging, and study. It is difficult to quantify such changes across the breadth of all insects because the relevant data, while existing, simply has not been collated in a usable form for most groups. One exception, and therefore an exemplar we will use to illustrate more general trends, is for ants.

Ants have received a great deal of study over the years, likely a result of their ubiquity, familiarity, interesting biology, and both economic and ecological impact. We have reviewed all of the literature in which new species-group taxa of ants have been described, and this forms the basis of the statistics we cite to support our interpretation of general trends. For this case, we see a dramatic increase in the number of persons and the number of publications in which new ants are described. In the sixty-two years between 1758 and 1819 a total of twenty-six authors published, largely in book form, descriptions of new taxa of ants. By the period between 1910 and 1920 this number had increased to sixty-eight. The number of publications increased from seven in the decade from 1820–1829 to 456 between 1910 and 1920. Finally, the number of species-group taxa described as new (including infrasubspecific taxa) increased from twenty-nine in the decade from 1820–1829 to 3,501 between 1910 and 1920.

While ants are a useful indicator for quantitative comparisons, the most extensive taxonomic activity clearly was focused on the Coleoptera. Beetles were among the most important early inspirations for arguably the century's most important biological theorists, Wallace and Darwin. Beetles are generally attractive animals, often spectacularly so, and are relatively easy to collect and preserve, being among the more physically robust of insects and, as a rule, relatively reluctant to take flight when pursued. Wallace became interested in beetles after meeting Henry Walter Bates in Leicester, when they were both about twenty. He continued his passion for collecting them for the next two decades, amassing more than eighty thousand specimens during his voyage in the Malay Archipelago (Shermer 2002). Charles Darwin was already collecting beetles at the age of eight, and started to do so seriously while neglecting, first, his medical studies in Edinburgh, and then his BA at Cambridge, with many of his collection records published when he was twenty (Stephens 1829). The activity of collecting beetles as children and young men started both Wallace and Darwin on a course that culminated in their independent discovery of the main cause of organismal evolution.

In terms of sheer numbers described, between 1820 and 1920 more than 120,000 species of beetles were newly recognized. Among the most prolific describers of beetles of the period were Francis Polkinghorn Pascoe (1813–1893) and Joseph Sugar Baly (1816–1890), these two being rivals not only for numbers of beetles described, but also for the strangeness of their middle names. George Waterhouse (1810–1888), George Charles Champion (1851–1927), David Sharp (1840–1922), and Martin Jacoby (1842–1907), all British with the exception of Jacoby (born in Hamburg) and the Swede Carl Boheman (1796–1868), are all members of the "1000 club," each having described more than one thousand taxa, in many cases several thousand. Waterhouse was responsible for describing most of the five thousand beetles brought back by Darwin from the *Beagle* voyage, a task later completed by his son Charles (1843–1917). Champion was a collector from the age of sixteen, traveling to Guatemala when twenty-seven under the sponsorship of Frederick DuCane Godman and Osbert Salvin. Four years later he returned to London with fifteen thousand species of insects. He described more than four thousand beetle species in the monumental *Biologia Centrali-Americana* series, also overseeing the production of the fifty-two volumes of that series. Sharp, Jacoby, Gorham and Bates also made major contributions to this series (Figure 1.3).

Pascoe and Baly together described more than three hundred beetles resulting from Wallace's collecting in Sarawak—all from a very small area around the Simunjan coalmine, during a nine-month period in 1855 (Polaszek and Cranbrook 2006). The Natural History Museum in London houses

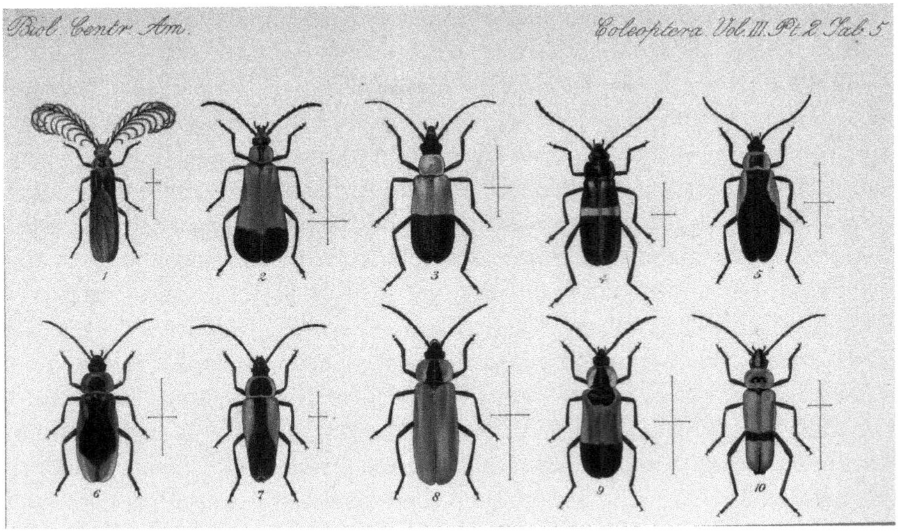

FIGURE 1.3 Beetles from the *Biologia Centrali-Americana* volume authored by the Rev. Henry Stephen Gorham in 1886. Biodiversity Heritage Library.

2,500 holotypes of Pascoe's beetles. David Sharp was another medical man and published more than 250 papers including several major revisions, as well as important work on male genitalia in beetles. Several thousand Sharp-type specimens are in the Natural History Museum, London. Sharp edited the *Zoological Record* for thirty years from 1892–1922, the year of his death. He was second only to Martin Jacoby for the number of beetle species described. Jacoby takes the first prize having described around five thousand species during his forty-year active career, including 347 species published posthumously.

There are more than a hundred thousand beetle species in the Natural History Museum (London) collection described during the century 1820–1920, by 1,200 authors, twenty-nine of whom described over a thousand species (Max Barclay, personal communication). Without doubt we can call this period "a century of beetles."

CULTURAL HEGEMONY

For most insects the greatest diversity in species is found in the tropical regions of the world. Conversely, historically the geographic concentration of specialists, those with the interest and capability of identifying that diversity, has been in the north temperate regions of the world. The period with which we are concerned embraces the height of the European colonial era, as well as US possessions in the Caribbean, Panama, Oceania, and the Philippines. However it was also the time in which the colonies of Spain and Portugal gained their independence. How was the cultural influence from the north reflected in the process of species discovery and description?

Although specimens were being collected in almost all corners of the world, it is currently impossible to provide an account of where they ended up. Natural history museums were established in Brazil (*Museu Nacional* in 1918, *Museu Paulista* 1890), Argentina (*Museo Argentino de Ciencias Naturales Bernardino Rivadavia* in 1826, *Museo de La Plata* in 1888), South Africa (South African Museum in 1825, *Staatsmuseum* in 1892), modern-day Indonesia (Bogor Zoology Museum in 1894) and Malaysia (Sarawak State Museum in 1891), and in Australia (Australian Museum in 1827, South Australian Museum in 1847, National Museum of Victoria in 1854, Queensland Museum in 1862) during this period. However, size and the extent of their respective insect collections in the early days is unknown. Certainly, the sponsorship of most collecting expeditions by European and American institutions and governments strongly suggests that the bulk of that material was sent back to those areas.

The disparity between the countries in which the biodiversity occurs and where the taxonomy was being conducted can be clearly seen in the publication output of those scientists. Our contemporary definition of what constitutes a "journal" should be taken with some caution. It was during this same period that learned societies devoted to the study of insects and their periodicals began to be established. The *Société Entomologique de France* was founded in 1832 and claims to be the oldest such group in the world. The Royal Entomological Society of London, in the form that continues to this day, was founded a year later. It is sometimes difficult to distinguish between short-lived journals and monographic series, particularly when the latter report the results of scientific expeditions. As a practical difficulty, journals can also change their titles over time, split into daughter series, etc. Therefore, the data we have concerning biodiversity publication should be viewed with appropriate caution.

From 1820–1920 a total of 791 scientific papers in which a new taxon of ants was described were published in what we today characterize as journals. There are 196 such journals, and only sixteen of these (8.2 percent) were based outside of Europe and North America (the US and Canada). Of the 791 articles, a mere 61 appeared in the latter set of journals (7.7 percent).

The language in which papers were published shifted from Latin to vernacular languages. Using, again, the primary taxonomic literature of ants as a reference point, publications appeared in a total of fourteen languages as their primary language. One-third of the publications in the 1820s were in the old lingua franca, Latin, and this language persisted well into the twentieth century, although in later years rarely as the primary tongue. German, English, and French were predominantly used throughout the period of 1820–1920 accounting for 84 percent of the papers (21.4, 30.4, and 32.1 percent, respectively).

Only a single paper appeared in a non-European language, Japanese, in 1912 (Matsumura 1912). One paper was also published in Esperanto in 1907 (Forel 1907)! The only other temporal pattern apparent is the first appearance of a paper in Russian in 1877 (Fedchenko 1877) and the continuation of Russian contributions through the subsequent decades.

A quick analysis of the scientists who produced the primary taxonomic literature of ants reveals additional demographic biases. A bit over one-half of the total number of publications describing new taxa of ants (55.8 percent) were the product of four persons: Auguste-Henri Forel (Swiss), Carlo Emery (Italian), William Morton Wheeler (American), and Felix Santschi (Swiss). A total of 138 persons appear as first authors of 886 publications. Of these, only ten (7.2 percent) appear to have been based outside of Europe, the US, and Canada. Their total output, twenty-one publications, is only 2.4 percent of the total scientific output on the topic. The gender disparity is even more striking. Of these 138 authors only one, Hazel Andrews, was a woman. She was a student of T.D.A. Cockerell at the University of Colorado. This situation persists to this day: over the history of myrmecology (study of ants), only 2.4 percent of the species-group taxa were described by women.

Insofar as the pattern we see in ants is generally applicable, on the one hand the number of languages in which papers were published increased from three in the 1820s (Latin, English and German) to nine a century later. This could be interpreted as an expansion of access to information about the world's insects. Other data suggest, however, that the effort remained similar to that of the days of Linnaeus and Fabricius, with a concentration of men from Europe and, to a lesser extent, North America.

THE (LACK OF A) PARADIGM SHIFT

Lindroth (1973), in his history of systematic entomology up until the time of Darwin, noted that the publication and publicity surrounding the *Origin of Species* (1859) had little practical effect on the description and naming of new species. This remained true to the end of our period of concern. Neither did the mechanics of the process of classification, putting species together in groups, change significantly. The major difference was simply the adoption of common ancestry as the underlying force accounting for the observed similarity. The philosophical struggles over the basis of classification—evolutionary taxonomy, phenetics, and phylogenetics—which would produce fundamental changes in the theory and practice of systematics, began only in the second half of the twentieth century. Without explicit, objective criteria with which to compare alternative classifications, an argument devolved to one based on authority. Authority, in turn, was based on power, and this was a function of academic pedigree, place of employment (or patron), publication record, number of disciples (and their power), and access to specimens, to name just a few factors.

Consensus on precisely what a species is eludes us to this day. Prior to the synthesis of evolution and genetics in the twentieth century, a species was simply one point on a continuum of morphological variability, and where that point lay seems to have been entirely at the discretion, experience, and ease of access to specimens of the individual taxonomist. The significance of differences in color, body sculpture, size, proportions, etc. was undoubtedly enhanced by the sparseness of collections anywhere outside of western Europe and eastern North America. The lack of samples between widely separated collectings disguised the fact that individual insects intermediate in those characters existed. The impression, then, was of discontinuous variability, fodder for recognizing and naming new taxa.

Since the rank of species was not thought to occupy any biologically significant position in the taxonomic hierarchy, there was a natural tendency to name not only new species, but ever finer variants within them. This was taken to the most extreme degree in groups like ground beetles (Carabidae) and ants. Lindroth (1973) briefly described this phenomenon in relation to the beetles. Thankfully for modern-day taxonomists, the names coined by the most egregious describers of these so-called infrasubspecific categories of beetles are no longer considered to be valid ("available," to use the specialists' jargon).

In ants, however, the process of formally naming minute, sometimes individual variants, was alive and healthy up to the 1920s. At its extreme the naming system had evolved from the binominal nomenclature introduced to zoology by Linnaeus in 1758 (which persists to this day) to a pentanominal system. That is, instead of a name of a taxon being composed solely of the combination of a genus name plus a species name, for example, *Homo sapiens* or, in the case of the honey bee, *Apis mellifera*, the name of the most finely separated groups was a combination of genus, subgenus, species, subspecies (or race), and variety: e.g., *Camponotus (Myrmocamelus) blandus* race *rosariensis* variety *mendozensis* described by Forel (1913). In modern systematics a subspecies is defined as a geographically restricted subdivision of a species. Other ranks that were used, sometimes basically as synonyms of subspecies and sometimes at ranks below that, i.e., within a subspecies, went by a range of names: variety, race, aberration, stirps, morph, etc. The practice of naming infrasubspecific taxa of ants had begun already in 1874 with the publication of the names (as *nomina nuda*) *Formica fusca* race *rufibarbis* variety *cinereoïdes* and *Formica rufa* race *pratensis* variety *rufoïdes* by Forel, and it reached its zenith in the early twentieth century. Modern practice, under the International Code of Zoological Nomenclature, deems such taxa and their names to be invalid. However, they were not expunged from the scientific record wholesale. Many have been "laundered" by using the same epithets at acceptable ranks and are considered to be valid names for species and subspecies.

TAXONOMIC PRACTICE

Distances and the trials of travel in the nineteenth century impacted the conduct of taxonomy. While today we, the authors, can collaborate in real time over a distance of 3,854 miles (6,202 km), both correspondence and physical travel took much more time. There was a transformation in the speed and comfort of travel in the early to mid-1800s due to the construction of canals, development of improved roads, and the expansion of the railroads. Nevertheless, an entomologist interested in studying specimens and understanding the taxonomic concepts of a colleague had two options. The first was to travel to the site of the collection, whether a personal cabinet, museum, or university. The second option was to rely upon the published descriptions.

The current rules of nomenclature stipulate that in order to validly name a new species that name must be accompanied by a statement purporting to distinguish it from others. This is a modern rule, but neither today nor in the past were there any regulations concerning how detailed, long or even accurate such a description must be. Linnaeus (1758) described the honey bee, distinguishing it from all other bees as: "pubescens, thorace subgriseo, abdomine fusco, pedibus posticis glabris utrunque margine ciliatis" (hairy, thorax grayish, abdomen dark, hind legs smooth, both margins fringed). Many descriptions, probably most, were not accompanied by any illustrations. The reader, therefore, had to rely on the verbal description in order to understand the author. The quality of a description is certainly not simply a function of its length. A perfectly functional diagnosis, that is a statement of how to distinguish the species in question, could in theory be as short as a word or two. The value of the diagnosis, rather, depends on the salience of the characters to the task at hand and the context of that task. Separating only two species—either because those are the only ones in that genus in an area, or because the author is ignorant of any others—can be done quite succinctly. On the other hand, recognizing one species out of five hundred others that it might be could require an extensive treatment. This difficulty in accurately sussing out the meaning of an author led to the situation where names were commonly misused and a name was often "qualified" by adding a *sensu* phrase, stipulating whose concept was being used.

Even by the end of the nineteenth century travel between America and Europe was relatively uncommon. For example, William H. Ashmead (1855–1908), hymenopterist at the US National Museum at the turn of the twentieth century, described over two hundred new genera. To do this he had to distinguish them from established genera described by European authors, one of the most important being Arnold Förster (1810–1884). Ashmead's understanding of what Förster meant his genera to be was based on the latter's powers of description and the single trip that Ashmead took to Berlin in 1890–1891. It is not surprising, therefore, that American and European entomologists often described the same taxa, particularly genera, in parallel. As an extreme and ironic case illustrating the impact of distance on communication among those describing new insects, in 1913 the Abbé Jean-Jacques Kieffer (1857–1925, a prolific worker in Hymenoptera and Diptera) described a new genus of parasitic wasps from West Africa and named it *Doddiella* in honor of the young Alan Parkhurst Dodd (1896–1981). Dodd was an avid taxonomist of insects in Australia, the son of another prominent entomologist, Frederick Parkhurst Dodd (1861–1937). He later achieved fame in a parallel field, biological control, in directing the work to introduce the moth *Cactoblastis cactorum* into Australia from Argentina to control the spread of prickly pear cactus. Kieffer, on the other hand, was the rector in Bitsche in Alsace-Lorraine. Kieffer's description was so inadequate for the purpose of identifying Australian insects that Alan Dodd could not recognize the insect that had been dedicated to him. As a result, in 1927 he redescribed the same genus from New South Wales, this time under the name *Aratala*!

By 1920 the inclusion of illustrations in taxonomic articles became more common, but they were still used only sparingly. Volume 22 (1920) of the *Proceedings of the Entomological Society of Washington*, an important organ of the researchers in the US National Museum included thirty-three articles, almost all of them taxonomic, in 246 pages. Seven plates of line drawings, one photographic plate, and nine pages with drawings or graphs were the limit of the illustrative material.

One of the pioneering influences of Latreille was the introduction of the idea stabilizing nomenclature through the use of type species of genera (Dupuis 1974). Instead of solely relying on imperfect wordsmithing to convey an author's taxonomic concept, a type species serves as an additional means to more concretely articulate one's thoughts. The type species would, ideally, be a species well known to others. The generic concept of the author is thus "anchored," that is, any genus in which this species was placed would also receive the name of the genus of which that species was the type. Note that, at least as this idea is used today, a type species is not constrained to be typical of the genus. This concept was then extended to apply to the level of species, and in this case the type of a species is a specimen. Explicit designation of type specimens by the authors of new species was gradually being absorbed as a standard operating procedure in taxonomy by the 1920s and, today, is a requirement.

ORGANIZING THE GROWTH IN KNOWLEDGE

The rapid growth in the number of species of all groups of insects inevitably led to specialization by researchers. Lindroth (1973) described this in general terms for the pre-Darwinian period. He cited John Obadiah Westwood (1805–1893) of Oxford University as the last of the "good" entomological polymaths. Thompson and Pape (2016) described how specialization occurred in the study of true flies, the Diptera. The sheer numbers of species and publications was too much for any single person to grapple with and led to an acute need for organization and structure.

Louis Agassiz (1807–1873), founder and first director of the Museum of Comparative Zoology, Harvard University, published a comprehensive list of all of the names of genera (1848) that had been used for animals, providing the name, the author of the name, and its place in the classification. In 1864 scientists from the Zoological Society of London and the British Museum founded *The Record of Zoological Literature*, aiming to catalog publications on animal biology, with a focus on biodiversity and systematics. Albrecht Günther (1830–1914) was the principal member, and editor of the first six volumes. He has been described as its "founding father and leading spirit" (Polaszek and Michel 2010). Albrecht Karl Ludwig Gotthilf Günther (anglicized as *Albert Charles Lewis Gotthilf Gunther*), was a German-born, later British, zoologist. He came to the British Museum in 1857, where he studied ichthyology, and was Keeper of Zoology at the British Museum from 1875 to 1895.

The *Record of Zoological Literature* was soon renamed the *Zoological Record*. Its aim was to serve as an annual index of all publications on animals, of which a large percentage were taxonomic in nature. In addition to the bibliographic data, the names of all newly described taxa were included. As the overall volume of publications increased the *Zoological Record* itself was forced to specialize, publishing separate volumes for each of the major groups of animals. For insects, these major groups largely corresponded to the rank of order. By founding *Zoological Record*, Günther and his colleagues initiated something apparently unstoppable. Some 155 years later, *Zoological Record* is still going strong, and currently indexes over seventy thousand items yearly published in over five thousand serials and non-serial sources, in many languages (Polaszek and Michel 2010).

During our century under scrutiny, 1820–1920, two major innovations profoundly affected entomology forever. From its genesis in the Stricklandian Code (named for Hugh Strickland, below), the first international code of zoological nomenclature, the "*Règles*" (as it came to be called, shorthand for *Règles Internationales de la Nomenclature Zoologique*), appeared in 1905. During much of the same period, though culminating in 1933, Charles Davies

Sherborn labored at, and published, the monumental *Index Animalium*, the foundation for modern biodiversity informatics (Polaszek and Michel 2010).

The process leading to the first zoological codes of nomenclature started in 1842, when the palaeontologist, malacologist, and ornithologist Hugh Strickland formed a committee to "consider of the rules by which the nomenclature of zoology may be established" (Melville 1995). This was the first attempt towards the establishment of a uniform and permanent set of rules (Strickland 1843). Charles Darwin and Richard Owen (first director of the Natural History Museum in London) were among Strickland's committee members. Later Kiesenwetter (1858), Lewis (1871, 1875), Sharp (1873), and Blanchard (1889) introduced alternative codes or changes to the Stricklandian code. While aiming to improve on previous versions, these changes inevitably caused disagreement among users. Such widespread disagreement regarding some of the foremost principles of the codes, e.g., priority, continues today.

The achievements of these successors to Linnaeus lacked authority; their codes were based on best practice and recommendations, and hence enforceable only by gentlemen's agreements. This situation persisted with some success, however, only until more taxonomists entered the fray. By the late nineteenth century the situation with describing and naming new taxa was becoming chaotic. This was changed drastically by the formation of the International Commission on Zoological Nomenclature (ICZN) in 1895. The formation of an international, democratically elected, body of experts to legislate and rule on matters of animal nomenclature was an important advance in zoological systematics. The Commission was founded at the International Congress of Zoology in Leiden in 1895, with Raphael Blanchard as the first president, continuing in this role until 1919. Despite much divided opinion within the Commission, in particular with respect to priority, the first international code, the *Règles* was published in 1905 (ICZN 1905). Two years later, the Congress of Zoology in Boston announced the intention of applying this newly published code to the scientific names of several thousand of the more commonly used animal names, in an attempt to effectively render them "set in stone." This has been regarded as the precursor of the ICZN's "Official Lists of Names and Works in Zoology" (Hemming 1958; Melville and Smith 1987; Melville 1995; Smith 2001) itself a precursor of the ZooBank register of animal names and nomenclatural acts (Polaszek et al. 2005; Pyle and Michel 2010; Dayrat 2010).

This double aim of trying to set in place a set of rules to stabilize past and future animal nomenclature, and to simultaneously provide an authoritative and lasting list of animal names, has been a constant feature of the efforts of generations of zoologists attempting to address this area. While Strickland, Blanchard, and their contemporaries were addressing the rules, others were also working towards a complete catalog of animal names.

The monumental work of cataloging all animal species, including insects, was that of Charles Davies Sherborn (1861–1942). *Index Animalium*, when completed in 1933, was the result of more than thirty years of ceaseless endeavor—eleven volumes (including the indexes) and more than nine thousand closely printed pages in small typeface, including all animal names from 1758–1850. For each name he cited, in abbreviated form, the bibliographic information needed to access the original description. Sherborn personally researched and checked information for *Index Animalium*, almost exclusively from the original sources (Polaszek and Michel 2010).

Sherborn's index documented only those names from the ninety-three years following Linnaeus, providing a firm foundation in the classical names, but leaving the subsequent years to the annual compilations of the *Zoological Record*. An attempt to catalog all insect genera was undertaken by the Belgian entomologist and ornithologist Philogène Wytsman (1866–1925). *Genera Insectorum* was intended to be a comprehensive treatment of all insects, with individual fascicles focused on a single group and authored by a leading specialist. Beyond a simple listing of all known species of all genera, a typical fascicle also included a classification, identification keys, descriptions (sometimes with new species), and fairly extensive illustrative material, often in color. A total of 219 fascicles were produced, beginning in 1902 and continuing, long after Wytsman's death, until 1970 (Evenhuis 1994).

The monographic series entitled *Das Tierreich* began publication in 1894 and continues to the present. Its scope includes all animals, with each volume focused on a single group. In the case of insects, these groups are often at the family level. Each such volume is similar in content to a fascicle of *Genera Insectorum*, except that descriptions of the individual species are provided.

Genera Insectorum did not cease publication because its original objectives had been achieved. Even those fascicles that were published did not become the endpoint of taxonomic research, but rather became the jumping-off point for even more species discovery. The goal of completely documenting insect diversity is still unfinished. The combined efforts of collectors, curators, engravers, editors, and of course the researchers themselves during the century between 1820 and 1920 vastly expanded our understanding of the scope of the insect fauna, increasing the number of known species ten to one hundred times. The actual number is not yet known as there still is no comprehensive catalog of what has been accomplished already. Taxonomy, as a science, no longer occupies the position of prominence that it had then, either. It must now compete for funding and professional positions with exciting new fields of inquiry. In this regard, the long history of taxonomic work probably works against it. Nevertheless, the task of documenting the plants, animals, fungi, and microorganisms with which we share the planet remains unfinished. The spirit of exploration and awe at the scale of the world's biodiversity that permeated the period of time that we have reviewed is still there today. The interrelated challenges of habitat loss, climate change, and human population growth add a sense of urgency to the goal of species discovery. It is still quite possible to discover something in one's backyard that no one has ever seen before. Those discoveries are the seed from which a better understanding of our world grows.

CHAPTER TWO

Insects and Disease

A Formidable War Alliance

MELISSA K. MILLER AND GARY L. MILLER

May He who has shielded you in the day of battle defend you from the Pestilence that walketh in darkness.

A father's letter to his son, in service,
Edward W. Bacon, 1862 (Burkhardt 2009: 74)

War pestilence has an ominous ring to it. The act of warring with all its misery and tragedy is as old as human civilization. But the addition of pestilence in the form of insects or insect-related disease adds another troublesome element to the mix. Insect problems and disease outbreaks within military operations were concerning enough but conditions were often ripe for civilian populations to suffer as well (Figure 2.1). Insects and their accompanying disease-causing organisms neither chose sides nor respected treaties. The opportunity of wartime conditions better served their basic biological needs.

The Age of Industry (1820–1920) is bookended by extremes. It begins nearly fifty years before proof of the germ theory by Louis Pasteur in 1867 and ends just prior to the discovery of penicillin by Alexander Fleming in 1928. Militarily, it begins shortly after the Napoleonic Wars (1803–1815) and ends after the First World War (1914–1918). Early in this period, military operations commenced with little or no knowledge of the cause or effective treatment for some of the major insect-borne diseases that would wreak havoc (Figure 2.2). For that matter, insects, if they were considered at all, were regarded only as

FIGURE 2.1 Petersburg, VA. Water-filled ditch on west side of Fort Sedgwick. Insects abetted the panoply of horrors in warfare. The confined spaces of the trenches with poor sanitation and drainage were bordered by a no man's land that held the dead and dying. The decomposing waste created prime breeding grounds for filth flies along with rodents and their associated fleas. This trench with its bombproof shelter (seen in background) was photographed outside of Petersburg, VA in 1865. But, the scene was unfortunately replicated during other wars. Library of Congress, LC-DIG-cwpb-02573.

an annoyance. But the insect ensemble and their arthropod kin (e.g., spiders, ticks, and crabs) that a soldier faced in the field could be daunting. The invading armies were themselves soon invaded. A member of the 7th Rhode Island Volunteers described the assault of his Mississippi bivouac during the American Civil War (1861–1865):

> By evening many tents were up and the camp was in fair condition but great was the complaint concerning vermin. It seemed as though all the insect life

FIGURE 2.2 Cholera "Tramples the victors & the vanquish'd both." The role of insects and disease was not fully recognized in the first half of the Age of Industry. Instead, mantles of miasmatic contagions were considered as the cause for many of the diseases. In 1831, Robert Seymour illustrated the specter of cholera inflicting death to all. US National Library of Medicine. US National Library of Health, NLM Unique ID: 101393375.

of the entire region had congregated here in anticipation of a glorious picnic. Necessarily our apartments were kept wide open hence they were free to all comers. Ants built catacombs beneath our couches, land crabs burrowed up through the fungus grown floor to inspect our resting places, woodticks climbed the tent walls, whence they could select the most favorable lodging place, flies covered our food as with sackcloth, and endeavored to rob us even of its scantiness, mosquitoes of unrivaled force and ferocity plied their lancets with merciless vigor, and, when they were appeased, rested on the ridgepole and mockingly barked at their victim until he went to sleep, great hairy spiders built nests in the peak, strange things whized [sic] and buzzed and boomed through the darkness, anon dropping on our faces with a sharp thud as if shot or alighting with sticky feet reluctant of dislodgment. All night long there was a rustling and a crackling of well nigh every type of winged and creeping abomination that earth produces.

(Hopkins 1903: 118)

By 1920, a transformation in attitude had occurred. Insects were viewed as more than a nuisance—they were another combatant.

With well over a million described species of insects, only a very small percentage are considered truly harmful to mankind. This chapter focuses on some members of this elite group (i.e., filth flies, fleas, lice, and biting flies) and underscores their role in human disease and suffering. We emphasize some of the experiences, conditions, and observations of those who encountered these six-legged legions firsthand. This chapter also underscores the indomitable human spirit, the ingenuity of those faced with great adversity, and how some persevered with optimism and a sense of humor.

FILTH FLIES

Here we have flies in regiments, in brigades, in divisions, and in corps, great armies of flies – big flies and little flies ...

Chaplin A.M. Stewart, 102nd PA Volunteers, Harrison's Landing, VA, 1862 (Stewart 1865: 208)

Filth flies are non-biting insects with a single pair of functional wings and sponging mouthparts. They defecate and regurgitate on their meal, lap up the soupy mixture, which may include disease causing organisms, and decamp on to the next meal. In addition, their feet (called tarsi) are covered with minute hair-like structures that can trap and transport pathogens from one landing stop to the next. Arthur Conan Doyle, of Sherlock Holmes fame, observed the flies in action when he served as a physician during the Second Boer War (1899–1902), "The air was poisoned by foul sewage and dark with obscene flies. They speckled the scanty food" (Conan Doyle 1900: 247). Conan Doyle's observation of fly-specked food had additional ramifications beyond the repulsiveness of fouled fare. Befouled food could have disastrous results in that filth flies spread diseases such as dysentery, cholera, and typhoid fever. These diseases are linked to unsanitary conditions including contaminated water, food, and feces (Figure 2.3). Lack of sanitation especially during wartime conditions, in turn, favored more filth flies.

There was ample opportunity for filth flies, such as the common housefly (*Musca domestica*), to thrive during military campaigns in the Age of Industry. Not only did they share the military fare, but the female flies found the associated detritus an enticing place for their offspring (maggots) to grow. The dung associated with cows, horses, and mules, which were often relied

FIGURE 2.3 Camp sanitation. Anti-fly campaigns infiltrated both the public and military awareness by the turn of the nineteenth century. This postcard, printed prior to the U.S. involvement in the First World War, overzealously underscored the importance fly control. R.R. Donnelley. Authors' collection.

on for food or transportation during much of the period, provided just such a nursery. To this add the daily excrement produced by humans as well as the offal related to encampments or the unburied dead from the battlefields and the breeding and feeding sites for filth flies was ripe for a fly population explosion. With overlapping fly generations, a single female fly can originate more than 5.6 trillion descendants in a single season (Howard 1911). This explosion is akin to compounding interest, except the payment would come in the form of annoyance, disease and debilitation, or death. The description of American Civil War camps would be appropriate for other theaters of war in the Age of Industry: "The emanations of slaughtered cattle and kitchen offal together with the noxious effluvia from the seething latrines and infested tents produced an olfactory sensation which has yet to be duplicated in the Western Hemisphere" (Brooks 1966: 108). Some would dismiss the smells as simply a "patriotic odor" (Steiner 1968: 17). However, the implications for attracting filth flies were huge.

It is hard to comprehend the obnoxiousness of the filth flies. A Confederate soldier during the American Civil War observed:

> When we open our eyes in the morning we find the canvas roofs and walls of our tents black with them [flies] ... It needs no morning reveille then to rouse the soldier from his slumbers. The tickling sensations about the ears, eyes,

> mouth, nose, etc., caused by the microscopic feet and inquisitive suckers of an army numerous as the sands of the sea shore will awaken a regiment of men from innocent sleep to wide-awake profanity more promptly than the near beat of the alarming drum.
>
> (Wiley 2008b: 248–9)

A few generations later and halfway around the world, a British medical officer confirmed the annoyance of "day flies" (filth flies) in First World War Mesopotamia.

> ... the whole roof of a tent at night might be uniformly black with them [flies] and eating was in the nature of a free fight. A couple of hundred or so to each tent was perhaps the average but they made rest a matter of difficulty.
>
> (Swayne 1917: 53)

It appears no aspect of military life was exempt—even military staging areas could be overrun with flies. The island of Imbros served as a staging post for the allied Mediterranean Expeditionary Force prior to the invasion of Gallipoli (1915). Flies were so prevalent and annoying that they drank fountain pens dry (Hamilton 1920). This would be a harbinger to the fly problems the Allied forces faced when they got to Gallipoli. Gunner Dudley Meneaud-Lissenburg of the Royal Field Artillery elaborated on his experience:

> We were invaded by millions of flies. There was no escape from these beastly insects. They swarmed around everywhere. Drinking and eating was a real nightmare and I avoided no matter how hungry I was rice pudding, which was served up frequently, mixed with currants and dehydrated fruit. It was difficult to distinguish currants from flies. They look alike in this repulsive mixture. Immediately the lid was taken off the dixie the flies would swarm down and settle on the rim in a cluster and many of them would fall into the pudding. The spreading of jam on to a hardtack biscuit was indeed a frustrating exercise. Driven by the pangs of hunger, the hated apricot jam was tolerated of sheer necessity. A concerted effort by at least three of us to transfer the jam from the tin on to the biscuit was necessary, one to open the tin, another to flick way the flies and a third to spread the jam and cover up. The ceilings of our bivouacs, a waterproof sheet, were black with flies crawling over each other and falling on top of one as you tried to rest.
>
> (Hart 2011: 229)

Large populations of flies were certainly an annoyance but diseases associated with them would have deadly implications as chances for disease transmission increased.

Recognition of the role played by arthropods in transmission of diseases at the turn of the nineteenth century focused attention on filth fly problems—especially related to the spread of typhoid. US Army training camps during the Spanish-American War (1898) were typhoid-ridden. The countless flies were breeding on exposed feces or in sewage pits that were often placed near the kitchen tents (Reed et al. 1904). Lime (calcium oxide or calcium hydroxide; used to reduce odor and accelerate waste decomposition) was spread on the pits, but Army cooks later noticed flies with whitened feet walking over food (Reed et al. 1904). Cavalrymen suffered even more from fly-related issues—not only from the proximity to their horses and associated manure but also from the visits from flies at the sewerage pits next to the kitchen area. "It was impossible to keep the flies from the already cooked food even if a man kept one hand over his plate and ate with the other" (Ealand 1915: 121). In at least one case, the fly problem was aggravated by the lack of tools to dig sinks (latrines) and the perceived threat of a louse infestation:

> When the camp was laid out, we had no tools for the digging of sinks. I personally visited the company streets and asked the men to use the ditch at our rear until we could get tools with which to dig sinks. Night after night the space between the Ninth New York and the Fifth Missouri was lined with deposits of fecal matter. Not having spades with which to cover this, it was allowed to lie there and dry.
>
> Flies became numerous, and as the kitchen was only 30 feet distant, it was only a question of time when the food would be contaminated. At last we obtained one pick and two shovels, and with these we dug a sink, but before we could finish and dig a second one the first one had been filled. Diarrheas now became common. Additional tools were obtained, and three battalion sinks were dug. Many men would not use the sinks on the plea that some of the men had body lice and left them on the seats. I obtained permission from the colonel of the regiment to arrest any man found defecating outside of the sinks. This order had some effect, but at night the nuisance was almost as bad as it had been in the daytime.
>
> (Reed et al. 1904: 222)

Walter Reed, who would also lead the US Army Yellow Fever Commission, raised awareness of sanitation as a major factor in fighting disease. Insect-borne diseases would sicken and kill many before adequate precautions were finally put in place.

Insect-borne diseases took a great toll on the fighting force, sometimes leaving entire units too sick to fight. Of Union deaths from disease during the American Civil War, 57,265 were attributed to dysentery or diarrhea and 29,336 were caused by typhoid (Clodfelter 2017). Deaths in the ranks due to dysentery or diarrhea reduced the fighting force by total elimination, but incapacitating a fighting force also had a profound impact on military effectiveness. One Union soldier recorded in his diary, "Sickest I ever was. My bowels moved 18 times in 3 hours" (Wiley 2008a: 136). Interestingly, typhoid fever progressively diminished as the American Civil War dragged on and disappeared almost entirely from veteran units (Wiley 2008a). Improved sanitation or fly control may have given some relief, but it is now understood that repeated exposure to *Salmonella typhi,* the causative agent of typhoid fever, infers a life-long immunity (Kakkilaya 2015).

While disease epidemics in theaters of war could spread to civilian populations, the reverse could happen as well. The prevalence of typhoid fever and dysentery outbreaks in troops during the Franco-Prussian War (1870–1871) was thought tied to their quartering in places where these diseases were already endemic (Prinzing 1916). Although the Spanish-American War lasted a little more than three months, more than 5,400 died of disease compared to 385 killed in action (Clodfelter 2017). Of total deaths during this conflict, more than 80 percent were attributed to typhoid fever; likewise, 30 percent of the deaths in the Boer War were attributed to the same affliction (Shipley 1915; Howard 1921). The human death toll from disease during war times was appalling but facilitated attention and research into the cause and prevention of insect related non-battle sickness and death.

British immunologist and bacteriologist Almroth Wright's development of an anti-typhoid fever treatment in 1895 led to experimental inoculations for British troops during the Boer War (Wright 1900). Although not without controversy, this would later have a direct benefit on British Expeditionary Forces (BEF) during the First World War, as the British Army were the only ones to enter the war being immunized against typhoid fever (Colebrook 1948). An effective anti-typhoid inoculum could have its downside in that fly control in camps became less important (Howard 1921) (Figure 2.4).

Cholera was another enteric disease with a fly connection. Generally associated with contaminated food and water, the bacterial agent *Vibrio cholera* can survive on the external surface of houseflies for five–seven days and inside the fly for its lifetime (Kotenok and Chicherin 1977). During the Crimean War (1853–1856), Allied (namely French and English) land forces, as well as the transport fleet, suffered from outbreaks of cholera. One physician diarist in the Crimean War interspersed his observations on

FIGURE 2.4 Hospital scene. American soldiers receive the typhoid vaccination during the Mexican Expedition (1916–1917). The vaccine was a breakthrough but entomologists such as the UDSA's L.O. Howard were concerned people would become complacent to the dangers of the house fly and poor sanitation. UCR Arts. Authors' collection.

flies between his passages on cholera stating, "it is difficult to avoid eating them [flies] in our food" (Robinson 1856: 133). Hospitals for cholera patients could be scenes of horror, "The cholera ward presented the most ghastly spectacle of helpless human suffering I ever witnessed. Poor wretches writhing in every posture indicative of agony their bodies covered with flies and no ministering hand to perform the slight office of humanity in keeping them off" (Robinson 1856: 114–15). The possible connection between flies and disease transmission was not fully realized but the revulsion was evident. Total Allied deaths attributed to cholera numbered nearly eighteen thousand.

So devastating was this disease that it was targeted for vaccine development. By the late 1800s a prophylactic vaccine was proven effective against this organism making it the first bacterial disease in man for which an effective vaccine was developed.

Having to endure fly problems was bad enough for healthy soldiers but those with open sores or wounds could soon be subjected to another horror—myiasis (fly larvae/maggots feeding on the body tissue of a live host). This condition, often encountered under war-time situations, is caused by the feeding preferences of blow flies—another type of non-biting fly. Adult flies are attracted to camp offal including the unburied dead, which during decomposition give off chemicals like putrescine and cadaverine. These chemicals and others associated with rotting flesh are a signal to adult flies that food and larval habitat are close at hand. In the scorching July heat of First World War Gallipoli, the following observation was made: "These flies were disgusting horrible pests, for they would come straight from the rotting corpses of the Turks, which lay in un-buried hundreds in front of our trenches ..." (Patterson 1916: 195). The female flies deposit their offspring in decaying flesh where the larvae (maggots) develop. Soldiers who couldn't be removed from the field could find their wound quickly infested or "fly-blown." Exposed wounds in triage units and hospital wards were also targets for the egg-depositing flies (Figure 2.5).

Medical staff were understandably concerned with myiasis. British doctors in the Crimean War found applications of creosote in water were most useful not only for deterring adult flies but also for killing the maggots in wounds (Shepherd 1991). During the American Civil War, both Union and Confederate doctors noted fly-blown conditions of exposed wounds. Years after the war Union surgeon, William Keen, remembered, "... I saw very many wounds swarming with maggots. They were very disgusting, but practically I do not think they did much harm" (Keen 1917: 135). He, like his fellow Union surgeons, would have used chloroform to remove maggots from wounds. Their Confederate counterparts did not always have access to those medical supplies. Confederate doctors accidently discovered the maggots were more efficient at digesting dead and decaying tissue than trying to remove it with scalpels or nitric acid. Besides digesting only necrotic tissue, maggots excrete allantoin which promotes wound healing. Confederate surgeon John Zacharias recorded:

> During my service in the hospital at Danville, Virginia, I first used maggots to remove the decayed tissue in hospital gangrene and with eminent satisfaction. In a single day, they [maggots] would clean a wound much better that any agents we had at our command.
>
> (Miller 1997: 231)

FIGURE 2.5 Wounded in a shelter tent three miles from Russian batteries. Wounded Japanese soldiers during the Russo-Japanese War (1905) shelter near Russian artillery. The photograph records not only head netting to deter bothersome insects but also documents what are apparently flies on the dressed wound and arm of one soldier and the legs and soles of footgear from another. Exposed wounds could become fly-blown if untreated or infected when exposed to microbes carried by filth flies. Hare (1905).

Some fifty years later, the battlefields of the First World War afforded another opportunity for myiasis. American physician William Baer observed wounded soldiers with huge maggot infestations but were asymptomatic of sepsis (Baer 2011). Baer's experiences led to his postwar pioneering work on maggot therapy for wound recovery and care. In the modern era of diminished antibiotics, the use of sterile maggots for wound treatment is deeply rooted in the suffering of war.

Non-biting flies of all types must have tried soldiers' patience and vexed their very souls. Yet there were some who evidently acclimated to the fly problem and epitomized the unquenchable human spirit.

> This morning instead of being awakened at 4.30 a.m. by a cluster of house-flies having a garden party on my nose I just opened one eye and looked at them running about outside my entrenchments then closed it and fell asleep again for an hour.
>
> (Hamilton 1920: 285)

At least one soldier could wreak havoc on annoying flies and gain a morbid form of entertainment at the same time.

> They [flies] settle in hundreds among the brush, which composes the roof of our mess house. We explode loose [gun] powder under them, which brings down a shower of dead and wounded [flies]. Still there are just as many more in a few minutes afterwards. I had two trees cut in front of my tent, on the stumps of which I made a sort of trough with stiff paper, [which] was filled with [gun] powder. The flies lit on the stump in the thousands to get at the running sap, [and] piled up on each other three deep, completely covering the top. With a piece of touch paper fastened to the end of a long switch or twig, I would blow these fellows up several times a day. But still there were just as many more there ten minutes afterwards!
>
> (Sneden 2002: 110)

Any type of distraction was a welcome relief in times of great hardship.

FLEAS

... nearly eaten up by the infernal fleas and other creeping things.

Diary of Pvt. J.J. Oswandel, Co. C., First Regiment PA Vols., Mexican-American War, 1847 (Oswandel 1885: 404)

Fleas, wingless blood-feeding insects, spend most of their adult life riding around on their host—often a rodent or an unlucky person on to which they may catapult themselves sticking tightly by way of their specialized legs and feet. As observed by a soldier in 1847, "Oh! I tell you, friends, they [fleas] stick to us like beeswax never letting go until squashed" (Oswandel 1885: 356). War correspondent Junius Browne wrote about the overwhelming flea problem during the American Civil War: "We were all covered with fleas. Fleas were everywhere. Tent, straw, books, blankets, valises, saddles swarmed with them.

The air scintillated with their blackness" (1866: 34). Fleas and their rodent hosts would be an integral part of soldiering during the Age of Industry.

In wartime conditions, rodents could reach high populations, each bringing uninvited guests to the party. Writing about his billet during the Crimean War, one British physician mused, "... mice in my room and fleas in my bed" (Robinson 1856: 110). During the First World War, rat populations reached epic proportions as the filth and fodder associated with trench warfare provided plenty of food and harborage for foraging rats.

> There was no proper system of waste disposal in trench life. Adding to the offal, empty tins of all kinds were flung away over the top on both sides of the trench. Millions of tins were thus available for all the rats in France and Belgium in hundreds of miles of trenches. During brief moments of quiet at night, one could hear a continuous rattle of tins moving against each other. The rats were turning them over.
>
> (Simkin 2017)

To compound the misery, along with the rats came their fleas.

> My fellow-lodgers and self have made a similtaneous [sic] and by no means agreeable discovery, viz.: the existence of certain little carnivorous animals, who elude discovery in the day time and at night dance over our unfortunate bodies and victimize us. Opinions are somewhat divided; one party who has effected the only capture, a bug asserting that *they* are the sole enemies; the other two on the strength of sensation only declaring they must be fleas.
>
> (Robinson 1856: 17)

The public health implications of this situation were not well understood until the transmission cycle of flea-borne diseases like plague and murine (flea-borne) typhus were worked out.

During the Russo-Turkish War (1828–1829), Russian troops quartered in Bucharest and surrounding villages exhibited symptoms of infection with the plague bacterium. With a combination of wartime overcrowding, subsequent troop movement and interactions with civilian populations, the disease was soon rampant. Misdiagnosis and lack of adequate hospital care left the spread of the disease unchecked. Like the patients they treated, most nurses and physicians died from the disease too (Prinzing 1916). One physician recorded the harrowing scenes:

> There was not a hospital not a quarter of the city not a division of troops not a family not a single place which had escaped infection and everywhere one came across victims of the pestilence some dying and some dead; for

> it spared neither sex nor age nor class. [He continued,] If the inside of the plague camp afforded a terrible sight, where sick men tossed about gasping in the burning summer heat, between the dead and dying the conditions outside of the plague-camp were no more pleasant to witness; for along the roads from the city to the hospitals, on fields and meadows, behind every shrub, in every ditch, dead and dying men were stretched out everywhere.
>
> (Prinzing 1916: 167–8)

Ignorance contributed to the chaos because the causative agent of plague was poorly understood prior to the end of the nineteenth century.

In 1894, Alandre Yersin observed the causative bacterium in human plague victims and in the lymph nodes of dead rats. Subsequently the whole picture came in to focus when in 1898, Paul-Louis Simond dissected fleas and observed the plague bacterium in the gut of fleas from dying rats. He later experimentally proved fleas as the intermediary in the disease process making the connection between the rat, the flea, and plague in man. The very mobile lifestyle of the flea and its propensity to leave a dead host and seek a blood meal from a living host makes it the perfect mechanism to transmit a zoonotic disease. Wilbur Hinman, an American Civil war veteran and popular writer, alluded to this very thing when he penned the following in 1887:

> This category would be incomplete without the nimble flea—the kangaroo of the insect world. The peculiarity of the flea is his jumping propensity, and the consequent difficulty of catching him. In this respect the flea is wiser and smarter than his fellows. Most of the bugs and insects that pester the human family become so absorbed in their biting and blood-sucking that they are wholly oblivious of personal safety. While they are gorging themselves they think of nothing else, till there comes a well-directed slap, and they are no more. But it isn't so with the flea—"put your finger on him and he isn't there." He is a believer in the Hudibrastic theory, that
>
> He who bites and runs away
> May live to bite another day.
>
> He keeps the danger flag flying when upon his forays, and if his quick eye detects a hostile demonstration he gives one of those jumps that have made his name a proverb.
>
> (Hinman 1887: 260)

The nimble flea was only one of many insects, by themselves or as a group, that were at the very least a topic of literature and conversation and at the worst, part of a disease cycle (Figure 2.6).

FIGURE 2.6 The Rebel Flea. A Union political cartoonist likened Confederate General P.G.T. Beauregard to the flea after Southern forces withdrew from Corinth, Mississippi in May 1862. The sub-caption reflecting the jumping abilities of the flea was also observed in other wars. Insects were universal in their ability to span reactions and associations. *Harper's Weekly* 6, no. 289 (1862): 435. Authors' collection.

Sometimes fleas attacked in concert with other insects. A Pennsylvania Volunteer soldier during the Mexican-American War (1846–1848) reflected in his diary,

> These pests [fleas] are getting so saucy and determined, that they and the *piojo* (lice) have regular pitched battles with one another on our bodies. But of the two pests, give me fleas in preference to *piojo*; but I am like all my comrades, troubled with both sometimes.
>
> (Oswandel 1885: 356)

Indeed, a combined infestation would be a force to be reckoned with.

LICE

I've slept in dugouts with the rats and the louse
And I've slept in France in a fairly good house.

Frank P. Jarvis, First Canadian Mounted Rifles,
part of a poem to his brother, 1918 (Brown 1919: 61)

Perhaps contemporary writer Emily Willingham (2011) best summarized the topic of lice and louse-borne disease when she wrote, "The history of lice and men is a long, itchy story that intertwines co-evolution not only with these blood-sucking little parasites but also with the microbes they carry." Unlike many other biting insects, human body and head lice are very much specialists, preferring to dine exclusively on human blood (Figure 2.7). While feeding, they cause an itching sensation that is almost unbearable. This leads to scratching during which unfortunate victims may infect themselves by rubbing a disease organism into the bite site.

> One of the worst features connected with cooties [lice] is that you cannot tell who had them last. They may even be of German origin after you have slept in captured German dugouts. While you soon learn to reach a state of toleration for allied cooties, you itch worse at the thought that they may be German.
>
> (Bettcher 1918: 223)

German or not, lice were a constant presence and source of irritation (Figure 2.8).

Lice were so common in the Age of Industry; they were part of the daily military lexicon. During the American Civil War, the body louse had many camp names but was most commonly called grayback (referring to its color), bodyguard, or vermin (Miller 1997: 231). Lice even crept into a Confederate artillerist prayer as he retired for the night:

> Now I lay me down to sleep,
> While gray-backs oe'r my body creep;
> If I should die before I wake,
> I pray the Lord their jaws to break.
>
> (Wiley 2008b: 251)

By the First World War, the list of aliases for lice also included chat, cooties, or the French term "totos" (Farrow 1918). Whatever body lice were called, at the very least they provided some level of consternation and discomfort, but

FIGURE 2.8 Läuse. This 1916 German postcard was postmarked from a "Fieldpoststation" probably on the Eastern front. The billeted soldier is surrounded by a family all busily scratching from their louse infestations. Military and civilians were all suffering from lice. The insect illustrations in the lower corners served to familiarize folks on the home front. Authors' collection.

at the very worst, were important vectors of diseases such as epidemic typhus, trench fever, and epidemic relapsing fever.

The inability to bathe or change uniforms regularly meant some means of delousing was necessary. So great was this problem in First World War Flanders that the following account was given:

> That vermin swarmed upon the bodies of all our boys who went into the trenches and tortured them. After three days they were lousy from head to foot. After three weeks they were walking menageries. To English boys from clean homes to young officers who had been brought up in the religion of the morning tub this was one of the worst horrors of war. They were disgusted with themselves. Their own bodies were revolting to them. Scores of times I have seen battalions of men just out of battle stripping themselves and hunting in their shirts for the foul beast. They had a technical name for this hunter's job. They called it chatting.
>
> (Gibbs 1920: 122)

A common method for individuals to delouse was to remove portions of the uniform and gradually work along the seams and folds of clothing to crush the hiding lice and their eggs (nits). This time-consuming exercise, called "skirmishing" during the American Civil War (Miller 1997) or "chatting" during the First World War (Gibbs 1920), has been employed by generations of affected military personnel. The task of physically delousing, also called "shirt hunting," often included conversation with fellow combatants and for many, it became a social event.

> During these hunts there are lots of pertinent remarks bandied back and forth among the explorers, such as, "Say, Bill, I'll swap you two little ones for a big one," or, "I've got a black one here that looks like Kaiser Bill."
>
> (Empey 1917: 22–3)

In the Boer War, delousing at one's leisure was one thing, but many could become adept at doing it on the march,

> ... as soon as a [British] regiment halted they stripped to the skin, turned their cloths inside out, and picked the Anoplura [lice] off. As a private said to me: "We strips and we picks 'em off and places 'em in the sun, and it kind o' breaks the little beggars' 'earts!"
>
> (Shipley 1915: 11)

Physical delousing gave context to soldier downtime but a more efficient method was needed.

When "chatting" or "shirt hunts" were just not enough, other swifter delousing methods were tried that often involved heat or fire. Some resourceful soldiers even used the company cook pot for the job with the following justification:

> So, in regard to using our mess kettles to boil clothes in it might be asked "Why not?" Were they not used to boil our meat and potatoes in, to make our bean, pea, and meat soups in, to boil our tea and coffee in, to make our apple and peach sauce in? Why not use them as wash-boilers? Well, "gentle reader," while it might at first interfere some-what with your appetite to have your food cooked in the wash boiler, you would soon get used to it; and so this complex use of the mess kettles soon ceased to affect the appetite, or to shock the sense of propriety of the average soldier as to the eternal fitness of things, for he was often compelled by circumstances to endure much greater improprieties.
>
> (Billings 1888: 84)

Others used a technique called "singeing," whereby garments were held over a fire until the lice were killed. This reminded one Confederate soldier whose clothes were "well stocked with big fat fellows [lice]" of popping corn (Wiley 2008b: 250). But campfires could draw enemy gunfire if they were too close to the front line or they were impractical in the confines of a trench. To get around this problem, a candle flame was used instead. An American serving with the BEF described the scene in this way:

> At night about half an hour before "lights out," you can see the Tommies grouped around a candle, trying, in its dim light, to rid their underwear of the vermin. A popular and very quick method is to take your shirt and drawers, and run the seams back and forward in the flame from the candle and burn them out. This practice is dangerous, because you are liable to burn holes in the garments if you are not careful.
>
> (Empey 1917: 21)

Heat treatments (i.e., flame or boiling) were proven effective for lice removal although often deleterious to clothing. Open flames could burn one's garment or if the uniform was made of wool (e.g., the standard Union uniform of the American Civil War), it might shrink to an unwearable size when boiled (Miller and Adler 2009). Being ever resourceful, eradication by cold treatment was also used by soldiers with varying degrees of success. During the Crimean War, one writer noted, "The only remedy available was to put the wet flannels in the snow for two days this killing all but the nits" (Pierce and Hutchinson 1921: 312). A similar technique was used by prisoners of war in the First World War but was met with less success (Depew 1918). Whatever the method used, personal delousing was hardly enough and soldiers were quickly re-infested.

Physical delousing remained a common technique but by the American Civil War chemical methods were coming of age. Soldiers painted their clothing seams with "blue ointment"—a compound that contained mercury and dusted their clothing or bedrolls using powders mixed with arsenic. These noxious concoctions were purchased from sutlers (traveling merchants) or sent from home in relief packages. British troops during the First World War had a favorite treatment:

> Recruits generally sent to Blighty [Britain] for a brand of insect powder advertised as "Good for body lice." The advertisement is quite right; the powder is good for "cooties" they simply thrive on it.
>
> (Empey 1917: 21)

By the First World War, louse control was part of governments' war research efforts. This resulted in new, if sometimes ineffective, methods being tried to reduce the louse problem. An American in the French Foreign Legion elaborated on the French issue:

> We were given bottles of strong liquid which smelled like lysol and were supposed to soak our clothes in it. It was thought that the cooties would object to the smell and quit work. Well a cootie that could stand our clothes without the dope on them would not be bothered by a little thing like this stuff. Also our clothes got so sour and horrible smelling that they hurt our noses worse than the cooties. They certainly were game little devils and came right back at us.
>
> (Depew 1918: 47)

It was rapidly becoming evident that mass delousing was necessary.

Recognition of the scope and importance of the problem with lice led to a proliferation of research among the combatant countries during the First World War and immediately afterwards (Grinnell and Hawes 1943). "It was not long before the German armies found the louse a very live problem and their scientific journals are full of papers on the control of the vermin" (Pierce and Hutchinson 1921: 312–13). The United States Department of Agriculture's (USDA) Bureau of Entomology explored the use of gasses of warfare as fumigants for louse control. Cooperative research between the US War Department and USDA developed dry cleaning methods for controlling lice in clothes and explored the use of oil for controlling nits. Using highly volatile chemicals for louse control sometimes had unintended consequences—like causing the delousing station and the soldiers' clothes to go up in flames (328th Field Artillery 1919). In a more novel approach, a high-frequency generator to control the pest was tried with little success (Howard 1919). War was fueling research into finding answers for louse control.

The First World War delousing stations sprang up in the theater of operations to include sites that had heat and steam sterilization (Pierce 1921) as well as mass showers. US soldier Walter Myers wrote home to his family in 1918, "Had a bunch of 'cooties' not long ago. However the Red Cross has places around here where you can steam your cloths. They [lice] only lasted on me about a day. Believe me, I sure got rid of them quick. Pretty damned uncomfortable." Unfortunately for "Soldier Bill" as he called himself, Myers was soon to become accustomed to having a constant louse infestation saying, "I have 'em most of the time. But they aint so worse after you have 'em awhile. I 'kinda' got used to 'em" (Myers 1918). Wary that transported troops could leave behind lice, the Germans even developed large vacuum tubes to hold entire railroad cars (Pierce 1921). But, for all the abatement initiatives, lice would remain entrenched for the war and for a time afterwards.

On the lighter side, lice provided some forms of amusement. As far back as the American Civil War, soldiers would fill their downtime with louse races and even made wagers on the outcome. A prisoner of war from the 40th Georgia Infantry observed this practice while held at Camp Chase in Columbus, Ohio (Ouchley 2010). An American Marine in the First World War described the ritual in detail:

> Some guys would pass the time with a louse race. You'd take a frying pan and make a two chalk marks in it, a starting line and a finish line. Then you'd pick two lice from your body and bet on which one would win. If you got the pan too hot they wouldn't run and they'd just pop. They didn't run so much as hopped anyway. We spent long hours picking lice and throwing them in a hot pan.
>
> (Eggleston 2016: 41, 43)

Lice occupied much of the soldiers' time. Whether soldiers were killing lice, suffering from the diseases they carried, fretting over their overwhelming numbers, or using them as a form of distraction or entertainment, lice were synonymous with soldiering.

BITING FLIES

> *... in the pestilential swamps of Louisiana, filled with alligators, mosquitos, and malaria. The regiment lost more men by disease than it probably would have lost in severe battles.*
>
> Union General I.H. Elliott (Elliott 1902: 56)

Continuing with the saga of insects that feed on man, we come to true biting flies, of which the mosquito is a major player. Battlefield experiences and the savagery of mosquitoes were indelibly etched in the memories of many. Fitted with a sharp stylet for mouth parts, the elegant and mobile female mosquito has tormented mankind for millennia. An Illinois regimental historian in the American Civil War commented on the hardships and supernatural abilities of mosquitoes in Louisiana bayou country:

> All suffered intensely from mosquitos. They were of enormous size, and the soldiers said they did not buzz, but barked. Their ferocity was beyond describing, and there was no living at night except under mosquito bars [nets], and the savage monsters sometimes tore holes in these.
>
> (Elliott 1902: 56)

The ever-present mosquito chooses her victim with the precision of a heat-seeking missile, lands on a patch of exposed skin, and settles in for a meal. She injects a specialized cocktail of chemicals that thins the victim's blood and may elicit a localized bite site reaction. Sucking up the blood, this meal nourishes her burden of eggs. If the mosquito has acquired a disease-causing organism in a previous blood meal, she may complete the disease transmission cycle by injecting the infectious agent into her new victim. The danger of the mosquito would be proven during the Age of Industry.

Early in the Age of Industry, human malaria was recognized but its cause was largely attributed to "night" or "bad" air (hence the name) (Figure 2.9). Scientists of the day had ever-improving tools and techniques with which monumental discoveries in the malaria disease life cycle were made. German physician Heinrich Meckel first visualized the malaria parasite, Ronald Ross in 1897 proved the mosquito as the vector (transmitter) of the parasite, and Giovanni Grassi transmitted malaria to a human volunteer in 1900 (Kakkilaya 2015). Understanding the transmission cycle of an important insect-borne disease affecting man had gone from noxious vapors to scientific proof.

Centuries of anti-malarial concoctions and cure-alls did little to thwart the disease. This changed in 1820 when two French pharmacists, Joseph Pelletier and Joseph Caventou, perfected a method to purify a yellow gum from the bark of the cinchona tree. They named it quinine after the Peruvian Indian name for the bark from which it was distilled. Quinine became the wonder drug of its day. It was effective in treating the symptoms of and preventing malaria. Pelletier and Caventou received many honors but very little monetary compensation because they published their techniques so that anyone could manufacture the compound (Kakkilaya 2015). The fact that anyone could produce quinine meant it was added to many patent medicines and potions. It became a dubious ingredient and panacea to treat other ailments including syphilis, rheumatism,

OUR NEW MAJOR-GENERAL.

"Our faithful old Ally of the North, GENERAL JACK FROST, shall come and clear away the Malaria of the South, and we shall march Southward from this place, and there shall be no footsteps backward until this Rebellion is crushed out of this Union."—*Speech of* Major-General BUTLER.

FIGURE 2.9 *Our New Major-general.* In 1861, *Harper's Weekly* illustrated General Jack Frost clearing away malaria for the Union's march southward. Cooler weather meant mosquito-borne diseases would diminish but the connection between mosquitoes and their associated diseases were not fully understood at the time. *Harper's Weekly* 5, no. 249 (1861): 640. Authors' collection.

neuralgia, diarrhea, and various fevers (Brooks 1966). Quinine was soon a much sought after miracle drug that was an important part of military medicine in its time.

Mosquitoes, malaria, and quinine became part of soldering in the Age of Industry with mosquitoes and malaria even impeding military campaigns (Steiner 1968). But sometimes mosquitoes and malaria also were used as intended weapons. During the American Civil War, Southern soldiers modified the environment to expose Northern soldiers to the "bad air" or miasmas.

> As soon as the swamps (the outlet to which had been dammed by the Rebels) were drained off dysentery of a most fatal form became prevalent amongst the soldiers. Its onset was of a decided malarial character and quinine speedily removed its periodicity but exercised no beneficial influence over the graver part of the disease.
>
> (Hunt 1867: 809)

While draining the swamps, the Northern soldiers likely exposed themselves to malarious mosquitoes that may have fed on their malaria-infected Southern brothers. The soggy, garbage-polluted area that the previous encampment left behind set the scene for transmission of vector-borne as well as filth-related diseases like dysentery. The fact that quinine did not completely alleviate the symptoms indicates either a poor product or some other pathogen was at work as well. By the First World War, the advantages of quinine as not only a treatment but also as a preventive were known:

> We had, for the men, quinine parades, when five grains were swallowed as a prophylactic against malaria every day. They were amusing affairs to watch – serried ranks with water bottles, standing to attention while the sergeant dispenser walked with proper dignity down the line handing a pink tablet to each man, who gulped it spasmodically, took a draught of water and returned to attention. It reminded one of a religious ceremony, of some strange communion service.
>
> (Swayne 1917: 112)

Quinine although effective was not always readily available and its treatment properties were not always appreciated.

Malaria was an unexpected adversary in the First World War (Brabin 2014). Hungry and mobile, mosquitoes knew no boundaries for spreading the malaria parasite from army to army and crossing into the civilian population at will. The effect, in terms of military readiness, soon became evident. By the end of 1916, 17 percent of the British forces were incapacitated by malaria and by 1918, this had risen to 39 percent (Withers and Craig 2003). The mosquito became a peacemaker, "leaving opposing forces too sick to fight" (Lockwood 2009: 69). In Macedonia during the First World War, British forces requested relief from nearby French forces only to have the French Commander respond, "Regret that my army is in the hospital with malaria." Fortunately for the Allies, the Central forces did not attack because they too were suffering from the disease (Spielman and D'Antonio 2001). But on other fronts in the First World War, mosquito control and establishment of non-malarial areas was part of the BEF offensive plan prior to full engagement of enemy forces (Alexander and Dunkle 2017). Some soldiers suffered from what physicians called "malaria defeatism" where the thought of being infected with malaria was preferable to facing battle. So horrible were the conditions in Italy that some avoided taking their preventive quinine in hopes they would become sick and sent home on convalescent leave (Snowden 2008). Mosquitoes were influencing both physical and mental wellbeing of fighting forces.

The mosquitoes' arsenal was not limited to malaria. Other mosquito-borne diseases also had war-time implications, with yellow fever being one of them. Named for the jaundiced appearance of some patients, the causative agent of yellow fever is frequently transmitted by the mosquito, *Aedes aegypti*. During the American Civil War, there were limited yellow fever outbreaks. But previous devastating outbreaks in various port cities indelibly etched fear into the military and civilian populations; "Yellow fever was dreaded because of its mysterious origins and high virulence" (Steiner 1968: 14). Many of the yellow fever cases entered from the Caribbean area. When outbreaks were discovered on ships, the ominous yellow signal flag or yellow jack was raised to warn others of onboard disease. A Union Navy Surgeon's Steward recorded in his dairy, "Up goes our yellow flag and no more intercourse with the outside world til we are adjudged free from contagion" (Earp 2002: 77). Quarantine helped control yellow fever outbreaks but the disease was not understood (Steiner 1968). Some thirty years later, yellow fever (along with malaria and typhus) ravaged American forces during and after the Spanish-American War. With yellow fever posing a continued threat to US forces after the war, the Army established the US Army Yellow Fever Commission and appointed Major Walter Reed to chair the body. Experimental findings from the commission resulted in breakthroughs toward the understanding, control, and management of yellow fever and its associated mosquitoes. Scientific research helped allay the mysteries of the disease but the human cost was felt for many generations.

Scientific discoveries aside, personal protective measures against mosquitoes were often less than optimal. Protection consisted of mosquito bars, messy ointments of dubious efficacy, as well as head nets and gloves that were not compatible with combat operations. Preventive measures during the First World War also often failed to protect soldiers from the relentless mosquito onslaught. Problems were confounded by negligence, skepticism, and often contradictory instructions given by officers (Brabin 2014). A soldier recorded the shortcomings of mosquito nets saying, "Inside the mosquito net you see various gorged little insects struggling to get out of the meshing through which they passed so easily when they were slim and hungry" (Swayne 1917: 53). It was probably not the mesh sizing at fault but, the failure of an improperly used or torn mosquito net to adequately cover a bivouac. Further compounding the failure were operational shortages that delayed replacement of torn or missing mosquito nets (Brabin 2014). Even with their shortcomings, mosquito nets were a valued commodity for soldiers and civilians alike (Figure 2.10). In the malarious regions of Macedonia during the First World War, General Milne of the BEF remarked mosquito nets were "as important as the rifle" but were in short supply (Withers and Craig 2003: 29). Local thieves in Mesopotamia robbed military personnel of their nets at knifepoint, "It was a good policy to remain motionless for a short time. It was better than remaining motionless for ever [*sic*]" (Swayne 1917: 133). Personal protective measures, like mosquito netting, was a valued

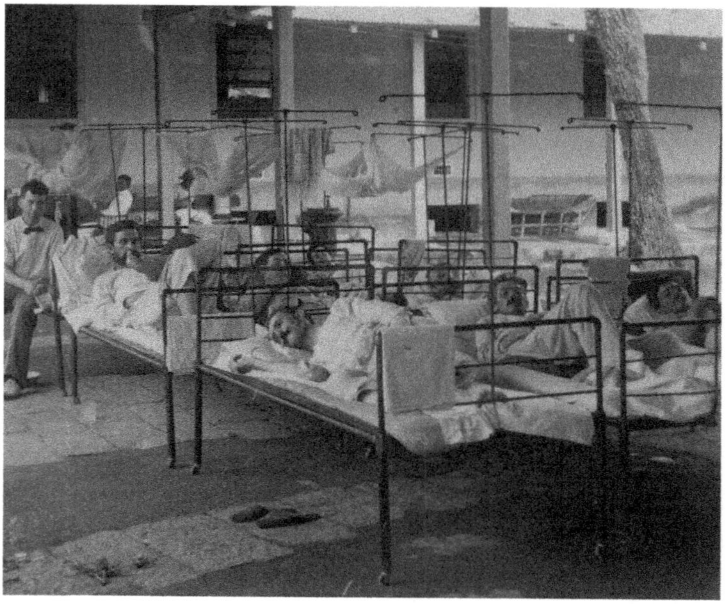

FIGURES 2.10a AND 2.10b Top: Washington, District of Columbia. Interior of ward in Harewood Hospital. Mosquito bars or nets shown in a soldiers' hospital ward in 1864 were used only as a relief from the annoyance flying insects. Library of Congress LC-B815-1006; Bottom: Stricken with fever. By the end of the Philippine-American War (1899–1902), mosquito bars were being used as an insect deterrent and a means to reduce transmission of insect-borne disease such as malaria and yellow fever. Here, some soldiers in a fever ward in Manila, Philippines are attended by both orderlies and netting in 1900. Authors' collection.

accoutrement and used as a physical barrier against larger flying insects, but did little to reduce insect populations.

As if other flies were not worry enough, in more tropical regions, like First World War Mesopotamia, soldiers had to contend with the almost invisible sandfly or biting midge. Writing under a pseudonym, Captain Maurice Nicoll of the Royal Army Medical Corps penned the following:

> Small almost impalpable colourless insects whose bite is like a red hot wire and who can penetrate the meshes of an ordinary mosquito net with ease began to infest the place. They are surely the most successfully maddening insect ever designed by the Lord of Flies. They give rise to a malady known as sand fly fever which is like influenza and drains the body of all vitality for many days.
>
> (Swayne 1917: 44)

He goes on to further describe the misery.

> For many days after this we moved about as it were in a vast furnace. The nights were broken by sand flies. Personally I found the only way of keeping them out was to wear socks on the feet and hands and smear the face and neck with some kind of ointment on which their feet slip, so that they cannot find a purchase when in the act of driving their sucking apparatus into the skin. In the morning, what with the sweat and the grease, and the tropical exhaustion, one looked like few things on earth. Oil of citronella is only of temporary use; paraffin and creosote are of little good. Butter muslin nets are out of the question, as the heat is stifling under them. The burning of aromatic or pungent compounds is useless, and as for killing them one might lie awake all night, scuffling and dabbing and slapping at the almost invisible forms without gaining the slightest benefit.
>
> (Swayne 1917: 109–10)

While a mosquito net or canopy could provide a physical barrier against some insects, the netting also blocked any cooling breezes and needed sleep could be even more difficult in sweltering conditions. Regardless of size, biting flies bedeviled soldiers during the Age of Industry, rendering some too sick or exhausted to fight and contributing to non-battle related injuries and deaths.

METAMORPHOSIS

The hum of the mosquito and the buzz of the house fly have become fraught with an entirely new significance.

Professor Glenn W. Herrick, Cornell University (1913: 1)

Great scientific and technological advances were made during the Age of Industry. On its outset, the Age was shrouded in miasmas and contagions of "typho-malarial" or "putro-adynamic" fevers. But cascading discoveries and advances throughout this period illuminated the relationship of insects, disease, and aspects of prevention. Patrick Manson's discovery of a filarial parasite associated with a mosquito set the pace in 1877. Within a brief time, other researchers were confirming arthropod's (i.e., ticks and insects) connections with many diseases. By the 1890s Theobald Smith and Frederick Kilbourne's discovery of the infectious organism and tick involvement in the transmission cycle of the organism causing Texas cattle fever played a major role in understanding how arthropods and disease were connected. Ronald Ross, Walter Reed, and Charles Nicolle produced major breakthroughs in malaria, yellow fever, and epidemic typhus, respectively. The creation of the USDA in 1862 highlighted strong public support for entomology where "These midcentury entomologists were the first to combine a viable entomological community with vigorous institutions" (Sorensen 1995: 259). Forged amid the American Civil War, the USDA named Towend Glover as an entomologist within the new organization.[1] Likewise, Francis Walker was appointed to The British Museum and Harold Lefroy was appointed as the first government entomologist to India. Advances in microscopy coupled with the way scientific exploration was conducted at the time opened the flood gates for monumental discoveries changing the way people viewed insects and disease. In 1913, American pioneer of home economics Martha Van Rensselaer included the importance of insects in her course for the farm home:

> Until within the last few years the economic importance of insects was due mainly to their indirect injuries to man through attacks on things that he produces. Suddenly, almost within the last decade, insects have assumed an entirely new and exceedingly important significance through their direct injuries to man himself.
>
> (Herrick 1913: 1)

The idea of "germs with wings" had taken hold in the public psyche (Tomes 1998: 145).

Understanding of the role of insects and disease was transformative to military culture as well as representing a pivotal point in terms of military sanitation. The Crimean War is acknowledged as the first major conflict for which data on the specific causes of military losses are available (Smallman-Raynor 2004). The efforts of Florence Nightingale, through her meticulous record keeping and nursing activities during the Crimean War, were responsible for focusing on this aspect (Christopher and Gill 2005). Preventive measures could be adapted to maintain or improve the health of the fighting force (Prinzing 1916). Even though insects abounded in most aspects of military life, entomologists were non-existent in military service and insect-related issues were tasked to military doctors (e.g., Major Walter Reed). By the First World War, the role of microbes, and vectors in disease, was recognized and the importance of entomology to the military was understood. The BEF added two entomologists to each of their sanitary units (Howard 1919). For the first time, the US Army commissioned eight individuals as officers and had several non-commissioned officers dedicated to insect control in military camps (Shultz 1991). Understanding the role of insects during this period was not restricted to commissioned or non-commissioned military officers; charity organizations, such as the Red Cross, also were involved. They assisted with insect-related issues ranging from sophisticated delousing stations to simpler relief methods from flies. The measures did not go unnoticed among the troops, "The Red Cross fortunately supplied us with instruments of fly destruction and later on fly experts were sent out" (Swayne 1917: 47).

By the end of the nineteenth century, it was also becoming clear that insect-related issues affecting military operations would require large government efforts. For example, in addition to its agricultural work, the USDA was also responsible for conducting "Investigations of insects in relation to disease of humans and animals and as animal parasites" as part of its mission as early as 1904 (Henneberry 2008: 16). In the First World War, L.O. Howard (Chief of the Bureau of Entomology, USDA) organized the Bureau's entomological focus toward insects to directly further the war effort. The Bureau of Entomology was to serve as a central clearinghouse on insect problems for state and federal authorities and would also publish on the topics. Howard specified, "With this information it will be possible to conduct a vigorous campaign against threatening pests" (Howard 1917: 229). The USDA's Bureau of Entomology worked in close cooperation with the Office of the Surgeon General of the Army and the National Research Council's Committee on Medicine. Research on louse-related issues, stored-product pests, and forest pests provided valuable information.

The importance of insect collections and their associated curators was also realized as part of this large government involvement. During the First World War, entomologists at the Natural History Museum (British Museum) played a critical role in the identification and education related to insects affecting both civilian and military health. Publications and popular posters on the life history and the diseases spread by houseflies, lice, fleas, mosquitoes, and sandflies were produced and distributed by the tens of thousands (Shindler 2018). In the Museum's exhibition halls, larger-than-life models of these insects also were made for both public display and military training purposes (Shindler 2018). Minute insects garnered enormous attention.

In the years immediately after the First World War, the importance of insects and entomologists had risen to the level where their work was considered an important part of camp sanitation and military operations.

> The work against vermin [insect pests] almost necessitates a specialist to take care of it alone. In fact it were best if three entomologists were located in each camp, one looking after the suppression of water and moist earth breeding insects, one looking after the suppression of fecal waste and manure breeding insects and the third handling the vermin of the person and the barracks.
>
> (Pierce 1921: 45)

The idea of even having a single entomologist would have been unthinkable a few decades earlier; it became a necessity by the 1920s.

Disease, adversity, and insects were intertwined with human conflict in the Age of Industry and there is little doubt that insects played an important, albeit often overlooked role on human culture. Insects were part of the heartbreak, hardship, and the humor. Perhaps the spirit of the Confederate camp song *Goober Peas* (peanuts) best illustrates this unique cultural mix:

> I think my song has lasted almost long enough,
> The subject's interesting but the rhymes are mighty rough,
> I wish this war was over when free from rags and fleas,
> We'll kiss our wives and sweethearts and gobble goober peas.
>
> (La Bree 1898: 554)

CHAPTER THREE

Insects and Food

A Global View of Edible Insects

FLORENCE VACCARELLO DUNKEL

Three essential understandings are important for readers as they begin this chapter. First, edible insect harvesting predates the evolution of *Homo sapiens*. This we understand from the fossil record of tool development and use by pre-Hominins (Lesnik 2018).

Second, the commercial silk production industry has provided a continual supply of insects for food for the past six thousand years. While the Chinese were industrially producing silk for millennia, they were also simultaneously producing the food insect, silk moth pupae, as a by-product on an industrial scale. Silk production originated in China in the Neolithic (Yangshao culture, fourth millennium BCE) and was confined to China until the Silk Road opened during the latter half of the first millennium BCE. China maintained its virtual monopoly over silk production for another thousand years. As we open the century 1820 to 1920, little has changed in six thousand years with this commercial production of food insects, mainly *Bombyx mori* pupae. The one exception during this period was that the industrial process of silk production ceased being a royal secret of the Emperor and Empress of China which, if leaked, was punishable by death. Eventually, commercial silk production spread from China to throughout Asia, to Europe and the US.

Third, while the culture of silkworms for silk production (sericulture) was providing a commercial source of human food in Asia, Native Americans, particularly in western North America in the Great Basin west of the Rocky Mountains, were harvesting insects for food on a near-industrial scale.

In this chapter we will move systematically around the world beginning in the Great Basin of North America and explore the collection, production, and use of food insects. At each geographical "port of call" we will reflect on the waves of the Industrial Revolution that were surging in leap-frog pattern globally. In some cases, this industrialization affected edible insect production and use significantly, in other cases it had no observable effect.

EDIBLE INSECTS AND THE US INDUSTRIAL REVOLUTION

As the nineteenth century began, several factors converged that created a favorable environment in the United States for entrepreneurs and American inventors: abundant natural resources; immigrants from Europe, providing a ready labor supply; strong government policies; and new energy sources. This juxtaposition of factors gave rise to US industrialization—the transformation from manual labor-based industry to more technical and machine-based manufacturing. Overall production and economic growth greatly hastened the shift from an agrarian national economy to an industrial economy. This shift brought with it changes in the use of edible insects.

The start of the American Industrial Revolution is often attributed to Samuel Slater, who opened the first US industrial mill in 1790 with a design borrowed heavily from a British model. Slater's cotton-spinning mill in Pawtucket, Rhode Island, run by waterpower, greatly increased the speed with which cotton threads could be spun into yarn (Takkanen 2019). The stage was set for adapting this technology to silk production, which would bring with it the edible by-product, silk moth pupae. Already the weavers of silk had come to Virginia from France via England, bringing with them mulberry trees (Figure 3.1) and silk moths. These were the French Protestant Huguenots seeking religious freedom. Virginia promoted sericulture by offering forty hectares (about a hundred acres) to every immigrant who agreed to plant mulberry trees on their gift of land. The mulberry silk moth (*Bombyx mori*) is a single plant feeder and requires mulberry leaves.

Silk, a high-value commodity, coupled with its nutrient-rich pupal by-product, seemed a formula for success, but by the late 1800s it was all a failure except for the spinning and weaving. Cottage-scale and industrial-scale production of silk moths in Virginia and elsewhere were abandoned. Failure was attributed to various factors, among them the inadvertent importation of diseases of the silk moth (*pébrine*) and mulberry tree rootstock. In addition, young silk moth larvae required intensive, round-the-clock labor that subsistence farmers adapting to a new land and unfamiliar climate were unable to provide.

FIGURE 2.7 *Le Poilu* (First World War French infantryman). This French postcard from the First World War whimsically illustrates how insects (and disease) can affect civilian populations too. The “petit souvenir” (little souvenir) of head lice from a soldier on furlough would make for a memorable visit. Since human lice are specific to humans, the scratching dog, cat, and bird would not likely be harboring human lice. However, the animals provide comic relief while emphasizing the degree of some infestations. Authors’ collection.

FIGURE 3.1 A 170-year-old mulberry tree and the Hudgins family, descendants of the French Protestant Huguenots, sitting under it survived, but not the Virginia sericulture industry that immigrated with the trees and the weavers. Author's collection.

Of all these factors and economic shifts, the greatest impact on food insects in the US came in the form of European immigrants, particularly from Ireland, Germany, Sicily, and various independent Italian states. These settlers traveled further inland, families joining with other families to form wagon trains with the goal of farming in the "wide open spaces" on the hunting and gathering lands of the Native Americans in the Western US. Young people seeking their fortune in gold mining journeyed rapidly westward on the first trans-continental railroad (1869) with little but the clothes on their back.

These settlers in unfamiliar lands faced a steep learning process, sometimes with life or death consequences. Those who chose to inhabit rural areas often needed to dip into the Traditional Ecological Knowledge (TEK) of the local Indigenous people, which included sustainable farming techniques, an intricate transportation and trading system, and a detailed knowledge of food insects. The Euro-American immigrants superimposed onto the TEK their industry-focused transportation system, connecting remote parts of the US. Looking retrospectively now for two hundred years, it seems that the intimate connections to the land combined with the TEK accumulated from hundreds of generations of Native Americans practicing Native Science on that land was a necessary resource, not replaceable by compartmentalized, reductionist Western Culture Science if that land, water, and air was to support generations of Euro-Americans.

Both during their journeys and once settled, immigrants needed nutritious, locally gathered food when their crops failed or they were unlucky panning for gold. Native Americans, especially in the area of the Great Basin, had the solution: pemmican, their traditional energy bar (usually in the shape of a patty or consistency of granola) made from dried local fruits, meats, nuts, and berries, proved to be just the right sustenance (Tallbull, personal communication to F.V. Dunkel). Pemmican's name reminds us of its most essential ingredient: "pimikan" is a Cree word which in translation is "manufactured grease" (Tannahill 1973: 228–9). Pemmican was the "daily standby" source of essential omega-3 fatty acids such as alpha-linolenic acid (ALA) and the omega-6 fatty acid, alpha linoleic acid, for Native Americans, early explorers, and pioneers. ALA is required for immune health, vascular health, and mental acuity, all important for living on the frontier, especially for the young children born in these rigorous circumstances. Omega-3 fatty acids are particularly high in grass-fed animals such as bison, deer, and orthopteran insects (e.g., katydids and grasshoppers), all of which served as the main meat and fat ingredients of pemmican, depending on which part of the West the pemmican was being processed. Pemmican also contains vitamin B12, essential for peak functioning of humans' innate immune system (Damayanti et al. 2018; Schmidt et al. 2019).

The classic story of edible insects in the nineteenth century in the US involves two communities: the Ute (Native Americans living around the Great Salt Lake) and Euro-American settlers moving westward. The first of these settler groups arrived in the Great Salt Lake Valley on July 24, 1847. They sought an isolated place after the violence they had experienced in the State of Illinois, and the wide, flat plains of the Great Salt Lake Valley seemed ideal. They immediately planted potatoes and turnips, and built a dam. Then they sent word back to their friends in the east describing the two-square-mile city they had settled. By the end of 1847, nearly two thousand of their colleagues had moved to the Great Salt Lake Valley.

These families planted crops as soon as they arrived. During the first summer, crops that didn't fail during the drought were ravaged by an outbreak of local insects. Realizing that their people would not make it through the coming winter without outside help, the pioneer leaders turned to the Ute, who prepared "prairie cakes" for the grateful pioneers. With the contribution of the prairie cakes, the families survived. Happy to be beginning their second summer, the settlers asked the Ute how to make the prairie cakes so that they, too, might develop winter provisions. The Ute replied that they used local nuts, seeds, berries, and also the insects that wash up on the edge of the Great Salt Lake and are easily collected. The settlers were appalled and said, "Well, we will have no more of this!" The settlers recorded their disgusting encounter with edible insects in their journals and told the story of disgust to their grandchildren (Dunkel 2010). Their legacy today includes the large settlement known as Salt Lake City, Utah thanks, in part, to the timely, nutritious resources provided by a food insect of the Great Basin.

The edible insect was a katydid, *Anabrus simplex*, but because of the adverse, disgusted reaction of the settlers, no statue commemorates either the Ute or the katydid that saved the settlers' lives. However, in respect for the life-saving role it has had as a food insect, the katydid species has since been given an official common name by entomologists: the Mormon cricket.

The Great Salt Lake is the final drainage destination of the Great Basin, an ecological region of what is now mainly in the US and in part of the southern areas of Canada's western provinces (Figures 3.2 and 3.3). It was an ideal area for the natural production of edible insects known to Native Americans in

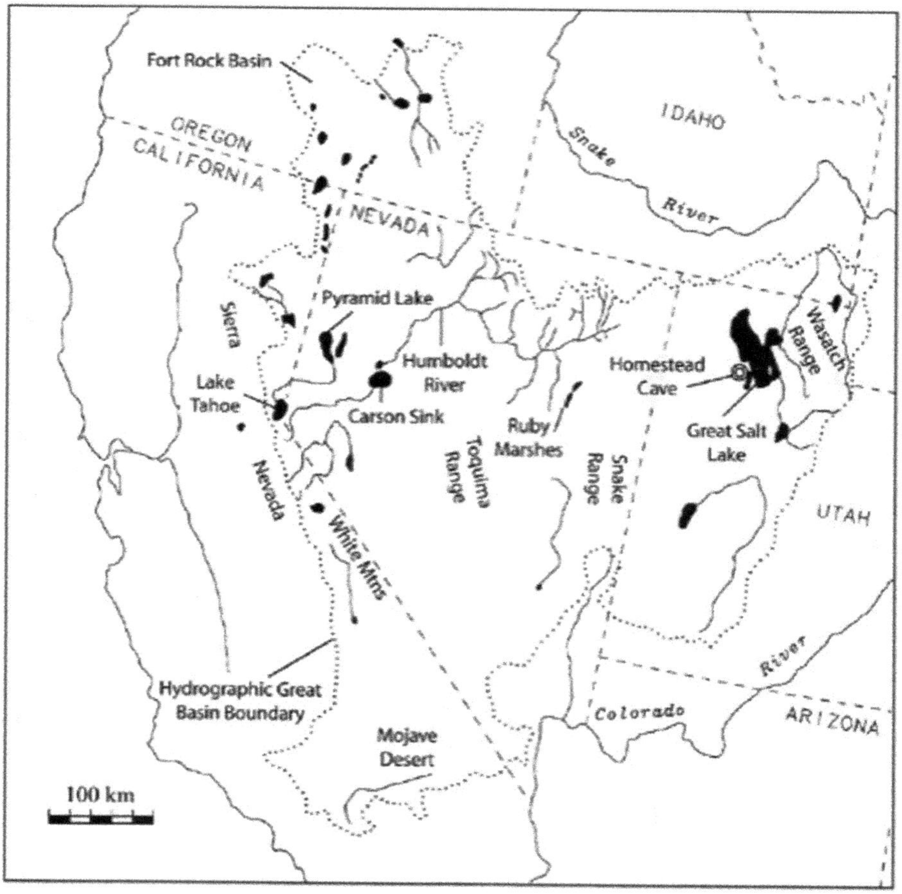

FIGURE 3.2 The Great Basin defined topographically. Grayson (2006).

FIGURE 3.3 The Great Basin defined by Native American nation and Traditional Ecological Knowledge base. These simplified ranges of nations are a generalization of the situation in the nineteenth century, with those of the Washoe, Northern Paiute, and Ute more for the first part of the century. D'Azevedo (1986). Redrawn by Gene Kritsky.

the area for over a millennium before the 1800s. By the nineteenth century, most Native American nations in the Great Basin (essentially all or parts of the western states of the US, excluding Texas) had developed traditional harvesting, processing, preparing, storage, and serving techniques for many of the local seventy-five edible insect species. Culinary uses for these Great Basin food insects included soups, pemmican, individual meat chunks, and simply as a powdered ingredient.

It is possible that the Ute practice of eating katydids had been serving as a natural population control for *A. simplex* (Bancroft 1889). When the Euro-American settlers drove the Ute out of the Valley, katydid (and grasshopper) populations exploded (Sutton 1988), but these insects were unappreciated as a food source by the Euro-Americans.

At first, food insects helped fuel Western expansion, but soon the European-style agricultural systems began to fuel insect outbreaks. The disgust of the new Western farmers led to an overreaction and likely contributed to the extinction of one of the food insects that became at times abundant over great expanses of the Great Plains (Lockwood 2004). This insect, the Rocky Mountain locust (*Melanoplus spretus* (Walsh)) was considered by some to be "one of the most spectacular and destructive natural phenomena in the American West" in the nineteenth century (Sorensen et al. 2018: 104). In 1874, Charles V. Riley, the state entomologist in Missouri and de facto entomologist for Kansas, observed swarms of locusts that "enveloped the plains from Manitoba to Texas" (Sorensen et al. 2018: 102). Reportedly "a housewife who frantically covered her garden with carpets watched in horror as locusts devoured the carpets before consuming the remaining vegetables" (Sorensen et al. 2018: 103). Riley advised farmers to plow their land to kill the locust eggs which he found died only if exposed to air.

Riley also designed the Locust Catcher or "hopper dozer" (Figure 3.4) to make industrial size harvests of the locusts, and invited government officials and citizens to taste the fine flavor notes of sautéed locusts. Riley did a complete

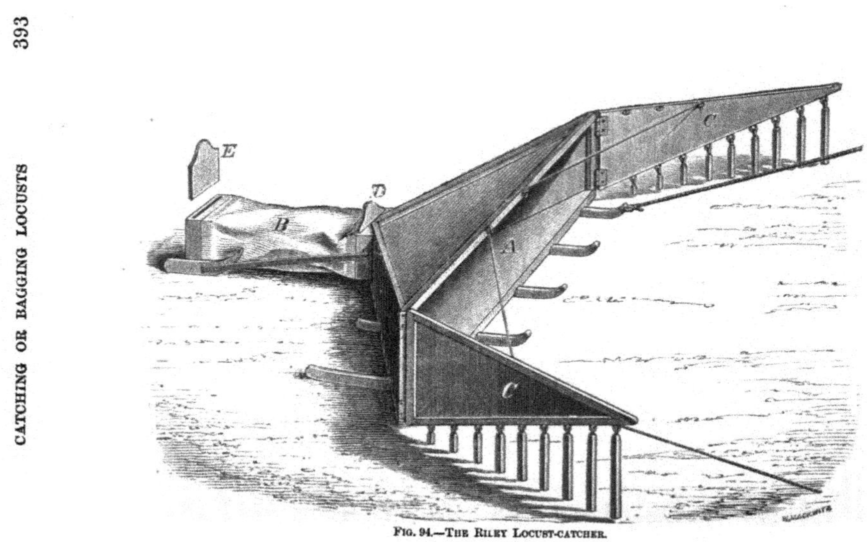

FIGURE 3.4 Riley Locust Catcher developed by entomologist Charles V. Riley to facilitate large-scale collection of the Rocky Mountain locust during outbreaks in the western US. United States Entomological Commission (1878). Hathi Trust Digital Library.

job of introducing food insects in Missouri and in Europe. First, he subsisted on locusts for twenty-four hours. Then he enlisted the help of a St. Louis restaurant to put locusts on the menu, and participated in the more difficult preparations. On a visit to London and Paris for their Entomological Society meetings, he brought a box of dried grasshoppers for tasting (Sorensen et al. 2018).

Afflicted farmers pleaded to state and national authorities for disaster relief. By the early 1900s the Rocky Mountain locust was extinct, due to disruption of the locust breeding ground by agricultural activities, especially plowing and flooding by European immigrants (Lockwood 2010).

In addition to orthopterans, western Native Americans consumed shore flies, cicadas, beetles, lepidopteran larvae, crane flies, salmon flies, bees, and yellow jackets. The White Knife Shoshoni (Harris 1940), the Western Shoshoni (Steward 1941), the Northern Paiute (Powers 1877), the Owens Valley Paiute (Chalfant 1933), and the Wahso (Lowie 1939) had a rich knowledge of how to use local insects in their diet. Two western nations even developed the technology of harvesting honeydew (aphid excrement) to use as local sugar (Heizer 1945). More recently Sutton (1988) reported seventy-five insect species used as food by Native North American tribes north of Mexico. These species comprise eight orders and twenty-eight families, mainly ants and grasshoppers. Other food insects included: seed beetles, long-horned beetles, diving beetles, scarabs, snipe flies, crane flies, giant water bugs, aphids, honey bees, bumble bees, gall wasps, paper wasps, yellow jackets, hornets, subterranean termites, tiger moths, tent caterpillars, giant skippers, noctuids, giant silk moths, sphinx or hawk moths, giant stoneflies, predatory stoneflies, darners, long and short-horned grasshoppers, sand or camel crickets, and other crickets.

Studies of the economic systems in the Great Basin during the 1800s and earlier mention piñon, a small seed, as an often-traded material of high value (Steward 1938). Food insects, despite their use in this area as a relatively major dietary constituent equal to game animals (mammals), are not well recognized. Anthropologists traditionally considered insects as famine food or backup reserves, usually taken individually in the course of other activities (Thomas et al. 1983). As many early scholars were Europeans or Euro-Americans, their biases and the dominance with which they applied them likely led them to diminish or even ignore the key dietary role that food insects played for European immigrants *and* Native Americans who thrived for millennia in the west.

In contrast Sutton, an ethno-historian and ethno-geographer who specialized in the Great Basin area, classified the edible insect procurement process in the Great Basin of the US as harvesting on a commercial scale during the 1800s, and for many centuries prior. He concluded that these harvesting events required considerable planning, travel, and effort and were not chance encounters (Sutton 1988). Furthermore, stored insects obtained from these large-scale insect harvests were combined with stored plant products. Altogether this nutritionally balanced diet provided food for a comfortable winter.

In his energy calculations Madsen (1989), a State of Utah archeologist, showed that this traditional use of insects as a food source "made a great deal of economic sense." He found that one person collecting grasshoppers or katydids from the water margin for one hour would yield 18.5 pounds. This gathering process would accomplish the collection of as many nutrients as collecting eighty-seven chili dogs, forty-nine slices of pizza, or forty-three Big Macs® (Madsen 1989). Crickets collected at 18.5 lbs/hr (23,479 cal.) compares favorably with deer and other large game (Madsen and Schmitt 1998). Thus, the food insect collecting and consuming during the nineteenth century in the Great Basin area not only made economic sense, but it also made a great deal of metabolic sense![1]

Northwestern Utah's Lakeside Cave provided another labor-efficient collection method, used for five thousand years (including in the nineteenth century) to collect the migratory grasshopper, *Melanoplus sanguinipes*. Ethno-entomologist Madsen (1989) returned to this cave and documented five rows, six feet in diameter and nine inches thick, containing up to ten thousand hoppers per foot (Madsen and Kirkman 1988). An estimated two hundred pounds of sundried grasshoppers could be collected per person per day (Madsen and Schmitt 1998). Although used as fresh meat, grasshoppers were mainly dried, often ground, and made into pemmican.

It is interesting, but not surprising, that a recent Google search of pemmican for Native American tribes of the Great Basin turned up no mention of insect meat as a component among the sources (Ngapo et al. 2021). Similarly, the original Native American pemmican was almost never mentioned in the diaries and records of migrants on the western wagon trails. Despite the fact that they often took mountain men as guides during the summer months, the respectable migrants seemed reluctant to eat Native American and wild men's food, sometimes with disastrous consequences—they perished.

A less dramatic transfer of the TEK about edible insects likely also happened as immigrants and other settlers encountered Native Americans in the northeastern, midwest, southwest, and Pacific regions of the US. European immigrants were curious about the ways of the people who had survived for millennia on this land that they themselves were now hoping would be the "land of milk and honey" where "money grows on trees." Some settlers actually went to live with the local Native American tribe. They "emulated the Native Americans, married them, were adopted by them, and on some occasions even fought alongside them" (Junker 2016). The opposite seldom occurred (Axtell 1979, 1981). In many of these locations, food insects were an important and celebrated part of the local diet. Certainly, this information on harvesting, processing, and eating insects would have been shared with the new settlers.

Similar stories may exist in the juxtaposition of European immigrants moving into land occupied for centuries by other Native American nations. For example, the Paiute in California harvested lake flies and shore flies as a sustainable nutrient source. Cicadas were sustainable protein sources for Native peoples in the northeast, midwest, and southwest US. One of the more unusual food insect stories from European immigrants is from Danish explorer Peter Freuchen (1961), who went to live among the people in Greenland in 1906. He shared that "A constant concern of the Eskimo [Inuit] household [near Thule, Greenland] is the lice, even though Eskimos get rather used to them. The wife delouses the husband, particularly his long hair when the plague gets too bad. And the husband, I have seen often delouses the children and eats their lice with great relish, only once in a while handing one to the tot so that he also could have a little pleasure from them." Likely lice supplied extra calcium, iron, and other nutrients not found in sufficient supply in seal, whale meat, or blubber that was the main nutrient source for these people.

Notwithstanding the preceding accounts, a general internet search relates quite a different story about typical foods in the US during this time period. Thus, we find in the north, cows provided milk, butter, and beef, while in the south, where cattle were less common, venison and other game provided meat. Preserving food in 1815, before the era of refrigeration, required smoking, drying, or salting meat. Vegetables were kept in a root cellar or pickled.

Why the discrepancy in information? Culture determines whether a plant, animal, or microorganism is considered food or not food (Dunkel 2017). Edible insects were a crucial dietary component for many of the 573 Native American nations. Given the more recent non-aggressive interaction and actual learning expeditions that have been documented by settlers who embedded themselves in these nations, it seems plausible that edible insects were part of the transfer of traditional knowledge. Settlers such as George Bird Grinnell (1849–1938) in Montana and Freuchen in Greenland espoused an ethno-relative world view (Grinnell 1972a, 1972b). They were curious, accepting, adapting, and ready to integrate these cultural practices into their daily lives.

While some of the European immigrants sought a good life in the wilderness, others settled in the growing cities. There, as well as in the nearby small farming towns, new processed foods were becoming available for the homemakers. Flour, particularly white flour, for cakes and cookies opened a new area of sugar-based foods. The culinary arts flourished with these new products. With the rise of Euro-American homemakers making more baked breads, cakes, and cookies, flour became a commodity stored in the home. Insects were competitors in this food supply and became unwelcome pantry guests in the US. In Europe, insects, except for honey bees, already had a vile reputation. Similarly, when dry cereals, cake mixes, dry milk, and other processed foods whose base was grain were introduced into homes in the early twentieth century, insects in the house were feared and certainly not considered food.

At the same time that the Industrial Revolution brought new foods and a growing transportation system to distribute them, other factors were operating. This period, the 1850s to 1920s, also saw the greatest period of faunal and floral introductions to the US via immigrants (Coates 2011). European cattle breeds were brought to replace the bison which, at the hands of the Euro-American immigrants, were slaughtered in vast numbers along with many Native Americans. Arrival of European cattle cinched the broad-spectrum abandonment of the food insect TEK encountered by Euro-Americans. Not until the twenty-first century would their descendants return to the recognition that our food systems must be sustainable and suggest that maybe the answers lie in the paleo-diets with edible insects that European immigrants had ignored in the nineteenth century.

FOOD INSECTS IN MEXICO AND OTHER LATIN AMERICAN COUNTRIES

While the Euro-American immigrants to the US were relatively quick to adopt the new processed foods and develop recipes based on them, Mexican rural and urban populations held tight to their traditional foods, particularly edible insects. As with Native American nations occupying what is now the US, the Mexicans preferred insects because they signaled more than food and nutrients. Edible insects were part of the social fabric and an integral food for celebrations. They brought back memories of family gatherings. Mexico thus differs markedly from the US in its edible insect history.

For Native Americans in what is now the US and Mexico, the use of insects as food dates to the pre-Columbian times. Mexican researchers documented more than two hundred species traditionally used in Mexico (Ramos-Elorduy 1993; Ramos-Elorduy and Pino 1993). More than sixty species of edible insects were reportedly used by Native Americans in the area that is now the US (DeFoliart et al. 2002, 2009) and likely it was far more, since in areas such as Montana, 125 species of edible grasshoppers occur. As with Native Americans in what is now the US, i.e., the eastern woodlands, and western Great Basin areas, the Indigenous people of Mexico, particularly in the area of Oaxaca, never lost their tradition of edible insect eating. At least two hundred species were part of Mexican cuisine compared to about 185 species documented in use in the US Native American cuisine in the nineteenth century.

Local markets were mainly supplied by wild gathering of edible insects, such as the *chapalinas* (small grasshoppers), stink bugs, honey of stingless bees, and *escamole* (ant eggs, larvae, and pupae). Silk moth pupae were available as a by-product of the cottage industry-scale production of silk. Local suppliers and families celebrated together in the nineteenth and twentieth centuries, as they did hundreds of years before, enjoying *chapalinas* and *escamole* in an unbroken chain to the present moment.

Agriculture in the Americas was well developed before the arrival of any Europeans. One of the historical triumphs in agriculture beginning 8,700 years ago was the domestication of the wild grass, teosinte, which Native Americans' impressive selective breeding programs developed into corn (maize). Along with corn came the corn earworm, a moth larva and one of the often overlooked food insects in Mexico and likely Central America. Corn earworm larvae supplied the two amino acids missing in corn, lysine and tryptophan, plus vitamin B12 and other essential nutrients humans cannot make metabolically (Martin 2014). The moth larvae are soft bodied and when boiled or roasted provided an additional boost of flavor. Thus, "association plantings" (several species planted in hillocks to give each other mutual support or sun shading, etc.) of corn, beans, and squash had a fourth component—the edible insect.[2] Aztecs favored corn cobs with insects, good sense that survived, despite European ingression, in the nineteenth and twentieth centuries.

EDIBLE INSECTS IN CHINA AND OTHER ASIAN COUNTRIES

We remember the first half of the nineteenth century in China for the Opium Wars, the resulting Treaty of Nanjing and establishment of treaty ports, and cession of Hong Kong to Britain as a colony. The second half of the century began with the Taiping Rebellion and entry of Taiping "Christian" ideology. Messengers of this ideology were mainly from cultures of European derivation and brought with them a disgust for edible insects that we noticed surfacing at the same time in these missionaries' ancestral homelands, e.g., England (Holt 1885) and their adopted homes in the US.

As the nineteenth century was ending, the First Sino-Japanese War began. China would experience political upheaval into the early twentieth century with the Boxer Uprising (1910), the Chinese Revolution (1911), and establishment of the Republic of China (1912) prior to formation of the Chinese Communist Party (1921). European influence in effect ceased.

During the many years of unrest, traditional Chinese foods were not significantly changed by the brief interaction with western cultures. The main food insects in north and central China, locusts (Donovan 1842), were so revered that temples were established in their honor (Hsu 1969). In the mid-nineteenth century hunting locusts was popular and profitable for young people (Oliphant 1860). Large-scale grasshopper and locust harvests were brought in baskets and boatloads to sell in Beijing markets, boiled, dried, with salt or sauce, ready for frying or storage (Sowerby 1925). Crickets were not reported as food anywhere in China. Traditional foods in southern parts of China (Guangdong) included: water beetles (Dytiscidae) and water scavenger beetles (Hydrophilidae) (Bodenheimer 1951), green-bottle fly larvae, giant water bugs, stink bugs, and caterpillars infected with fungus of the genus *Cordyceps* (Simoons 2014), particularly silkworm pupae produced by the usual cottage industries (in the home, in the countryside). Commercial production of silk continued to grow during this period.

Silkworms are the caterpillars of moth species raised for silk. In the nineteenth and twentieth centuries, 90 percent of the world's silk production derived from cocoons of domesticated mulberry silk moth (*Bombyx mori*). Silk was also produced from other domesticated or wild Saturniidae moth species, notably the Eri silk moth *Samia cynthia ricini*, the Assam silk moth *Antheraea assamensis*, the tussore (or tussah) moth *Antheraea mylitta*, and the small tussore *Antheraea paphia* (Mishraa et al. 2003; Longvah et al. 2011). When these silk moth larvae enter the pupal stage, they build a protective cocoon with a single thread of raw silk. At the end of pupation, the pupa releases an enzyme that severs the silk thread in a circular fashion, creating a trap door in the cocoon for the moth to emerge. Once that enzyme is secreted the silk is worthless. Thus, for silk production, pupae are killed by boiling, drying, or soaking in sodium hydroxide before the enzyme is produced (Datta and Nanavaty 2007; Jintasataporn 2012). Resulting dead pupae, or spent pupae, are produced in large quantities and are a major by-product of silk production. Each 1 kg of raw silk yields 8 kg of wet pupae (2 kg dry pupae) (Patil et al. 2013).

Silkworm pupae have for centuries been eaten by humans in Asian silk-producing countries and are considered a delicacy in regions of China (Luo Zhi-Yi 1997), Japan (Mitsuhashi, 1997), Thailand (Yhoung-Aree et al. 1997), India (Longvah et al. 2011), and elsewhere. Due to its high protein content, silkworm pupae meal has also been a livestock feed, mainly for monogastric species (poultry, pigs, fish) but also for ruminants (Trivedy et al. 2008).

Historians of food and of the silk industry during this period paint a mixed portrait of how this significant, readily available protein resource was used. Simmonds (1885) contends it was simply food for the poor. Others (Gray 1878) observed that pupae were considered delicacies, sought after for their "strengthening and refreshing" properties. Others noted pupae were a typical every day meal, fried with onions eaten with a sauce or put in an egg omlette. Women unwinding cocoons in cottage factories were observed simply snacking on pupae (Simoons 2014). When available at certain times of year pupae were sun-dried and ground (Wijayasinghe and Rajaguru 1977; Usub et al. 2008; Jintasataporn 2012).

Degradation of mulberry-fed silkworm pupae produces a foul smell attributed to mulberry leaf chemical compounds sequestered by silkworms, e.g., flavonoids, terpenoids. Donovan (1842) reported this issue was circumvented in the nineteenth century in central China by combining pupae with mulberry leaves, frass, molting residues, and other silk industry wastes and using them as farmed fish feed or as fertilizer (YaoWang 1989).

Silk production during the 1800s went through many changes and only in China did traditional handcrafted or hand-reeling survive through the century and into early 1900s. Compared to the best technology of the 1920s, Chinese hand-reeling still lagged in productivity per worker. It is possible that the 1920s were the watershed of Chinese handicraft silk. Yet for all the quality advantage that filature (factory-reeled) silk enjoyed over hand-reeled silk, traditionally reeled silk was crucial to the hand-weaving industry for more reasons than price. Raw silk produced by filatures is actually more brittle than hand-reeled silk since the natural gum in the filament is removed during the steaming process.

While the Chinese had exported silk for hundreds of years, exports rose dramatically in the mid- to late 1800s (Xu 2012). The first step towards formation of the global silk markets occurred in the early nineteenth century, when rapidly increasing manufacturing needs of France and northern Europe contributed to a steady demand for raw silk (Eng 1986). The food resource as a by-product remained in China.

It is always tempting when comparing traditional and modern techniques of production to disparage traditional techniques and means of food preparation as backwards and inefficient. In reality, traditional techniques of silk production in China were surprisingly advanced (Xu 2012). While industrial silk reeling did eventually surpass traditional reeling in productivity and quality, traditional manufacturing continued to be a viable, resilient enterprise (Xu 2012). In terms of economic viability, handicraft silk benefited from the relatively small efficiency gap that existed between filature and hand reeled silks. During periods of social uprisings and political turmoil, cottage industries such as silk moth production provided not only a source of cash from the raw silk, but a ready source of essential nutrients from the pupae, particularly for children who are usually the first to suffer during political upheaval.

EDIBLE GRASSHOPPERS IN THAILAND, KOREA, AND JAPAN

In the nineteenth and early twentieth century before broad-based use of strong, synthetic chemical pesticides, grasshoppers were a favored addition to daily meals in Thailand, Korea, China, Japan, and throughout east Asia. "Metdugi" is the rice-field grasshopper. This Korean food insect literally disappeared from commercial markets in the later twentieth century because of pesticides, but the deliciousness was not forgotten. Luckily, consumers began demanding organic rice and metdugi are now once again a common snack in restaurants and bars in Korea (Pemberton 1994).

Silk moth pupae or "Beondegi" (번데기) were also a part of daily meals in Korea in the nineteenth century. Still today, if you are in Korea, finding beondegi for a snack is not difficult. Beondegi are now a street food served in paper cups with toothpick skewers (Pettid 2008). Canned preparations are available in grocery stores and even convenience stores, but not as easy to find and with not as appetizing flavor notes has been my experience (F. Dunkel personal communication 1994, 2016).

WASP CULTIVATION IN JAPAN

To fully understand edible insects in Asia, one needs to appreciate the intimate relation of family traditions and local festivals developed over the past hundreds of years, centered around an edible insect. For illustration, I have chosen the ancient Japanese tradition of raising and eating wasps, specifically, *Vespula flaviceps*, or *hebo* in Japanese. This set of traditional practices, which spans five months of the year, occurred in the nineteenth and early twentieth centuries and continues into the twenty-first century.

Preparation for the festival begins in summer in Kushihara, a community in central Japan. Nests of the *hebo* are located in nearby forests. Each nest, about the size of a tennis ball, is carefully brought back to the home by the wasp hunters and put in a sheltered hut in the garden. Over the summer and fall, the *hebo* are fed a steady diet of sugar water, honey, and raw chicken meat. When the larvae are large and bursting out of the nest they are ready to harvest. It is similar to harvesting apples at the end of the summer along with the other garden vegetables and fruits.

The festival happens in early November. Once, it was a competition of who was able to harvest the most *hebo* at the end of the season. Now it is a festival bringing together family and friends to gather and eat traditional foods, and of course, wasp tempura and wasp *onigiri* rice balls.

Most of the over two thousand species of edible insects have similar traditions associated with them. These festivals, it is important to note, are not based on making a powder of the insect, but of eating the whole insect, making the eating an adventure in texture and the actual flavor specific to that insect species. The impulse to "hide the insect" by incorporating it into food in powdered form makes it seem like it is not a food and more like a vitamin that you take (Joost, qtd. in Ho 2019).

EDIBLE INSECTS IN AUSTRALIA AND PACIFIC ISLAND COUNTRIES

As the nineteenth century opened, food insect collection processes were practiced in Australia as they had been for centuries. Witchetty grubs or witjuti grubs were a staple in the diets of Aboriginal Australians (Isaacs 2002). These large, white, wood-boring larvae of several moths, particularly *Endoxyla leucomochla* and other cossid moths, continue to be an important food of the desert. Grubs are dug from the roots of the witchetty bush, which is widespread throughout Northern Territory and parts of Western Australia and South Australia (Cavendish 2003). Roots of the *Acacia kempeana* shrub are another source of the grubs.

In other parts of the Pacific Island countries, bamboo borer larvae and palm weevil larvae were a standard dietary item for Pacific Islanders, particularly in the Philippine Islands, and continue to be so. This unbroken pattern of edible insects in human diets is encouraging for meeting future food security challenges. An important lesson, which originated well before the nineteenth-century harvesting of food insects, comes from these islanders, and persists to the present. It is that taboos and cultural use stories evolve in various Indigenous cultures to regulate use of a limited or diminishing resource. When such stories are not in place, edible insects may cease to be harvested sustainably and run the risk of becoming endangered and eventually extinct species. It is important today to monitor the wild populations of these favored food insects, especially after Pacific Islanders' contact with Europeans or visitors of European derivation.

A most encouraging outcome has been the sharing by the Aboriginals of their TEK of witchetty grubs with the other cultures within Australia and with visitors to the continent. The sharing of their knowledge of this insect—crispy on the outside with a smooth and creamy almond-flavored center—is key to engaging other cultures in exploring their own local opportunities for delicious, nutritious local foods.

FOOD INSECTS IN AFRICAN COUNTRIES

Similar to the industrial scale gathering of orthopterans (grasshoppers, crickets, katydids) in the Great Basin of North America, locusts and other orthopterans were commonly dried for food throughout Africa. Technologies for gathering, processing, and storing of locusts and other orthopterans were well known. Sun drying and salting (but not always) was the main processing technique in the nineteenth century and into the twentieth century. The arrival of locust swarms was seen as a blessing as much as a scourge. It was only when the Europeans and Euro-Americans arrived and announced that locusts were not food that the Africans began to abandon this most nutritionally complete food of all foods in their entire diet. Throughout sub-Saharan Africa, insects, particularly larvae and grasshoppers, were considered the special snacks of children. Considering the high nutritional needs of children, this practice had adaptive significance. It is interesting that in 1984 the top concern of the nutrition section of the Food and Agricultural Organization (FAO) of the United Nations was children's weaning foods, when at the same time FAO was not recognizing insects as food and especially not a food of particular nutritional significance. In the twenty-first century, this recognition is changing. On May 13, 2013, a watershed moment occurred with e-publication of the FAO Forestry Report *Edible insects: Future prospects for food and feed security* (van Huis et al. 2013). Within twenty-four hours after release, one million copies had been downloaded.

Mali has at least thirty-five separate languages and eight major ethnic groups for whom insects were once an important and diverse part of their diet. Now the Bambara villagers with whom the author interacts mention only limited numbers of insect species that they recall using as food. For example, villagers in Sanambele, Mali recognized only subterranean crickets *Brachytrupes membranaceus*, termites, katydids, and grasshoppers as good for eating.

At least one food insect did not meet the same fate. Mopane or madora "worm" cultivation in Botswana and other southern African countries has remained a strong, unbroken tradition from likely before the 1800s, throughout the 1800s to the present. Mopane is the large, edible larva of *Gonimbrasia belina* (Family: Saturniidae), the emperor moth, native to the warmer parts of southern Africa. This nutritious caterpillar is usually found on leaves of the mopane tree, *Colophospermum* sp. (Dreyer and Wehmeyer 1982).

FOOD INSECTS IN THE MIDDLE EAST

In 1800 and for millennia before, both the Moslems and Jewish populations considered insects as acceptable food. Acceptability was based on Moslem food laws (Halal) and Jewish food laws (kosher) and on their individual cultural histories. According to biblical accounts, John the Baptist survived his fast of forty days and forty nights with locusts (and honey). The Jewish community earlier had survived on "manna from heaven," likely a fungus that grew overnight on the sugary honeydew (frass or excrement) of aphids. The literature provides no mention of the ideal environmental conditions for orthopteran populations (crickets, grasshopper, locusts, katydids) to thrive in this area, especially around the Dead Sea, similar to the area in what is now the US around the Great Salt Lake and in the Great Basin.

The late 1800s and early 1900s were busy times in the Middle East filled with immigration, wars, and major epidemics of malaria. Large tracks of the area around the Dead Sea were found devoid of humans by visitors such as Mark Twain (1869). Thomas Cook and Son, the travel agent, in its 1876 *Handbook for Palestine and Syria*, said of Palestine,

> Above all other countries in the world, it is now a land of ruins. In Judea it is hardly an exaggeration to say that ... for miles and miles there is no appearance of present life or habitation, except the occasional goatherd on the hillside, or gathering of women at the wells, there is hardly a hill-top of the many within sight which is not covered with the vestiges of some fortress or city of former ages.
>
> (Cook & Son 1876: 35)

Indeed, the areas were considered uninhabitable by many due to the existence of anopheline mosquitoes carrying malaria, yet in order for malaria to continue, humans must have had to have been present (Alexander and Dunkel 2017).

Modern Political Zionism, the movement for Jewish self-determination, arose in the late nineteenth century as a reaction to anti-Semitic and exclusionary nationalist movements in Europe. Anti-Jewish pogroms in Russia (1881–1884) stimulated growth of Zionism, resulting in formation of pioneering organizations and the first major wave of Jewish immigration to Palestine (Palestine Department of Health 1941).

In 2019, a headline in the online newspaper *The Times of Israel* reported a phenomenon that occurred in the Middle East that was also common in the 1800s: "Grasshoppers swarm into Israel from Syria. Unusually large numbers decimate vegetables in Moshav Alonei Habashan on Golan Heights, lay hundreds of thousands of eggs in soil" (Surkes 2019). Then, as now, Middle Eastern habitats were ideal for producing edible insects. These orthopterans are a more nutritious source than the irrigated vegetables that now cover the once uninhabitable mosquito marshes and xeric grasslands of Israel, Palestine, Syria, Iran, Iraq, and the Arabian peninsula.

EDIBLE INSECTS IN EUROPE

One highlight of entomophagy is the nineteenth-century treatise by British author V.M. Holt, titled *Why Not Eat Insects?* (1885). Holt begins his work acknowledging the "difficulty of battling against a long-existing and deep-rooted public prejudice" (1885: 5) but argues that certain insects, like some mollusks and crustacea, are edible and palatable. Apparently fascinated by nutrition, Holt tried to convince his Victorian era readers that rather than confining the poor to workhouses to earn a square meal, these workers could assist farmers in the countryside by gathering agricultural pests. This would aid in pest management and also be the harvest of an ignored, but healthy food: "How can the farmer most successfully battle with the insect devourers of his crops? I suggest that these insect devourers should be collected by the poor as food. Why not?" (Holt 1885: 14–15). A touch of ethno-relativism, acceptance of other culture's practices without judgment, was another feature of Holt's writings:

> If I bring forward examples from ancient times, or from among those nations, in modern times, which are called uncivilized, I foresee that I shall be met with the argument, "Why should we imitate these uncivilized races?" But upon examination it will be found that, though uncivilized, most of these peoples are more particular as to the fitness of their food than we are, and look on us with far greater horror for using, as food, the unclean pig or the raw oyster, than we do upon them for relishing a properly cooked dish of clean-feeding locusts or palm-grubs.
>
> (Holt 1885: 32–3)

Holt had that rare ability to place himself in the "other's shoes" and then look back at himself, or, in this case, look back and reflect upon his own culture. It would be another 135 years before England would espouse the ideas of Holt. In 2018, Sainsbury became the first grocery store in the UK to offer insects (Giordano 2018).

CASU MARZU

Another highlight of the nineteenth century's European stories of edible insects comes from the Mediterranean area, from Sardinia, Corsica, France, Tuscany, and Sicily. It is live fly larvae that make *casu marzu*, one of the most amazing, yet controversial cheeses—"la rivincita di un grande formaggio" (Manicardi 2018). *Casu marzu* is in the pecorino cheese group along with other cheeses not produced by insect larvae including feta of Greece, Roquefort of France, Manchego from Spain, and pecorino romano and ricotta of Italy. Sheep's milk is ideal for all of these cheeses, especially for *casu marzu,* created by digestive action of the fly larvae of the cheese fly, *Piophila casei.* Larvae are deliberately introduced to the cheese. In the nineteenth century, *casu marzu* was created by removing part of the rind, often positioning the whole cheese round outdoors. With her good olfaction skills, the female cheese fly finds the cheese and lays more than five hundred eggs at one time. Eggs hatch and larvae begin consuming the cheese. In the process, acid from the larvae's digestive system breaks down the cheese fats (Overstreet 2003). The cheese becomes very soft, even seeping liquid. When the cheese is ready for eating, a typical *casu marzu* will contain thousands of *P. casei* larvae (Hegarty 2006). Larvae are white and cylindrical in shape, about 8 mm (0.3 in) long.

Casu marzu has a strong, challenging, but actually very enjoyable flavor, with a hint of gorgonzola and black pepper (Ward 2013). Even for some gourmet cheese buffs, *casu marzu* is a difficult cheese because of the live fly larvae. It's the larvae, however, that give this cheese its greatness. The larvae give a special burst of flavor as they squish on one's molars, creating the extraordinary culinary experience (A. Guisti, personal communication to F.V. Dunkel). The negative side of consuming this insect delicacy is that when disturbed, the *P. casei* larvae spring out of the cheese for distances up to 15 cm (6 in.). If this is a problem for diners, those in these Mediterranean cultures developed a way to collect and kill the larvae by suffocating them while in the cheese in a tightly closed bag (Frauenfelder 2005; Bethune 2006). Of course, without the live larvae, that extraordinary burst of flavor is absent when eating the cheese. In the late twentieth century, with the rise of food sanitary laws, *casu marzu* became illegal to sell or serve in a public establishment. However, new foods rules of the European Union may provide a way to put this insect cheese back on the public market. In the past decade, scientists at the University of Sassari in Sardinia, working cooperatively with local sheep farmers, have developed a hygienic way to raise *P. casei* for cheese production and still meet EU restrictions (Dessi 2008). In Sardinia, this traditional cheese is called *casu modde, casu cundídu* and *casu fràzigu* and is believed to be an aphrodisiac. A variation of the cheese, *casgiu merzu*, is also produced in some Corsican villages as well as in a number of Italian regions (Manunza 2018*)*, such as: *bross ch'a marcia* in Piedmont; *cacie' punt* ("*formaggio punto*") in Molise; *casu puntu* in Salento (Apulia); *casu du quagghiu* in Calabria; *gorgonzola coi grilli* in Liguria; *formag punt* in Apulia; *furmai nis* ("*formaggio nisso*") in Emilia-Romagna; *marcetto* or *cace fraceche* in Abruzzo (Formaggio marcetto 2021); *salterello* in Friuli-Venezia Giulia (Prodotti tipici 2011). Goat herders also made *casu marzu* with goat milk and added it to white wine with grapes and honey (Comuni italiani 2011). In the nineteenth century and for hundreds of years before, *casu marzu* was a traditional food of the region. It is still consumed, mostly at weddings and celebrations. In Germany and France, cheese mites, the close cousin of insects, worked in the nineteenth century to create deliciously aged cheeses, Milbenkäse and Mimolette, respectively (Mite Cheese, Mimolette) (Mite Cheese n.d.).

Had the Englishman V.M. Holt traveled to the Mediterranean area of Europe, especially the islands, and other areas of the mainland, he would have been pleased to find that while not part of British cuisine, insects and their arthropod "cousins," the cheese mites, were being deliberately eaten by people in Europe.

CONCLUDING COMMENTS

During our time period of interest, edible insects in the US encountered two opposing forces. The invention of trains, railroads and other forms of transportation opened the western US to Euro-Americans, who came face to face with edible insects and their role as a survival food. Traditional ecological knowledge (TEK) was a cultural wealth that these pioneers learned from Native Americans, including where to find edible insects, how to gather them, and how to prepare them to last for a year or more. However, a series of genocide actions taken by the US government and perpetrated on the Native Americans, coupled with Euro-American cultural tendencies, were impediments to settlers' development of a favorable response toward edible insects. Industrialization brought added hindrances including food opportunities, such as artificial packaging, finer controls on stoves and ovens, the ice box, and milling of white flour. With food manufacturing and new home processing techniques came a disdain for the place-based, Indigenous-culture-based TEK associated with locally abundant edible insects. These forces sounded a death knell for the acceptance of entomophagy.

Industrialization in primarily western cultures gave them a dominance and in some Indigenous cultures a reversal occurred: the old culture was shed and "modern" cultural practices adopted. Along with this reversal, edible insect practices were not just shed, but buried. Scholarly sources summarize current edible insect use in the US as being near zero (Costa Neto and Dunkel 2016).

The industrialization that arose in a few locations in the nineteenth century world was accompanied by the awakening of revolutionary action in East Asia and the colonization of Africa by Europeans. The nineteenth century also witnessed the Euro-American takeover of much of the land in what is now the US, land that had an active transcontinental trading system, an industrial-scale food insect harvesting processes, as well as various successful, sustainable, agricultural systems. As the twentieth century opened, large-scale use of silk moth pupae provided food in Asia, as did other large-scale harvesting and field husbandry, e.g., rice field grasshoppers in Korea, wasps in Japan, the mopane in southeast Africa, and grasshoppers and other orthopterans throughout Africa and Mexico. Indigenous peoples of Thailand, Japan, Mexico, Southeast China, and Brazil successfully integrated their TEK of edible insects into their societies even as they adopted many aspects and technologies from western cultures. In these countries now in the twenty-first century, edible insects are available commercially, served in upscale restaurants and eaten when families and friends gather in their homes. For Native Americans and others who have learned the place-based TEK from their Elders, the now-needed transition of the world's population to local foods with a low environmental footprint will be easy.

Western civilization, from its origins in Greek and Roman times, was a dominant entity focused on conquering and colonizing other civilizations. With the Industrial Revolution arising in Europe and blossoming in the US, the industrialized nations gained worldwide dominance. In the process, the wealth of traditional, place-based knowledge was put aside, subsumed by technologies developed during and in the aftermath of the Industrial Revolution. In the twenty-first century it behooves us to recover this ancient knowledge of how and where to raise what food crops sustainably. Edible insects are a complete source of the essential nutrients. Localized pockets in western cultures, its people and its commercial sectors, are awakening to edible insects. Western civilization as a whole must reacquaint itself with edible insects, their resources, their production, and their deliciousness.

We can act more wisely in the twenty-first century. We have tested the positives and negatives of monoculture. We are well aware of the finite energy resources we manage on Earth. Transportation that developed and flourished in the nineteenth century opened the way to dependence on carbon-based energy to supply us with abundant food in such variety that we forgot to eat seasonally. Indigenous people in each location hold the secrets of what grows best in each micro-region. These colonized people that Western cultures tried to subsume had knowledge of how to survive and thrive without intercontinental transportation or use of non-renewable resources. This means now in the twenty-first century new recipes that draw on the rich culinary history of Indigenous people can expand the culinary arts of western civilization. For example, as a dietary source corn by itself is missing some essential nutrients for human growth and development. The Indigenous people of Mesoamerica and the southwest US knew to include in their corn-based diet the corn earworm and corn smut. With these two additions, corn became a crop with all the necessary essential nutrients.

For humans to survive and thrive equitably this next century, we must learn to decolonize our methodologies. We need to relearn and teach to our grandchildren and seven generations of our grandchildren the interconnectedness of all life, including the water, the land, and the air on Earth and in our entire Universe. Specifically, we need to continue to strengthen the serious pursuit of adapting TEK about edible insects to meet our culinary and nutritional needs as well as to promote environmental sustainability.

CHAPTER FOUR

Insect Products

Apiculture, Sericulture, and Other Insect-Derived Commodities

WALTER S. SHEPPARD AND CAROL M. ANELLI

Humans have exploited insects and insect-derived products from time immemorial. Two of the best-known examples, with perhaps the longest and most romantic histories, are apiculture (beekeeping) and sericulture (cultivation of silkworms to produce silk). The origin of removing honey from honey bee colonies is decidedly prehistoric, with cave paintings made more than seven thousand years ago depicting such activity in Spain. By the time of the Egyptian Pharoahs, humans had graduated from simple robbing of natural hives, to bona fide "beekeeping," where honey bees were housed in mud tube hives and combs of honey were removed as desired by the keeper of the bees (Figure 4.1). The origin of sericulture is shrouded in ancient legend, although we know that by 2700 BCE, the practice of growing silkworms and spinning thread made from their silken cocoons into cloth was an established practice in China. Our chapter opens with a discussion of apiculture, focusing on nineteenth-century innovations in beekeeping, queen rearing, honey production, and commercial beekeeping. We then turn our attention to silk and other insect-associated products.

FIGURE 4.1 Ancient Egyptian beekeeper in a tomb near Thebes. Gene Kritsky.

APICULTURE: HONEY BEE HIVE DESIGNS

Beekeeping of the early 1800s had not changed substantially for centuries. It mainly consisted of providing honey bees with domiciles in the form of baskets, boxes, hollow tree trucks or pottery/stone enclosures (Figure 4.2). Beekeepers usually destructively harvested honey from selected hives in the apiary by killing the bees therein and removing all the honey and wax that could be found. Beekeepers sustained the apiary from season to season by excluding a portion of the hives from the annual destructive harvest. Empty hives were placed in or near the apiary following harvest and, with luck, were refilled through the actions of bees themselves as a result of reproductive swarming. Over the course of the next century, advances in beehive design, beekeeping equipment technology, transportation (especially steamship travel) and an increased awareness (and use) of honey bee biodiversity led to systems of managing honey bees that, by 1920, would be thoroughly familiar to apiculturists today.

From the early skep or box hives that were in use by European and North American beekeepers in 1820, a myriad of hive designs was developed and

Dutch swarm specialists inspecting bargains at the bee market in Holland.

FIGURE 4.2 Honey bee skeps being inspected in a market in Holland, 1920. Author's collection.

promoted over the next few decades. In 1865 Moses Quinby, who would later devise the forerunner of the modern bee smoker (used when opening a hive to calm the bees), commented wryly on the many hive designs being reviewed and promoted by state fair committees for certification or patents: "[W]hen a man comes along who has nothing to offer in favor of his hive, further than the unprecedented amount of honey secured by it and a favorable notice from a stupid committee, I feel very much like dismissing him without ceremony; it is evident he has chosen a sorry vocation" (Quinby 1865). He further clarified, "I will guarantee that more honey will be stored in a barrel, box or hollow log, just large enough to hold all that is gathered, than any patent fixture ever presented for this special purpose." Despite his disparaging remarks, Quinby went on to describe a number of the improved types in use at the time, evidencing the perceived need among Quinby's contemporary beekeepers for a better beehive. There was the "Chamber Hive," which had a lower box for brood and an upper one for honey, the "Suspended Hive," a design promoted to keep the combs away from pesky mice, and the "Inclined Bottom Board Hive," in which falling pestiferous wax moth larvae (that Quinby referred to as "worms") were supposed to roll away from the hive (described by Quinby as an "utter failure"). Quinby covered additional patent hives such as the "Changeable Hive"—a stackable design to allow the beekeeper to remove the oldest section of combs each year—but dismissed it as actually just three "box hives" stacked one upon the other. While the list continued with a number of other designs, a notable patented design that attempted to employ bee biology in service of the beekeeper was the "Dividing Hive," used to multiply (bee) stocks "at pleasure." The hive was made of two parts with a central divider and when full (with combs containing brood, honey and pollen) could be separated into two parts, with a new empty half added to each. The idea was that the queenless half would produce a new queen and the beekeeper would then have two hives. Quinby noted that bees would often put brood in one half and honey in the other and the subsequent division thus would not lead to a new queen in the split.

Quinby's 1865 book was a rewriting of his original book of nearly the same title first published (with various intervening editions) in 1853. It was important for Quinby to write this revision because of a true breakthrough in hive design that occurred in the early 1850s. In 1852, L.L. Langstroth patented a practical moveable frame hive design that remains the primary hive design in widespread use today, almost 170 years later. Langstroth is generally credited with the novel incorporation of "bee space" into his hive design. "Bee space" is the space left by bees between the combs they build and between combs and the walls of their nesting cavities, generally around 0.5–0.9 cm (1/4–3/8 inch) (Root 1920). By incorporating a hive design with dimensions that maintained a "bee space" distance between frames (the wooden rectangles that surround each wax comb), between the frames and the hive walls, and between the frames and the beehive top and bottom, the bees did not glue together frames with either wax comb or propolis (bee-collected tree resin) and, thus, frames remained truly "moveable" by the beekeeper. This allowed frames to be moved around the hive and, importantly, also between hives. While Langstroth could not have foreseen the worldwide impact of his design (and realized little financial reward from it in his lifetime due to legal battles related to patent infringement), he was rather immodest about his design: "... I shall enumerate certain very desirable, if not necessary, qualities of a good hive ... I prefer inviting the attention of beekeepers to the importance of these requisites, some of which as I believed, are contained in no hive but my own" (Langstroth 1853). Perhaps unappreciated at the time, Langstoth's development of a practical moveable comb hive changed the beekeeper-honey bee relationship. No longer was it necessary to destroy bee colonies to harvest honey and beeswax. Instead, honey combs could be removed, wax cappings removed and the honey extracted in a centrifugal machine. Afterwards, the combs could be returned to the hive for reuse. This opened new management possibilities for improved honey production (it is costly in terms of honey for bees to secrete wax and build fresh

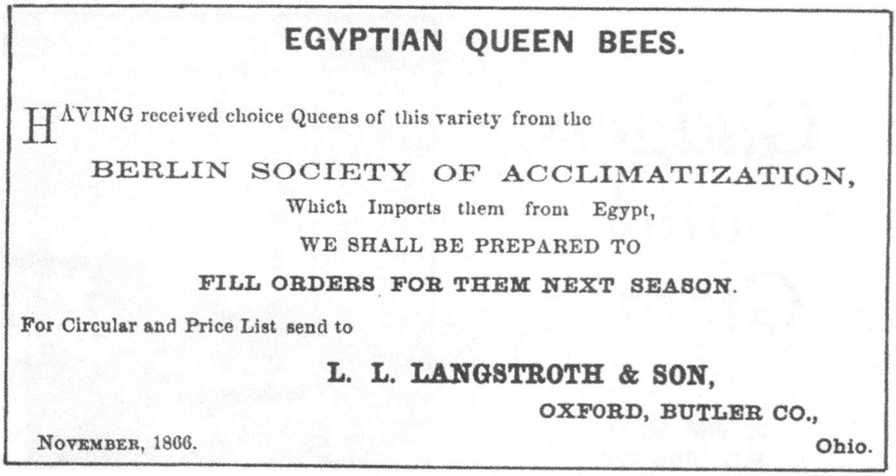

FIGURE 4.3 Advertisement of honey bees for sale, 1866; Berlin Acclimatization Society. Author's collection.

combs). Moreover, given a growing understanding of bee biology, beekeepers could now make "splits" from their hives and include frames containing correct-aged larvae to assure that the split would be able to rear a new queen. Although Quinby remained a strong supporter of a simple box hive design, due to its low cost and simplicity, homage to the Langstroth hive is evident in the full title of his 1865 edition, *Mysteries of Bee-Keeping Explained. Containing the result of thirty five years' experience, and directions for using the movable comb and box-hive together with the most approved methods of propagating the Italian Bee.*

Reference to the Italian bee in Quinby's title was remarkably topical in 1865, given that this honey bee was imported and introduced into the US in 1859 and first offered for sale to beekeepers a year later. An awareness of the remarkable biodiversity of the honey bee throughout its native range and the desire to use and evaluate "exotic" honey bees had been growing in Europe, in part due to the activities of the Berlin Acclimatization Society (Figure 4.3). This group and others were active in importing honey bee strains from the Middle East and Egypt into Europe.

HONEY BEE IMPORTATION, PROPAGATION, AND DIVERSITY

The honey bee (*Apis mellifera*) is endemic to Europe, Africa, and portions of Asia. In 1820, the only honey bees available in North America were descendants of a subspecies imported from northern Europe known scientifically as *Apis mellifera mellifera.* Commonly known as the "Dark Bee of Northern Europe," and also as the Black Bee, German Black Bee, and German bee (Cook 1894; Phillips 1928), this honey bee was introduced into Virginia in the 1600s (Crane 1976; Oertel 1976). It remained the only strain available in the New World until the introduction of the Italian honey bee in 1859 (Anonymous 1859). It is difficult to exaggerate the acclaim that Italian honey bees found awaiting them upon arrival in the New World. Advertisements for the new golden yellow honey bee spoke of its beauty, gentleness, vigor and productivity. Prices as high as $100 for an Italian queen were reported, equivalent to over $3,200 in 2021. The importation of the Italian honey bee was only the beginning of a relatively active period of importation, promotion, and distribution of honey bee subspecies from Europe, Africa, and the Middle East to the Americas. Technological developments in transportation led to widespread transatlantic steamship service in the latter half of the 1800s and reduced crossing time to as little as two weeks. Shortened shipping times meant less mortality in transport. From 1859 until 1922, seven currently recognized Old World honey bee subspecies were introduced into the United States (Sheppard 1989). These included subspecies with endemic "homelands" ranging across Europe, the Middle East, and Africa.

The bee shipping boom would change two decades after the turn of the twentieth century. As early as 1904, honey bees on the Isle of Wight began dying from an unknown malady (Imms 1907) and, soon thereafter, widespread colony deaths were reported from continental Europe. The causative agent was a mite that lives within the breathing tubes of honey bees (Rennie 1921). While surveys conducted after Rennie's discovery revealed no evidence of this mite in US honey bee populations, Congress passed the Honey Bee Act of 1922, restricting further honey bee importation into the United States. At some level, the Act was successful, as tracheal mites remained outside the US until 1984. Thus, aside from the initial importation of the "Dark Bee," a few specific importations of honey bee genetic material made with USDA oversight and the arrival of Africanized honey bees from South America (bees derived from an African subspecies initially introduced into Brazil in 1957), the vast majority of honey bee genetic material that now exists in the United States derived from introductions made during the Age of Industry.

Of the eight honey bee subspecies that were introduced (including the Dark Bee), descendants of only three European varieties found "favor" with US beekeepers, as measured by persistence beyond 1920 as commercial "strains" available to beekeepers. These include the wildly popular Italian honey bees and two dark strains of bees from mountainous origins, the Carniolan (homeland in the Alps) and the Caucasian (homeland in the Caucasus). The yellow Italian honey bee remained by far the most popular strain in the US, although the Carniolan was more widely used in Europe (Ruttner 1987). In fact, in Germany the Carniolan honey bee has now almost entirely replaced their endemic Dark Bee—an ironic happenstance, in view of the common name "German Black Bee" that was widely used for the Dark Bee in the US. The popularity of the Italian honey bee after its introduction into the US, and the growing demand for more and more Italian queens to replace the widespread Black bee, led to another technological beekeeping milestone that occurred during the Age of Industry: large-scale queen production.

ECONOMIES OF SCALE: INDUSTRIAL PRODUCTION OF HONEY BEE QUEENS

As the development of a practical moveable frame hive changed the relationship between beekeeper and bees from one of destructive honey harvesting management to one of agricultural "husbandry," it also provided the means for finer management of the inner workings of the hive. Frames could be moved about, within, and among colonies and larvae of different ages could be accessed and handled to make colony divisions. For the first time, the widespread adoption of moveable frame hives made it possible for beekeepers to closely observe the combs and detect brood diseases or comb-destroying insects. This advancement would be vitally important for large-scale honey bee operations on the horizon.

Various strategies to produce queens were developed that used comb manipulation and modifications to produce larger numbers of daughter queens than could be made by simple colony division. In 1888, Doolittle published a method of moving young larvae to specially prepared wax cups and placing the cups into frames in queenless colonies. This revolutionized queen production and made possible the growth of a commercial queen production industry (Doolittle 1888) (Figure 4.4). The basic principles that Doolittle presented (some of which he admitted appeared to him in dreams) are still used today by queen producers who produce tens or hundreds of thousands of daughter queens annually. The full title of Doolittle's contribution, *Scientific Queen-Rearing as Practically Applied being a Method by which the best of queen-bees are reared in perfect accord with Nature's ways*, gives some idea of how an accumulating empirical knowledge of honey bee biology and trial and error testing led to major advances in beekeeping technology and output.

Doolittle's technological innovation allowed a beekeeper to "graft" hundreds of worker honey bee larvae from a single queen, place them in specially prepared hives, and rear them all as new queens, yielding a never before imagined level of queen production. Doolittle made use of a growing understanding of bee biology at the time. In this case, the determining factor of whether a female honey bee larva develops into a functional queen or a

FIGURE 4.4 Beekeeper holding frame of newly capped queen cells. Photo: Walter S. Sheppard.

sterile female (worker) depends on a physiological response to different feeding regimes by nurse bees. In the absence of a queen, a strong colony without access to other young larvae will feed correct-aged larvae with a special diet that leads to their development into queens. By manipulating the brood and bees in such a system, up to a hundred or more queens could be made in a single hive at a time. Using Doolittle's published principles, beekeepers could specialize as queen producers and make money through the sale of queens, rather than relying on honey production alone.

Specialized operations for queen rearing became common in warm areas with an early spring, such as the southern US and California, and such operations continue to this day. Augmented by progress in aviation, Hawaii is now also a center of extensive queen production. Queen producers graft from their strongest, most vigorous colonies and sell their products as being derived from "selected" breeding stock. However, the fact that virgin queens fly in the air and mate with multiple flying drones presented a special difficulty for producers to actually control matings in a breeding program, a problem that was only beginning to be solved in the late 1920s with the development of instrumental insemination (Watson 1928). Nonetheless, the vast majority of commercially produced queens, up to one million annually in the US, are the result of "natural mating" of virgin queens from grafted stock, fulfilling the dreams of Doolittle (Delaney et al. 2009).

The vision of a modern queen production operation, including the manual manipulations employed to graft larvae, the movement of frames to set up "queenless cell builders" (specially prepared colonies for rearing new queens), and the use of "mating nuclei" (small hives) to emerge new virgin queens and hold them until mating flights are completed, would be unsurprising to Doolittle. In contrast, he would find quite unfamiliar the employment of current-day technologies, such as forklifts, palletized migratory beehives, and large trucks used in an agricultural system that moves more than a million beehives annually across the US to pollinate specific agricultural crops. However, during Doolittle's lifetime (1846–1918), a period covering a major portion of the time frame of this volume, no less a set of sweeping changes occurred in beekeeping as the result of industrial production during the Age of Industry.

INDUSTRIAL MECHANIZATION BRINGS TECHNOLOGICAL ADVANCES IN BEEKEEPING EQUIPMENT

The widespread adoption of Langstroth's 1850s moveable frame beehive subsequently saw the development and improvement of a number of other essential tools that impacted beekeeping tools and honey production. With the use of moveable frames, and the promise of "reusability" of combs following the removal of honey, mechanical centrifugal honey extractors soon followed. Major Francesco de Hruschka, a retired military officer living near Venice, Italy, is credited with the invention of "Il Smelatore," the honey extractor (Cook 1894; Comstock 1905; Dadant 1915). According to most accounts, in 1865 Major Hruschka observed his son placing a piece of honey comb on a plate, setting the plate in a basket and swinging the basket around him on a string. As a result of his son's play, the honey drained from the comb and, thus, the Major came upon the idea to use centrifugal force to remove honey from combs. Two years later, the first American honey extractor was reportedly made by Langstroth and Wagoner (Cook 1894). By the late 1800s, extractors were built in many configurations including with or without gearing. Years after the Hruschka discovery, Charles Dadant (1915) noted, "as soon as we heard of the discovery, we had a machine made." The present-day Dadant and Sons, Inc., who trace the start of their company to 1863, still produce hand-cranked and electric centrifugal honey extractors.

FRUITS OF THEIR LABORS—A NEW ERA IN HONEY PRODUCTION AND OPPORTUNITIES IN THE ORCHARDS

The development and improvement of machinery to remove honey from combs with centrifugal force led to an immediate change in the output of beehives related to honey production. Anna Comstock (1905) asserted that "almost double the honey may be produced," given that the emptied combs could be reused, the bees did not need to completely rebuild the honey storage combs, and beekeepers could even extract honey from brood combs. She also noted that hive management, especially swarming control, was easier "because the bees more readily accept the enlargement of their quarters when the supers contain fully made comb, [and] also larger hives may be used." She also identified a potential disadvantage: "unless great care is given, the bees will be starved through this convenient way of pilfering their stores." The handling, storage, and shipment of liquid honey was found to be simpler and more economical than handling comb section honey (small squares of honey-filled comb built by the bees in special wooden frames); and technologies to re-liquefy honey that had crystallized in storage containers developed, including hot water baths and jacketed storage tanks. End-user consumers could buy clear liquid honey in glass containers and avoid the need to squeeze combs to get liquid honey in the kitchen or to eat beeswax along with honey on their toast.

While honey extractors modified the production, handling, and consumption of honey, at the end of the nineteenth century beekeeping was still arguably circumscribed by the production of honey (and to a lesser extent) beeswax. Products of the hive remained the prime motivator of the art of beekeeping, both philosophically and economically. However, by the end of the Age of Industry, inklings of the future appeared regarding what we now think of as the primary "mission" of managed beekeeping—the pollination of plants used in agriculture. In the 1920 edition of the encyclopedic tome *ABC and XYZ of Bee Culture*, the author refers to the historical conflict between beekeepers and orchardists, whereby "fruit-growers ... asserted that bees injured the bloom, punctured the fruit, and interfered with the packing" (Root 1920). In some instances, beekeepers were even asked to remove bees from the orchard due to the perceived nuisance. Yet some modicum of harmony was materializing: "[At] the present time the two factions understand that their industries are mutually dependent" (Root 1920). Thus, in the chapter on fruit blossoms the author cites a number of orchardist testimonials on the value of honey bees, including a comment by a large California fruit grower, "Bees and fruit go together. I can't raise fruit without bees"; and by the owner of an 800-acre New Jersey fruit farm, "I could not do without bees. I never take a pound of their honey. All I want them to do is to pollinate" (Root 1920). These remarks, reflecting grower recognition of the importance of managed bees to large-scale agriculture, were very prescient, considering both the current economics of commercial beekeeping and the value of pollination compared to hive products. The primary source of income for many commercial beekeepers today derives from payments by growers to locate honey bees on their crops during bloom, with almond pollination service fees accounting for over 82 percent of all paid US pollination service fees in 2015 (Ferrier et al. 2018). The relative value of hive products compared to the value of pollination shows how far things have come since 1920. While few records exist to estimate the annual production of honey and beeswax at the end of the Age of Industry (1920), Jordan and colleagues estimated the recent value of honey and beeswax from honey bees to be around $250 million and $30 million, respectively (2021). In comparison, the economic value of pollinating insects in the US (with honey bees as the main contributor) was estimated to be $34 billion per year. Similar numbers are reflected in the European Union, where an estimated 80 percent of 264 crop plants require insect pollination (Chauzat et al. 2013). Overall, the annual global economic value of pollinating insects has been estimated to be €153 billion (Chauzat et al. 2013), far in excess of any estimates for the beehive products of honey and beeswax.

During the Age of Industry, four other insect-derived products—silk, cochineal, shellac, and gall-based ink—were widely traded commodities. Locally produced, each traveled along circuits of global commerce. Over time their production and consumption fluctuated, subject to the law of supply and demand, which affected the livelihood of individuals and impacted the economies of nations. As the nineteenth century progressed, cheaper synthetic substitutes were developed and manufactured for each product—nylon for silk, aniline dyes for cochineal and ink, vinyl for shellac-based phonograph (gramophone) records. Each synthetic commodity grew in popularity and eventually surpassed the demand for its natural alternative. Below we explore the origin of these insect products, their various uses, and their significance on the world market.

MOTH TO CLOTH: THE SILK INDUSTRY

Dubbed the "queen of textiles" for its sheen, luster, and luxuriousness, silk is the most expensive natural fiber. Soft yet resilient, warm yet breathable, absorbent without feeling damp, silk also has great tensile strength and elasticity. These extraordinary attributes explain its great desirability through the ages. Many insects (and all spiders) produce silk, but the product used most in textile manufacturing originates as two very thin, long filaments—each up to half a mile in length—secreted by the paired salivary glands of *Bombyx mori* (Linnaeus) caterpillars, commonly called silkworms. Numerous species of feral silkworms exist, but *B. mori* are fully domesticated. Prior to pupation, the caterpillar spins a silken cocoon within which it metamorphoses to the pupal stage before emerging as an adult moth. For silk production, pupae must be killed before adult emergence to obtain unbroken, silken cocoon filaments that will be reeled and twisted together to form threads of a desired thickness. This raw silk contains a gummy protein, sericin, removed by boiling in soapy water before dyeing, spinning, and weaving.[1]

The cultivation of silkworms to produce silk is an ancient Chinese practice, traditionally the purview of rural women. For millennia, sericulture remained shrouded in secrecy, per imperial rule that forbade sharing of the insects and techniques under penalty of death (Cloudsley-Thompson 1976). Once the secret was revealed, the practice spread, first to Japan and later to the Mediterranean basin, Europe, and beyond. In pre- and post-colonial US, raw silk production occurred as a cottage industry, safeguarded for decades by tariffs, and by 1820 various states had enacted laws promoting sericulture (Klose 1963). To further encourage its citizens to "grow and manufacture" silk, the US federal government issued a lengthy "how-to" manual. This publication also bemoaned the five-year values (1821–1825) for US silk imports ($35 million) versus exports ($8 million) and chided, "What a bounty is paid by us to support the agriculturist and manufacturer of other nations, on articles which our country ... might supply!" (Rush 1828). The suggested recourse? If every farmer would plant an acre of mulberry trees to feed silkworms, five hundred–six hundred pounds of raw silk could be produced (presumably per farm), enriching both farmer and nation. While some valiant souls did rise to the challenge, success and wealth from this venture proved elusive. Developing caterpillars require staggering amounts of their preferred food—fresh leaves from *Morus alba*, white mulberry, native to Eurasia.[2] Red mulberry (*M. rubra*) found in the eastern US, is a substandard food source, as is *M. multicaulis*, the species that speculators imported and resold at preposterously high prices to people hoping to get rich quick (Ryland 1939). Those hopes never materialized. In the final analysis, Americans could not or would not muster the resolve required to rear silkworms, nor could the industry compete with the relatively low wages paid to Asian workers.

Sericulture was arduous and exacting work. Cocoon reeling, the process of manually unwinding the cocoons' silken filaments and combining them to yield a single thread, required years to master. As the cocoons bobbed about in simmering water the worker had to locate each end of the filament from several cocoons (the number determines thread thickness) and blend them together to produce a single, smooth thread of uniform thickness (Figure 4.5). Only the middle portion of the cocoon's filament was used, the outer and inner portions being unsuitable for reeling. Such discernment required

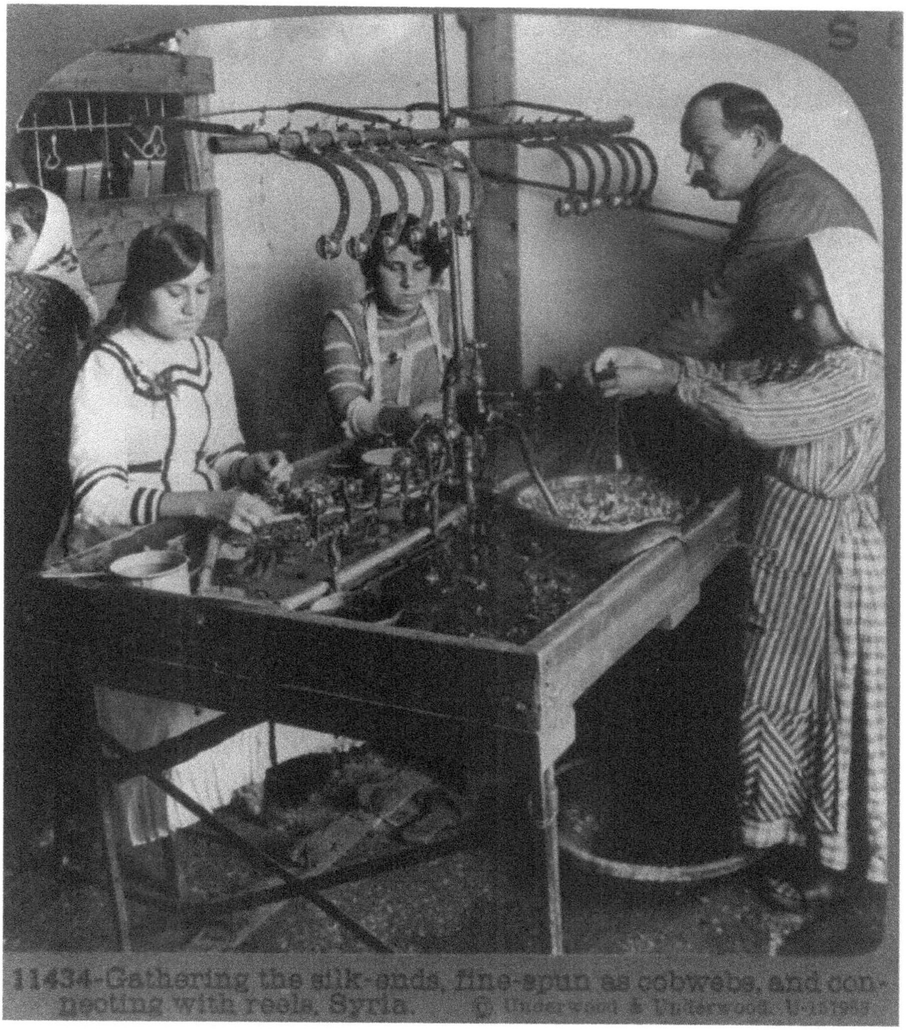

FIGURE 4.5 "Gathering the silk ends," silk cocoon reeling. Syria, Mount Lebanon, 1914. Keystone View Company. Carol M. Anelli.

competent judgment, gained through experience over time. Variably sized threads, while practical for hand sewing, were incompatible with machine looms, and American silk was badly reeled. Ultimately the US would excel at silk production using imported raw silk and its own power looms, with various silk dyehouses also in operation (Heusser 1927 provides a comprehensive view of the industry).

For much of the Age of Industry raw silk formed the basis of the greatest trade nexus between the industrialized West and the Far East. Silk traversed the globe along two major circuits: westward to the Mediterranean, Europe, and across the Atlantic to New York City; and eastward to Vancouver, BC, and US ports in Seattle/Tacoma, San Francisco, and Portland (Ma 1996, 1999; Vanterpool 2010). In 1820 Italy dominated the raw silk export market (66 percent of world total), well ahead of India (17 percent) and China (12 percent)[3] (Federico 1997). After the first Opium War (1842), Britain imposed a treaty port system that ensured traded goods would cross China's borders unfettered and tariff-free. As a consequence, new east–west silk trade systems developed that impacted global commerce of the commodity.

In addition to armed conflicts and unequal treaties, silk production and trade fell victim to the vicissitudes of nature. In 1845 *pébrine* (also known as "pepper disease"), an infectious protozoan disease of silkworms, first appeared in France (James and Li 2012). Renowned since the Renaissance for their woven silk (particularly from Lyon), the French saw cocoon production plummet from about twenty-six million kilos in 1853 to four million in 1865 (Li 1981). *Pébrine* soon devastated European sericultural operations, causing a fall in silk output worldwide and supply-side shifts. As a result, by 1860 China commanded 51 percent of export shares, eclipsing Italy (27 percent) and third-ranked India (9 percent) (Federico 1997). France, whose sericulture industry never recuperated, became reliant on China for its supply of raw silk. Two additional, concurrent events also affected world silk production: hostilities of the Taiping Rebellion, which reached the sericulture areas around Shanghai in the late 1860s, decreasing the global silk supply; and the Meiji Restoration in Japan, after which the country's feudal system was transformed along the norms of western industrialized societies (Li 1981; Federico 1997). The Japanese adopted modernized silk-reeling and implemented scientifically based sericultural methods, including breeding programs to improve silkworm stock, sanitary cocoon warehouses, and precautionary methods against silkworm contamination and disease transmission. These efforts had far-reaching effects, uniquely positioning Japan to meet the increasing global demand for silk in the coming decades.

Determining the origin of the mysterious affliction that had devastated sericulture in his native France fell to the illustrious scientist, Louis Pasteur. After years of meticulous observations and experimentation, he showed that *pébrine* and *flacherie* (flaccidness), another silkworm malady, were caused by completely different microorganisms. Pasteur established simple, effective techniques for rearing silkworms disease-free (which Japan implemented) that not only saved sericulture,[4] but also proved foundational for the germ theory of disease, a tenet of modern biology and medicine (Debré 1998).

During the latter half of the nineteenth century innovations in communication and transportation, and unprecedented technological advances, transformed the silk industry. Transatlantic telegraph cables enhanced communication and international trade, and the opening of the Suez Canal in 1869 (an enterprise

Britain opposed) provided a direct Silk Road connection between Europe and East Asia via the Mediterranean and Red Seas. This route obviated the need for France to purchase British re-exports of raw silk, and Britain's commercial dominance in this arena declined (Ma 1999). Subsequent innovations and improvements in transoceanic steamship design and fabrication further reduced transit time of Far East exports. But it was completion of the first North American transcontinental railway (1869), which connected the west and east coasts, that gave rise to the phenomenon of special silk trains (Field et al. 2007).

Operating from about 1875 to the 1930s, silk trains, freighted only with bales of raw silk from Japan and China, enthralled the public and became the subject of legends (Iseminger 1994; Vanterpool 2010). Departing from major west coast ports in BC and the US, these trains sped eastward, "as fast as humanly and mechanically possible,"[5] reaching New York City within five days. No passenger trains approached that record and the few people aboard a silk train were those paid by the railroad. From the port of embarkation in the Far East to the cargo's final destination, each transit step was painstakingly choreographed for speed and efficiency.[6] A week or more in advance, the trans-Pacific steamer wired the railway of a ship's expected arrival; stevedores and customs officers awaited dockside. Raw silk entered the US duty-free as "immediate transportation entry goods" (Vanterpool 2010). Thus crews worked posthaste to transfer silk bales, each about 133 pounds, into special railroad cars with smooth interiors and sealed, protective doors. Once loaded, silk trains stopped briefly, only to change engines about every 150 miles (an operation faster than transferring cargo), and about every forty miles to replenish the water supply needed to produce the engine's steam (whence the term, "jerkwater"). Along the way dispatchers cabled ahead to ensure that nothing—passenger and mail trains included—would impede the train's progress.

Speed was paramount as the stakes were high and delays costly. Insurance premiums, charged hourly, mounted quickly. The premium on $1 million raw silk cost about $600 per hour; thus, an eighty-hour transit portside to New York City cost $48,000 (Vanterpool 2010). To guarantee delivery of the train's cargo the railroad furnished a bond, typically around $2 million, that carried an interest rate of over $300 per day (Vanterpool 2010). Delays also increased the chance of theft. The railroad hired armed guards to protect the costly cargo, although thefts did occur. In 1919 about $45 million of raw silk was looted from silk trains on US railroads (Vanterpool 2010).

The silk market was unstable and notoriously risky. To survive financially, raw silk importers needed to turn a profit in a highly volatile market with fluctuating exchange rates while absorbing the cost of product, brokerage fees, transportation, etc. The manufacturers' milieu was subject to the caprice of fashion, which encompassed a dizzying array of silk weaves and fabrics (charmeuse, crepe, chiffon, satin, velvet, etc.), colors and hues, styles and trends. The wedding dress worn by President Teddy Roosevelt's daughter and gowns selected for the inauguration of President Woodrow Wilson by his wife and three daughters had an enormous effect on what was considered *de rigueur* (Li 1981). Hemlines rose, dress styles became simplified, and the manufacturing industry was impacted. In 1906 a typical dress required thirteen yards (more than ten meters) of silk fabric compared to just two yards in 1929 (Vanterpool 2010). But unquestionably the wildest swings in the industry occurred in the aftermath of the First World War. Raw silk prices on the New York market stood at about $5–7 per pound in early 1919 and spiked to $16 per pound in January 1920 before plummeting in December to $5–6 per pound (Silk Association of America Annual Report 1921). Industrialists who bought high to meet silk orders lost money when prices nosedived. Newspapers reported attempted, and successful, suicides: for example, one wealthy silk merchant took his life after losing $2 million (*New York Times* 1920). Reflecting upon the industry, one US silk manufacturer lamented "… the year 1919 might be described as one long golden dream of mounting (paper) profits and the year 1920 as a horrible (but unfortunately too real) nightmare" (Silk Association of America Annual Report 1921). Such dreams and nightmares were likely unimaginable back in 1874, when the US imported just 5 percent of the world's raw silk, valued at $4.5 million (Federico 1997; Vanterpool 2010). In the industry's boom year of 1919, the US imported nearly forty-five million pounds, valued at nearly $330 million dollars (Silk Association of America Annual Report 1921; Federico 1997). Between the world wars, the US imported greater than 75 percent of the world's raw silk (Iseminger 1994).

The US silk industry had undergone a transformation. In the 1860s it comprised a handful of factories and workshops that produced only sewing silk and trimmings on imported surplus machinery, which American entrepreneurs had purchased at rock-bottom prices from inactive British silk textile centers (Margrave 1975).[7] In a few decades the US industry had expanded rapidly and was flourishing. From its large-scale factories came broad silks (for dresses, underskirts, and blouses—known as "waists"), velvets, laces, pajamas, lingerie, knitted goods, hosiery, ribbons, men's ties and shirts, upholsteries, and much more (Allen 1902; Silk Association of America Annual Report 1921), often fabricated on American-made machines. The impetus for this dramatic boom in the industry can be traced to various political, economic, and social forces abroad that led to a diffusion of silk manufacturing skills and technology from Europe to the US (Ma 1996, 1999). The British silk weaving industry had collapsed following the country's 1860 treaty with France, whereby French silks entered Britain duty-free. Faced with operating under a competitive disadvantage, a growing number of British silk mill owners and jobless, highly skilled laborers immigrated to the US, where the silk industry was entering a major growth phase. Numerous immigrants would also arrive from southern and eastern Europe, prompted by a variety of factors and circumstances (discussed below).

Technological advances represent another critical element in the US silk industry boom, reflected in a growing number of mills to house the new machinery. Improvements and innovations had produced larger, and greater labor-saving US looms compared with European looms. Moreover, US manufacturers had become more adept at fabricating machines to produce the array of specialty silks that characterize the industry's heyday. Small towns and rural areas in the northeast accommodated ever more silk factories and complexes. In 1876 about thirty mills were in operation (Vanterpool 2010), burgeoning to 483 mills across sixteen states within twenty-five years (Allen 1902). John Ryle, a British émigré from a family of silk manufacturers, had established the first successful US silk factory in Paterson, New Jersey. Paterson would become the foremost US silk manufacturing center, producing the most ornate weaves and patterns owing to its many skilled immigrant workers who had trained in European silk mills (O'Donnell 1985). The city was known as "Silk City, USA" and Ryle earned the moniker "father of the US silk industry."

By 1900 the US had catapulted to the top of the global silk import market, a trend that persisted into the 1930s. Li (1981) regards this as the "single most important development in the international silk market of the late nineteenth and [early] twentieth centuries." The prodigious amount of raw silk entering the US required a labor force to convert it to consumable goods. Driven by a host of reasons—escape from crop failures and severe economic conditions, freedom from religious and political strife, the promise of a better life for themselves and their families—between 1880 and 1920 more than twenty million people emigrated from southern and eastern Europe to the US.[8] Many found work in factories, including the silk industry, where women and children typically toiled ten hours per weekday plus a half-day Saturday. Child labor laws, which established a minimum employment age and prohibited "oppressive child labor," would not be enacted until 1938 during President Franklin D. Roosevelt's second administration.[9] Mill owners increased their profit margins by hiring women and children and some even moved operations to different states with more lenient labor regulations. In 1913, gender-based daily wages for males (silk weavers $3.00, dyers $2.50) were substantially higher than those for women (silk winders $1.40), girls (silk spinners $1.50), and children (tasks unspecified $1.00) (Silk Association of America Annual Report 1913). Clearly, very few associated with the US silk industry shared in the wealth and prosperity of its meteoric success. Paterson, known internationally as "the Lyons of America," acquired widespread notoriety for a labor strike by more than twenty thousand silk workers that lasted nearly five months, and which claimed the lives of an innocent bystander and a striking worker. In the end, virtually none of the workers' demands were met. The Second World War would effectively end raw silk imports from the Far East, and competition from synthetic fibers, first rayon and later nylon, would sound the death knell for US silk manufacturing (Scranton 1985).

Toward the end of the Age of Industry the supply side of the global raw silk market also underwent significant changes. As noted, during the later decades of the nineteenth century China supplied over 50 percent of the world's silk, and by 1887 silk had surpassed tea as the nation's top export, valued at about twice its tea trade (Li 1981). But as China diversified its exports, its share of the global silk export market declined appreciably. Meanwhile Japan, having modernized its industry, assumed the top slot among the world's suppliers of raw silk, with some individual silk factories employing more than twenty thousand workers (Federico 1997). These represent phenomenal transformations when one considers that, just thirty years prior, Japan laid claim to less than 10 percent of the world's silk export share. Yet as was the case with the US silk industry, Japan's spectacular success came with a human cost. As the industry grew and with it the demand for labor, many impoverished rural families sent their young daughters off to jobs in silk factories, unaware of the dire working and living (factory dormitories) conditions, excessively long hours (upwards of eighteen hours daily), and low wages that typically awaited them (Hunter 1993).

COCHINEAL

Long before the advent of synthetic dyes, people had been extracting cochineal from gravid female scale insects to produce dyestuff of brilliant crimson and scarlet (Figure 4.6) (Cardon 2007). As one of the most expensive of all dyes, cochineal often added more to the cost of luxury textiles than did the silk, wool, or linen used to create them. Used internationally for centuries in textile production, cochineal also found applications, which persist to the present, in the food and cosmetic industries and as a dye for visualizing tissues under the microscope (Lee 1913; Dapson 2009). During our period of interest most cochineal traveled along trade routes that originated in the New World, where the long history of cochineal traces back to cultures of Indigenous peoples of pre-Columbian Mesoamerica, Mexico, and South America. The domesticated variety of the dried insect, termed *grana fina*, is the most valuable, producing the highest grade of dye. The insects, which feed on prickly pear cacti, required laborious and careful cultivation by peasants; approximately seventy thousand dried female cochineal insects yielded one pound of dye (Salinas 2018). Cochineal gained favor with European painters, who employed it as a red lake pigment (Anderson 2015), and modern analytical techniques

FIGURE 4.6 "Insectes Utiles." Nineteenth-century French trade card depicting cochineal insects on cactus, produced by Aralette and Co. Courtesy of James P. Strange

confirm its presence in several paintings in the National Gallery, London. For example, Delaroche (1833) used cochineal as a red glaze for certain articles of clothing in *The Execution of Lady Jane Grey*, which depicts the seventeen-year-old "Nine Day Queen" moments before her beheading at the Tower of London (Kirby and White 1996). Oaxaca remained the global source of New World cochineal for two centuries, with peak production of about a million pounds annually (Donkin 1977). But Guatemala supplanted Oaxaca during the 1850s by nearly doubling that amount (Anelli and Prischmann-Voldseth 2009). After its New World colonies gained independence, Spain sought to recoup some of its losses by importing cochineal insects to its Canary Islands. There, aided by government incentives, cochineal exports skyrocketed, from an initial eight pounds in 1831 to more than six million pounds in 1875 (Donkin 1977). Yet as fate (and human invention) would have it, the cochineal market all but collapsed with commercialization of the world's first artificial dye, which also had an entomological connection. William Perkin, a student at the Royal College of Chemistry, London, had been endeavoring to synthesize the antimalarial drug, quinine, for British soldiers in the Crimean War, when he serendipitously produced the colorfast aniline dye "mauveine." The new shade of purple triggered a fashion sensation, and other synthetic dyes soon appeared.

LAC PRODUCTS

Used together, cochineal and shellac, another insect-derived commodity, produced an enduring legacy in the exceptional finish characteristic of Stradivari's violins of the "Golden Period," celebrated for their inimitable acoustics (Echard et al. 2010). Shellac comes from lac insects, a type of scale insect native to India, China, and parts of southeast Asia. After settling on the twigs of their host trees, these soft-bodied bugs suck sap for nutrients and secrete a reddish-brown resin that hardens and protects them. Commercial use of lac to fashion toys and bangles and make a dyestuff, lacquer, and sealing wax dates to antiquity. But for centuries, lac insects were mainly cultivated to obtain lac dye, as the value of lac resin to produce shellac did not gain prominence until the nineteenth century (Hicks 1961). Of greatest economic value are female lac insects, which aggregate and make great quantities of encrusted "stick-lac" that is harvested and processed to produce shellac flakes. Flakes dissolved in alcohol yield liquid shellac, used as a protective finish by luthiers (Berenbaum 1995) and on fine furniture such as was crafted by Chippendale, Sheraton, and Hepplewhite in England and Duncan Phyfe in the US (Hicks 1961). The centuries-old method of French polishing requires the laborious and expert application of thin layers of shellac to produce a high-gloss finish on furniture and wood floors.

Shellac is a bioadhesive polymer, and during the Age of Industry enterprising inventors used it as a binding agent to manufacture thermoplastics (a substance that is moldable when heated and hardens when cooled), and thus arose a variety of lucrative commercial ventures. The increasing prevalence of daguerreotypes, which were easily damaged by scratches and fingerprints, drove the need for miniature cases to protect and display these early photographs. At first, wooden boxes were fashioned to meet the need, but soon patents were filed that described hard, decorative picture cases, molded from a mixture ("union") of shellac and sawdust plus a coloring agent, such as Japan Black. By the 1850s thousands of so-called Union Cases (Figure 4.7) were being manufactured by factories in the northeastern US. These would prove popular with American Civil War soldiers, who safeguarded their cherished images of loved ones within such thermoplastic cases[10] (Berg 2016). Around the turn of the twentieth century the demand for shellac increased greatly with its use as a binding agent to manufacture phonograph cylinders and 78 rpm phonograph discs. In 1920 the US alone imported more than 11,500 tons of shellac, valued at nearly $25 million (Demillo 2020). Much of this product was undoubtedly pressed into records, to

FIGURE 4.7 Shellac thermoplastic union case, Berg #1-158, "Chasing Butterflies." Berg (2016).

the wonder and delight of countless music lovers, whose phonographs audibly transported them to orchestral concerts, operatic performances, and popular dance halls. Britain's gramophone industry is estimated to have consumed more than one-third of the global shellac produced at the outset of the twentieth century (Roy 2021). Shellac's high insulating properties also found favor with the electrical industry (Hicks 1961), and its waterproofing properties proved useful in munitions manufacturing (Parry 1935). When combined with other materials, shellac was used to attach silver to the backs of mirrors (Hicks 1961).

GALL-BASED INK AND DYE

Humans have long benefitted from the intimate relationship between a particular tiny wasp and the oak species it inhabits. When a female of the cynipid wasp (*Cynips gallae- tinctoriae*) inserts her egg into tissue of the Aleppo oak (*Quercus infectoria*) she also injects chemicals that induce the formation of an abnormal growth termed a gall, rich in tannic acid. Within the gall the developing larva finds a safe haven and all the nutrients it needs to reach pupation, after which the adult wasp chews a hole in the gall, emerges, and locates a mate to propagate the next generation (Figure 4.8). Used medicinally by the Ancient Greeks and Romans, during the Age of Industry both the British

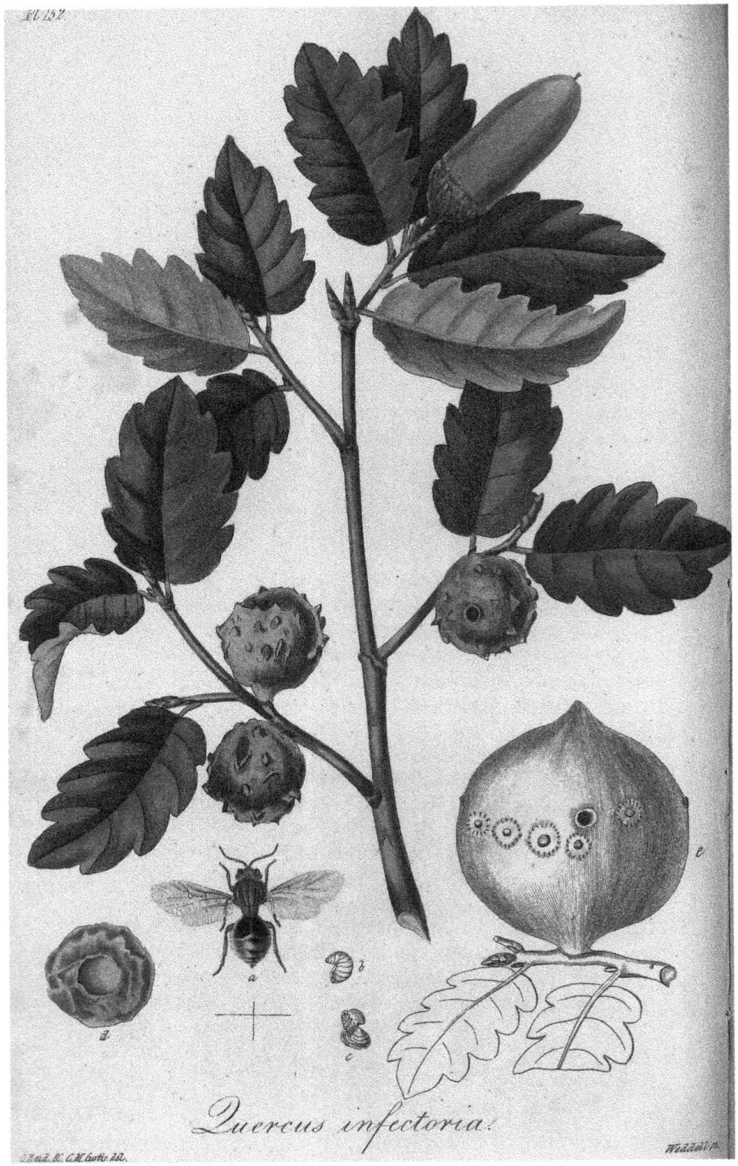

FIGURE 4.8 Cynipid wasp, oak host, and resultant gall. *Medical Botany* by John Stephenson (1831), pl. 152. From the Collection of The Lloyd Library and Museum.

and US Pharmacopoeia designated galls, specifically Aleppo galls (a.k.a. Turkey, Levant, or gallnuts), in recipes for astringent formulations (Fagan 1918). Gall extracts also were used in leather tanning and dyeing, and combined with iron sulfate to manufacture nonfading, indelible black ink for legal deeds, official records, maps, and musical scores. *The Spatula Ink Formulary*, a compendium of recipes, specifies Aleppo galls for inks used to produce banknotes for the US Treasury and Bank of England, among others (Oyster 1912). Unfortunately, those who strive to preserve historical documents, drawings, and silk must often grapple with the corrosion and degradation inherent in iron gall ink and dye preparations (Traveling Scriptorium 2013).

INSECTS AS MODEL ORGANISMS

In addition to insect-related products, scientific contributions from insect studies to the emerging field of genetics were demonstrably transformative. Mendel laid the groundwork in the mid-1860s by publishing his now-famous garden pea experiments, but the significance of his findings lingered in obscurity for thirty-five years until scientists, working independently, unearthed and confirmed his fundamental laws of heredity. Major breakthroughs in genetics soon followed, notably from a series of remarkable discoveries by T.H. Morgan and his students with the humble fruit fly (*Drosophila melanogaster*) (Moore 1993). Today Mendel's laws are taught to elementary school children, and the role of the fruit fly in clearly elucidating Mendelian inheritance and launching modern genetics is indisputable, memorialized in introductory science texts. *D. melanogaster* remains the insect workhorse for fundamental research in biology and medicine, yielding to date six Nobel Prizes in Physiology or Medicine.

IMPORTED SPECIES DEBACLES

In pursuit of commercial products from insects but unaware of the dangers of importing non-native species, humans unwittingly unleashed two exotic organisms during the Age of Industry, with disastrous results. In the first instance, an amateur entomologist living near Boston imported egg masses of the spongy moth (*Lymantria dispar* (formerly known as the gypsy moth)), hoping to cross the species with the silk moth to obtain a disease-resistant, silk-producing hybrid. Some of the *L. dispar* larvae escaped and by the end of the nineteenth century, the species had become established throughout New England, defoliating great swaths of forested areas and urban landscapes (Berenbaum 1995). In the second instance, prickly pear cacti were introduced in an ill-fated attempt to establish a cochineal industry in the British colony in Australia (Queensland and New South Wales) (The State of Queensland, Department of Agriculture and Fisheries 2020). The cacti spread rapidly, as detached pieces of cacti reroot readily and the plants produce an abundance of seeds favored by birds, aiding their dispersal. By 1900, ten million acres had been overrun with the spiny cacti, and by 1920 an area greater than the US State of Minnesota had been rendered virtually useless. Eventually a South American moth, whose larvae feed on cacti, was imported and released as an effective biological control agent, reducing cacti populations to manageable levels with other control methods.

CONCLUSION

Arguably, the day-to-day relevance of some products discussed in this chapter may have diminished as we enter the second century past the Age of Industry. A plethora of alternative materials now addresses some of the basic needs once filled by honey, beeswax, silk, cochineal dye, lac, and gall-based inks. But these alternatives have both advantages and limitations relative to the original insect-derived products and therein can reside the space for argument. For example, even with the wide assortment of synthetic fabrics that have replaced silk in many applications, the beauty and properties of silk continue to ensure its reputation as the queen of textiles and demand stays high. Similarly, some artisans prefer cochineal as a natural dye source to synthetic alternatives, and shellac is favored over synthetic substitutes by fine furniture builders and restorers.

Honey was highly prized as the sole sweetener for Europeans and people in North Africa until the early sixteenth century, when cultivated sugar cane from the New World was introduced. Although the prominence of honey has waned with the widespread availability of alternative sweeteners, the vastly more important role of honey bees and apiculture in providing food security looms ever larger. The recent worldwide attention given to the plight of honey bees related to widespread colony losses, parasitic mites, and new viral impacts drives home that point. Advances made during the Age of Industry stand as testimonials to human innovation. More than a century later, the need for innovation is undiminished and perhaps greater than ever, as we face challenges to feed an ever-growing population within a finite landscape and fragile agroecosystem.

CHAPTER FIVE

Insects in Mythology and Religion

Natural Theology

MEGAN E. MEUTI AND SUSAN W. FISHER

In Western culture, religion and science have vied for primacy in explaining nature for centuries. Following the Enlightenment of the eighteenth century, scientists and philosophers increasingly expected science to provide material explanations for natural phenomena and that religious explanations would cede to science (Porter 2003). Indeed, the notion that all natural phenomena would yield to evidence-based scientific explanations became prevalent. As we are still debating the relative merits of scientific vs. religious explanations of nature, this begs the question: why didn't the Enlightenment view of science prevail in modern times even though our daily existence depends heavily on scientific advances? An important waystation in the journey of competing viewpoints resides in the Age of Industry during which time the philosophy nominally called Natural Theology came into full bloom. This chapter examines the history of natural theology and the forces that led to its dominance in the nineteenth century using insects as a focus. We also examine the influence of natural theology on Charles Darwin and how his view of natural theology changed as the elements of his theory of evolution by means of natural selection solidified. Finally, we explore how the views of scientists and clergy towards natural theology and science changed in Europe and the US following the 1859 publication of Darwin's *magnum opus*.

NATURAL THEOLOGY AND ITS ANTECEDENTS

Natural theology argues that the complexity and majesty of nature both proves God's existence and reveals His nature. Natural theology is distinct from other ways of knowing God, such as divine revelation found in scriptures, although most natural theologists would argue that natural theology is consistent with sacred texts. Because of its naturalistic basis, natural theologists were often clerics, lawyers, or doctors as well as highly skilled naturalists, many of whom made significant contributions to both theological and scientific works of their times. Reflecting their religious training, it was typical for natural theologists to profess a theistic understanding of God's role in creating and sustaining nature. That is, they believed that a supernatural God actively maintained a personal relationship with His creations. In contrast, deists believed that a deity created the universe, but no longer interacted with humankind.

Natural theology can be traced to various ancient cultures including the Greeks (Thompson 1913). However, the modern understanding of natural theology is rooted in the work of three parson-naturalists with significant interests in entomology. The first was John Ray whose 1691 book, *The wisdom of God manifested in the works of the creation*, is considered the rock upon which modern natural theology is founded. The second parson-naturalist, Friedrich Christian Lesser, published a book specifically devoted to insects in 1740, the short title for which is translated as *Insecto-Theology*. In 1802, the third and arguably most famous parson-naturalist, William Paley, published *Natural Theology or Evidence and Attributes of the Deity, Collected from the Appearances of Nature*. From the views espoused by these progenitors arose the version of natural theology that was prominent during the Age of Industry. Because their work was published prior to the Age of Industry, they are not detailed here, but please see Fisher and Meuti (2018).

THE CHANGING RELATIONSHIP BETWEEN RELIGION AND SCIENCE

Scientific progress in the early nineteenth century was hindered by social unrest. The upper class felt threatened by *hoi polloi*, creating a milieu in which evolutionary ideas were dangerous. After all, if organisms evolved, couldn't both humans and society also change? As a result, public protests among the lower classes became linked to antitheological heresies that emerged from scientists and intellectuals. As Appleman (2001: 4) commented: "It is not easy to enter fully into the mindset of the early Victorians, in which both religion and science presupposed the glory of God to be manifested in nature. Nor is it easy to comprehend the fixity of species, or to imagine the intellectual or between readers and nature, many portrayals of insects and their surroundings were figured as if the reader was actually set at the miniature scale of the insects themselves, and functioned to revive, and indeed to magnify, a sense of proximity to nature that had been largely lost with life in the city. Unavoidable urban insects, such as the fly, the cockroach, and the flea, were often portrayed in ways that transcended their mundane ubiquity to reveal a plethora of detail and activity of which readers had hitherto been largely ignorant, framing them as symbols of nature, alive and well, within the cities themselves.

Within this modern, mediatized environment, in which images of insects were directly encountered more often, or perhaps more consciously, than insects themselves, the figure of the insect as denizen of an otherworld took on the qualities of an apparition. This malleability of the insect image has made it an effective vehicle for the symbolic promotion of ideas about society and the self in the modern era. Hogue observes that "the use of an insect image to suggest or denote something other than itself, an idea, or quality, often abstract is to be recognized as true symbolism and distinguished from purely linguistic, artistic, or pragmatic representation or venerations of the insect" (1975: 3). Yet, the ways in which "purely linguistic, artistic or pragmatic representations" might be distinguished from symbolic ones are not entirely clear. Although the accurate, objective and non-symbolic representation of insects has become an essential component of entomology, the idea that such representations are essentially non-symbolic underestimates the symbolic potential of the image, while privileging the dubious "purity" of the image-maker's product. It also overlooks the blending of insect symbolism and natural theology apparent in many nineteenth-century scientific texts.

While Hogue's definition of symbolism is useful, a more extensive appreciation of insect symbolism is possible when it is acknowledged that, as Jaffé observes, "the history of symbolism shows that everything can assume symbolic significance: natural objects ... or man-made things ... or even abstract forms" (1964: 257). The earliest known representations of insects, dating from twenty-five to thirty thousand years ago, suggest that insects were afforded symbolic meanings by humans very early in our cultural lives. The scientific or non-symbolic representation of insects, however, is distinctly modern, and indicative of modern culture's "visual literalism"—the idea that images may function as direct, objective representations of their subjects. It remains unclear, though, how any insect image could break entirely from symbolic heritage of insect representations. As Hogue has observed, it is in anthropomorphic portrayal of insects and their behavior that the ongoing symbolic significance of insects becomes most evident. In the case of early entomologists, this usually extended to the naming of insects themselves after the anthropomorphic deities of Roman and Greek mythology, a practice prompted by the classical education received by many early naturalists (Yancey, in Kritsky and Cherry 2000: 40).

emotional upheaval that Darwin's work would cause." Much of this conflict played out under the banner of natural theology.

Indeed, some long-standing ideas were beginning to change in the early years of the Age of Industry. Until this time, the biblical narrative of the natural world was dictated by the events described in Genesis, the first chapter of the Bible. The world and all its inhabitants were created in six days by divine fiat. As God's work was perfect, no change was permitted. The Bible was likewise used to pinpoint the precise time that the Lord's work was finished. Counting the many "begats" in Genesis 5–11 gave an unbroken record of the male lineage from Adam and Eve to Abraham that accounted for the time between Creation and 1922 BCE. Thereafter, other biblical passages were used to estimate the time before Jesus' birth. Using historical records after this point, creation was reckoned to be roughly six thousand years before present, although Archbishop Ussher, head of the Church of Ireland, was more precise: creation took place in 4004 BCE on October 23.

As historian John Hedley Brooke (1991: 299) noted: "By the close of the eighteenth century, it had already become difficult to defend one of the assumptions on which much of natural theology had been based—that God had created every creature it was possible to create and that they had all existed together since the original days of creation." Indeed, several lines of evidence prevented acceptance of this formerly canonical view. First, Victorian naturalists were venturing far afield, collecting biological specimens across the globe. These included fossils—organisms that were once living but which were no longer present. Natural theologists reconciled fossils by rationalizing that these organisms might still be alive somewhere else on Earth. However, after analyzing discrete geological strata, it became clear that extinct and modern species were not all alive at the same time, discrediting the biblical narrative. Nonetheless, enthusiasts of natural theology worked hard to reconcile the new reality with the static biblical account, and many argued that the fossilized remains were of organisms that perished during the Great Flood.

Another old work horse that fell by the boards during the Age of Industry was the idea that species were immutable, fixed and unchanging over time. Jean-Baptiste Lamarck, notable for his discredited evolutionary theorem, critically argued that living organisms could change (Lamarck 1809, 1815). Lamarck deduced this heretical idea by comparing fossil forms with living relatives. It seemed reasonable to Lamarck that organic change could take place over time and lead to the diversity of related forms in the present. Ever nimble, natural theologists adjusted by arguing that a changing environment might necessitate a change in form in living organisms and that the wise Creator, of course, foresaw this necessity (Brooke 1991). Although Lamarck did not correctly deduce the mechanism by which organic change occurred, he directed attention to the idea that transmutation was ongoing and continuous.

In 1807, the Geological Society of London was founded. It provided a place where gentlemen scientists could convene to discuss current geological research and argue the merits of theological and scientific worldviews. A rising star among them was Charles Lyell, a lawyer by trade in addition to being a student of geology at Oxford. According to Janet Browne (2006: 33): "Without Lyell, there might not have been any Darwin: no intellectual insights, no voyage of the *Beagle* as commonly understood."

How did Lyell exert this profound influence? Notably, Lyell advanced an objective scientific view of the age of the Earth. He argued that geological features on the Earth were not formed by cataclysmic events but were the result of slow continuous processes acting over time. Further, Lyell believed that only observable forces in the present could be presumed to have operated in the past and with the same force and intensity (Moore 1986). These ideas were the basis for Lyell's three-volume set of books entitled *Principles of Geology* published in 1830–1833, which were widely read in scientific circles and by Charles Darwin aboard HMS *Beagle*. In fact, Darwin was deeply impacted by Lyell's view of gradual change, and it is perhaps no wonder that Darwin commented regarding his own writing: "I always feel as if my books came half out of Lyell's brain ... the great merit of *Principles* was that it altered the whole tone of one's mind" (DCP-LET-480).

Thus, by the 1840s, slavish devotion to a biblical view of the origins of the Earth and its inhabitants was beginning to yield to a scientific and materialistic understanding of creation. The conversion was critical in allowing Darwin to publish *Origin of the Species* in 1859 and for the reception of this work among the public. Yet even at this time, Victorian society was still negotiating the proper place of science in a life that was deeply religious. There is perhaps no better evidence of this than a perusal of the textbook that served as the basis of entomological thought for a century: Kirby and Spence's *Introduction to Entomology*.

The authors and their book will be discussed in greater detail in a subsequent section as it embodies natural theology as it applies to insects. For the moment, it is sufficient to note that in the preface to the first edition, the authors make the case that the book was written with the goal of "making converts among their countrymen to a study equally calculated for promoting the glory of God and the delight and profit of man" (1815). To that end, the textbook is littered with allusions to the Will of the Creator, compliments to the Wisdom of the Creator, and even verses extracted directly from the Bible. The magisteria occupied by science and religion overlapped considerably during the Age of Industry.

The Age of Industry, thus, became a time-period when religious and scientific thinking initially mixed freely and later became starkly separated. This is largely because science must be based in material rather than supernatural forces. As Browne (2006: 5) argued, the changes that took place during this conversion were inextricably linked to the "spread of middle-class economic and political values through the era." Those who favored religious explanations of natural phenomena, including most of the clergy who acted as parson-naturalists and Victorian stalwarts of science, were usually gentlemen of means who came through an educational system that was infused with religion. For example, before 1840 Cambridge students had to swear allegiance to the Anglican Church's thirty-nine articles of faith upon matriculation (Anelli 2006). As the Age of Industry progressed, social forces loosened the grip of religion on science by allowing those who were not independently wealthy, and therefore required a salary to pursue their passion, to become scientists. This led to the professionalization of science and the establishment of objective measures of performance that were devoid of religious overtones. As these forces played out, religiosity took a back seat to empiricism.

NATURAL HISTORY AND NATURAL THEOLOGY IN THE UK

Because of the limited ability of the upper classes to do much but be served by the other classes, they were with left many hours in the day with simply nothing to do. Barber (1980: 20) notes: "Apparently, this boredom was widely admitted, so that when natural history writers came forward and offered an instant panacea they were sure of an attentive audience." Thus was born a series of successive crazes for various organisms in the natural world which took off between 1820 and 1860 (Barber 1980; Anelli 2006). Victorians went to the garden, woodlands, and seashores, vigorously collecting specimens as they went and proudly displaying their collections at home. Even Victorian ladies, usually reticent about matters of science, could wax eloquently about the wonders of ferns, and the beauty of butterflies. Even more remarkable is that the natural history craze swept across all Victorian society. Of course, the upper classes could fund trips to exotic places, pay to have their specimens curated, and display their collections in beautifully adorned cases. But one of the wonderful things about the natural history craze was that even those who did not have means could participate. An insect collection could be made by anyone with a net and bit of curiosity.

One might question the descriptor "craze" to describe anything in staid Victorian society, but the term seems apt considering the reports of its reach. According to Owen (1894), when the zoologist Sir Richard Owen visited Prime Minister Robert Peel, Owen led a fascinating examination of the lunch buffet with a microscope that he had in tow. Indeed, evenings at the microscope became a form of entertainment for those wishing to expand their knowledge of natural history.

The many benefits of pursuing natural history were ably chronicled at the time. These benefits included exercise in the outdoors (Wood 1862), curing disease (Swainson 1834), finding solace during bereavement (Lewes 1860), acquiring new themes for needlework among women (Barber 1980) and, generally, increasing the level of happiness (Allen 1976). Still, Victorians believed that even leisure time must be occupied with high-minded pursuits and valued "rational" amusement that was also morally uplifting (Barber 1980). Many writers claimed studying natural history met this criterion, but the case was truly made by uniting natural history with natural theology.

As eloquently stated by Barber (1980: 23), "Natural theology made the study of natural history not only respectable, but almost a pious duty, and there was nothing the Victorians liked so much as a duty." Even if the primary motivation for studying natural history was relief from boredom, a pivot to the tenets of natural theology eliminated any concerns about not having a higher purpose in mind. The clergy of the time were already using nature as a text for sermons, and even made insects and other arthropods the subject of entire homilies. For example, William Wilberforce Newton (1878) published a series of sermons entitled *Little and Wise: Lessons from the Ants, the Conies, the Locusts and the Spider*. This four-part sermon series was based on Proverbs 30:24–8:

> There be four things which are little upon the earth, but they are exceeding wise:
> The ants are a people not strong, yet they prepare their meat in summer;
> The conies are a feeble folk, yet make their houses in the rocks;
> The locusts have no king, yet go they forth all of them by bands;
> The spider taketh hold with her hands and is in kings' palaces.

In his first sermon, Newton stated that people, like ants, should not be discouraged by a lack of strength or size, and urged his followers to engage in spiritual works now, in the summer of their lives (Newton 1878: 7–29). Similarly, Newton used locusts to illustrate that self-control is superior to obedience and that to be truly good Christians his congregants must know themselves and act independently (51–70). Finally, Newton recommended that his followers emulate the spider by using their God-given talents to work in the best ways and in the best places (73–94). While these moral lessons are indisputably estimable, the biology underlying them was sometimes questionable. For example, Newton operated under the misconception that "I suppose the little baby ants grow up into what they call big ants" (14), when in fact every ant that has ever walked on six legs is a fully grown adult and little baby ants more closely resemble maggots. Additionally, Newton includes a picture of what he calls a migratory locust, but is actually a katydid (Figure 5.1). Newton was also prone to anthropomorphize his arthropod subjects; for example, he claimed that spiders "tremble when they see a chambermaid with a broom, for that always means death to them" (78). Yet in spite of these slight inaccuracies, these sermons clearly illustrate the interest that Victorians had in the natural world, and the lengths that they would go to connect what they observed in creation to the wisdom of the Creator; a step that barely required explanation for Victorians. In the end, Victorians wholeheartedly believed that natural theology made pursuit of natural history a worthwhile pastime.

The fact that science and religion could be joined into a unified theory under natural theology protected the religious view from deliquescing into irrelevance

THE MIGRATORY LOCUST.

FIGURE 5.1 What Newton labeled as a migratory locust in his *Little and Wise* book of sermons is actually a katydid. Migratory locusts have short antennae and females lack the long sword-like ovipositor, both of which are present in the illustration that Newton included. Newton (1878: 54).

as scientific advances marched on. Natural theology satisfied both religious and scientific needs, and provided an argument for those who wanted to defend religion as well as those who wished to attack it. It was difficult, for instance, to falsify or test the default explanation of natural theologists (e.g., "God did it and His ways remain obscure"). Eventually, however, the religious burden being asked of science became too great (Brooke 1991); accordingly the shared path between science and religion began to diverge in the second half of the nineteenth century.

WILLIAM KIRBY AND WILLIAM SPENCE: FOUNDING ENTOMOLOGISTS AND UNEASY PARTNERS

Two British entomologists, William Kirby and William Spence, provide probative value in understanding the intermixing of social, political, religious, and scientific work in Victorian times. While nominally colleagues who studied entomology, these two eminent scientists possessed different worldviews that illuminate the forces at work in Victorian science. William Kirby was both a Tory and a high churchman; William Spence was a merchant capitalist. Their contrasting worldviews led to clashes over religion and politics which, in fascinating ways, were interjected into their classic book, *Introduction to Entomology*. This book marked an important milestone, allowing any layperson to become a bona fide amateur entomologist. Indeed, the book was hailed as the "single greatest impetus to production of a general taste for entomology" (Wilson 1835) and the "most entertaining account of the habits of insects that has ever been put into popular form" (Anonymous 1842).

Kirby (1759–1850) was a consummate entomologist. Indeed, he is called the "founder of entomology" (Armstrong 2012: 128). He graduated from Cambridge in 1781 and took Holy Orders in 1782. He spent the next sixty-eight years ensconced as a curate at Barham in Suffolk (Moore 2004). As such, Kirby is yet another in the long line of Parson-naturalists who preceded him. In addition to his religious credentials, Kirby was highly regarded as a scientist; he was an original member of the Linnaean Society, a member of the Royal Academy of Science, and a founder the Entomological Society of London. Kirby wrote several scientific books, including *Monographia Apum Angliae*, published in two volumes in 1802. This taxonomic work on the bees of England established Kirby as a leading entomologist.

A devout natural theologist, Kirby believed that "insects were evidence of God's beneficence, their unique and vital roles giving humanity a glimpse of the intricate and harmonious design that allowed us to make sense of our place in the universe" (Endersby 2010). Kirby eschewed materialistic explanations of nature because he feared it was the first step on the road to deism. In the years following the French Revolution, Kirby feared that atheism was being promulgated by figures as diverse as Napoleon Bonaparte and Jean Baptiste Lamarck (Clark 2006). Embracing natural theology allowed Kirby to combat that threat while promoting social stability. Ultimately, Kirby's religiosity in matters of science put him at odds with his co-author, William Spence.

William Spence (1783–1860) is described as an economist, his vocation, and as an entomologist, his avocation. His interest in entomology apparently began when he was twenty-two and started corresponding with William Kirby. The relationship grew into a lifelong partnership that brought the science of entomology to the masses (Moore 2004).

Kirby and Spence's *Introduction to Entomology* is in many respects a rational, competent piece of science, and is likely among the world's most successful entomology texts. However, most scientific information was used to support the existence of God and to demonstrate His beneficence. In particular, Kirby's writing was laden with religious references. Ostensibly a work of science, the preface to the 1815 edition reads:

> And here it may be proper to observe, that one of their first and favourite objects has been to direct the attention of their readers "from nature up to nature's God." ... To see all things in God, has been accounted one of the peculiar privileges of a future state; and in this present life, "to see God in all things," in the mirror of the creation to behold and adore the reflected glory of the Creator, is no mean attainment; and it possess this advantage, that thus we sanctify our pursuits, and, instead of loving the creatures for themselves, are led by survey of them and their instincts to the love of Him who made and endowed them.
>
> (Kirby and Spence 1815: xii–xiii)

It is unlikely that there is a more heartfelt explication of natural theology anywhere and, yet, it is oddly placed in a science text.

Scholarship of many types abounds throughout the textbook, including long passages in Latin, Greek, French, or German. Experts are quoted from many disciplines of science as might be expected, and high-quality illustrations of insects enhance the text (Figure 5.2). But, one also finds a dazzling selection of poetry in the book and references to Herodotus, Shakespeare, Pliny the Elder, and innumerable other authorities. Clearly the authors were exceptionally erudite and well-learned. However, what is disconcerting for a science textbook, even one drawn from Victorian times, is the appearance of biblical passages and religious commentary on virtually every page of the book. Sometimes, the need to find good in all God's creations requires great forbearance such as in the appeal to the reader to find the will of God in disease-causing insects. After

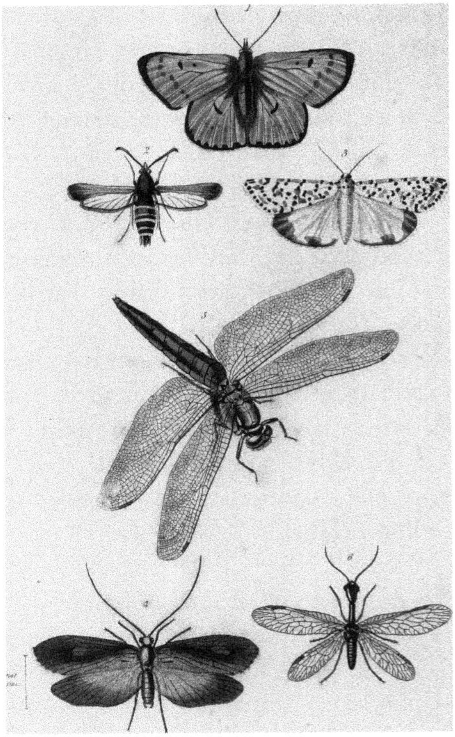

FIGURE 5.2 Illustrations from Kirby and Spence's *Introduction to Entomology* (1826: pl. III).

citing Joel 2:25 for a list of "instruments of punishment" that the Almighty can choose to send against offending nations, Kirby describes "numerous tribes of insects" as "[God's] armies marshalled by him, and by his irresistible command impelled to the work of destruction; where he directs them they lay waste the earth and famine and the pestilence often follow in their train" (Kirby and Spence 1815: 85).

Kirby and Spence's collaboration was often fraught with conflict owing to their different worldviews. In contrast to Kirby, Spence's professional endeavors and life experience imparted him with a secular view of science, divorced from religious influence (Clark 2006). Apparently, discord arose over the injection of religious notions into the text, particularly in the area of animal intelligence and instinct: Kirby believed that these were gifts from God whereas Spence maintained that they arose naturally and without intervention from the Deity (Endersby 2010).

Ultimately, Kirby allowed Spence to write about animal instinct, although Kirby insisted on telling readers of *Introduction to Entomology* that the chapters on animal instinct were written by Spence alone. Fortunately for Kirby, he had the opportunity to share his views on animal instincts through his contribution to the *Bridgewater Treatises*. In fact, Kirby clearly distinguished his own views from the godless materialism of Spence and announced his decision to "take the field" against his "worthy friend and learned coadjutor" (Kirby 1836).

THE *BRIDGEWATER TREATISES*

The *Bridgewater Treatises* were established by a bequest from the Reverend Francis Henry Egerton, Earl of Bridgewater who, upon his death, directed his trustees to fund a series of treatises to comment on the "power, wisdom and goodness of God as manifested in creation." Nominally based on Paley's *Natural Theology*, the authors were chosen by the President of the Royal Society of London (with assistance from the Archbishop of Canterbury and the Bishop of London) and given a stipend to write a treatise which would then be published and made widely available. Seven of the authors selected for the eight original *Treatises* were distinguished men of science (all being Fellows of the Royal Society of London); the eighth author was a Scottish theologian, moral philosopher, and political economist. Four were in holy orders; all subscribed more or less to Christian orthodoxy (Brock 1966).

The *Bridgewater Treatises* were widely read and were hailed as "scientific best sellers of the nineteenth century" by historian Jonathan Topham (1992: 17). The volumes appealed not only to scientists but to readers in the "landed, mercantile and professional" classes as well. Among the religiously devout, the treatises were valued as nontechnical religious commentaries on contemporary science. They also penetrated the educational market, filling the gap caused by inadequate science textbooks in nineteenth-century England but were not likely read by the lower classes as they were comparatively expensive.

William Kirby's *On the Power, Wisdom and Goodness of God, as Manifested in The Creation of Animals and in Their History, Habits and Instincts* comprised treatise VII of the *Bridgewater Treatises*. In the beginning of his work, Kirby (1836: xxi) asserts: "it has always been the habit of the author of the present treatise to unite the study of the *word* of God with that of his *works*." Shortly thereafter, Kirby cautioned that those who wish to venture into this area "should always bear in mind ... that all depends upon the *right* interpretation, either of the *written word* or *created substance*" (1836: xxii). Kirby then admonished those scientists whose knowledge led them to assume that their own minds were the source of their insight, not God, and were, therefore, in danger of self-worship. He had two scientists in mind: Pierre Laplace, often referred to as the French Newton and author of the five-volume series *Mécanique Céleste* (Celestial Mechanics; 1799–1825); and Jean-Baptiste Lamarck, the aforementioned French naturalist who advanced the idea that species change according to natural laws. Both Laplace and Lamarck were likely deists. However, according to Kirby (1836: xxii), both scientists "disregard of the word of God ... and have fallen into errors of no small magnitude." The particular trespass that both Laplace and Lamarck committed, according to Kirby, was to ascribe natural phenomena to secondary (i.e., materialistic) causes while ignoring the fact that a primary cause (i.e., God) is at work. Kirby even brings the honey bee in to illustrate the utter inanity of these two prominent French scientists:

> If we select, as a well-known instance, the *Hive-bee* for an example. Its long tongue is specially formed to collect honey; its honey stomach to receive and elaborate it either for regurgitation, or for the formation of wax; and other organs or pores are added, by which the latter can be transmitted to the wax pockets under its abdomen; connected with these, are its means and instruments to build its cells, either for store cells to contain its honey and bee-bread, or its young brood, such as the form of its jaws, and the structure and furniture of its hind legs.
>
> (Kirby 1836: xxvi)

Kirby went on in this vein for several more pages so that Laplace's and Lamarck's great error can be made very plain: all of these wonderful contrivances, so exactly suited to specific purposes, simply could not be the result of a physical cause. According to Kirby, the only reasonable way to understand these wonders was to believe that all life was created and sustained by the Deity. Charles Lyell is given similar treatment for a few pages until the topic changes to a discussion of all organisms in the world, starting with Cherubim and including all animals. In every case, the biblical narrative as described in Genesis is rendered into scientific terms to assure the reader that both biblical and scientific accounts are consistent with one another.

Kirby also addressed biodiversity in his treatise. He wanted to understand the current geographical distribution of all of the living animals on the planet. Kirby's starting point for this exploration was how each species got from Noah's ark on Mt. Ararat to its current location. He acknowledged several problems, but in the end, decided that all could be explained by periodic, seasonal migrations, help from humans and, of course, Divine intervention.

In Chapter III, Kirby turned to his main thesis: the functions and instincts of animals. Recall that this section of *Introduction to Entomology* was written by Spence because the two could not agree on a unified approach. Frequently using insects as examples, Kirby's religious views were on full display in Chapter

XVIII, entitled "Instinct." A few examples of Kirby's love of arthropods and his gratitude to the Creator will suffice:

> Of all animals, *insects* afford the most numerous instances of instinctive proceedings with this sole end in view; the pit-falls of the ant-lion; the webs and nets of various sorts of spiders spread over the face of nature; and many more, furnish instances of stratagems to secure their daily food; while an infinity of others acquire it, aided only by their senses and natural weapons .... Thus, we see the Almighty and All-wise manifests his *goodness*, as well as his wisdom and power, in providing for the wants of all the creatures that he has made.
>
> (Kirby 1836: 332–3)

Writing on the functions and instincts of arachnids in Chapter XIX, Kirby states: "I shall select an instance or two from animals of this Class, which show the care of the Creator for these little beings, apparently so low in the scale of Creation; His foresight of every circumstance in which they would be placed; and His adaptation of their structure to their assigned station" (1836: 350). Kirby ably made this case using trapdoor spiders (Figure 5.3). He first addressed the care that the female invests in constructing her burrow. Kirby noted the walls of the tube, how the tube was bored, the mortar used to solidify the tube, and how the web is integrated into the design. These elements, Kirby deduced: "shows that she was directed in her work by a Wise Master" (342). But Kirby was far from done connecting the contrivances of this spider to the work of the Deity. Indeed, he devoted another ten pages detailing how the Wise Creator endowed the female spider with the ability to exclude intruders from her burrow by constructing a door, attaching it to a frame and integrating the "tapestry of her web" into these structures. Kirby then concluded that the Wise Creator had spared no extravagance in providing for the humble spider.

DARWIN AND NATURAL THEOLOGY

Alison Winter (1997: 25) described scientific thought in Britain as entering an era of "heady optimism" by the 1840s. According to Winter, this was the time when the term "scientist" was first used to describe the work of those engaged in biology or physics, and new scientific societies and journals were established (Morrell 1976). While "fringe" scientific movements such as phrenology and mesmerism blossomed during this time (Davey 1854), mainstream science was on the march, enjoying increased acceptance by Victorian society. However, these items were simply preparatory to the main event that would transpire in

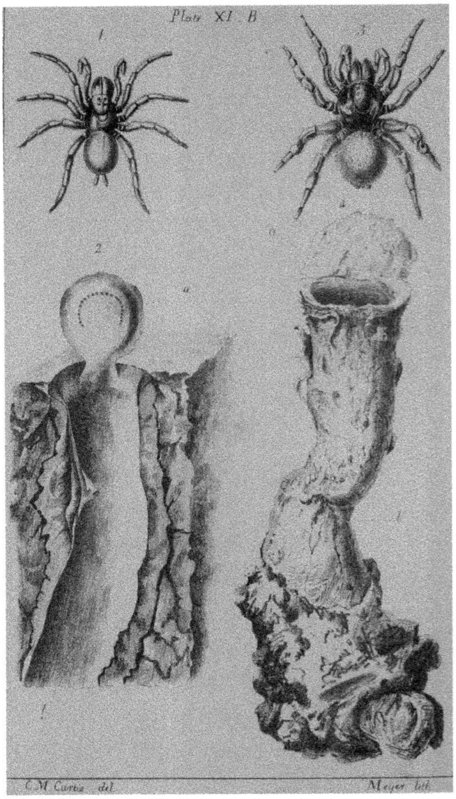

FIGURE 5.3 Illustrations of trapdoor spiders and their burrows from Kirby's *Bridgewater Treatise, On the Power, Wisdom and Goodness of God, as Manifested in the Creation of Animals, and in their History, Habits and Instincts* (1836: pl. XI B).

1859: the publication of Charles Darwin's *On the Origin of Species by Means of Natural Selection.*

It is generally known that Darwin was a devout Christian during his early life. His time at Cambridge was steeped in Christianity and Darwin averred that he could even read the Greek Testament with some facility. In his autobiography (Barlow 1958: 59), Darwin claimed that reading Paley's *Natural Theology* "gave me much delight as did Euclid," and that he could recite much of the book by rote. Notably, Browne (2006: 18) commented, "Natural theology, indeed, was commonly regarded by the British cultural establishment as one of the strongest bulwarks against social unrest because it reinforced ideas of stable hierarchy, a powerful antidote against civil insurgencies and rebellion."

Although Charles Darwin was not formally trained as an entomologist, he was nonetheless an avid insect collector with considerable expertise. In his autobiography, Darwin gave a charming description of his passion for beetles: "One day, on tearing off some old bark, I saw two rare beetles and seized one in each hand; then I saw a third and new kind, which I could not bear to lose, so that I popped the one which I held in my right hand into my mouth. Alas it ejected some intensely acrid fluid, which burnt my tongue so that I was forced to spit the beetle out, which was lost, as well as the third one" (Barlow 1958: 62). Clearly, Darwin had an intense fondness for insects! It is not surprising that his interest and later rejection of natural theology would, therefore, significantly revolve around arthropods.

Darwin was enormously influenced by the works of Lyell and Paley, both of whose volumes accompanied Darwin on the *Beagle* voyage. As previously mentioned, Lyell's writing taught Darwin that the earth was not six thousand years old, but likely many millions of years old and that small forces working over long periods of time could create large changes. While Paley argued that contrivances confirmed the beneficence of the Creator, Paley's writings also opened Darwin's eyes to the variation in nature. Darwin became very interested in how variation could arise and be maintained.

Part of the mechanistic question about how favorable variation was preserved was resolved by a paper that Darwin read after he returned to England. Reverend Thomas Robert Malthus wrote *An Essay on the Principle of Population as it Affects the Future Improvement of Society* (1798). The basic thesis was a mathematical one that populations increase geometrically while the food supply increases only arithmetically. This led to a chronic mismatch between the number of organisms and the available food. To Malthus, it meant that society should do nothing to help the poor, as their wretched living conditions actually improved their lot by suppressing population growth and increasing moral constraints among survivors (von Sydow 2005). Darwin, however, reasoned that because of the mismatch, there is a *struggle for existence.* He then deduced that if a particular organism had a variation that improved its ability to acquire or use limited resources, it would be more likely to survive and produce offspring. Over time, favorable traits would therefore increase in the population and this became the mechanistic explanation of how favored traits were preserved.

Ospov (1981), thus, credits Malthus with convincing Darwin to think in relativistic rather than absolute terms. Darwin read Malthus in 1838 while he was working on his "*Transmutation*" notebooks. Variability, how it was created and preserved were very much on Darwin's mind. As an adherent to the Paley school of natural theology, the young Darwin believed that God worked in absolute terms: God's plan was perfect, and organisms were immutable. However, by studying transmutation in the light of Malthus, Darwin slowly changed his view to an understanding that perfect adaptation was not the way nature worked. Instead, nature could mechanistically adjust to external change; Darwin called this mechanism "natural selection." Organisms that currently existed did so because they were more successful in producing offspring under current conditions than potential competitors. If conditions changed, organisms better-suited to the changed environment would prevail.

Yet before Darwin could bring his theory to fruition, he had to decisively divorce himself from the religious narrative of his upbringing. Darwin's theory of evolution was materialistic. It did not require the intervention of a deity. During the *Beagle* voyage, Darwin began to question the legitimacy of religion. He alluded to the "manifestly false history of the world, with the Tower of Babel, the rainbow as a sign, etc." in the Bible (Barlow 1958: 71). But Darwin's biggest problem with God seems to come from his inability to reconcile the idea of a beneficent God with theodicy: "... it revolts our understanding to suppose that His benevolence is not unbounded, for what advantage can there be in the sufferings of millions of the lower animals throughout almost endless time?" (Barlow 1958: 75). That reality, according to Darwin, was much clearer proof that "organic beings have been developed through variation and natural section" (Barlow 1958: 75), not from a loving God. It appears that Darwin's observations on insects may have been persuasive in leading him to this conclusion.

In a letter to his colleague Asa Gray in 1860 (DCP-LETT-2814), Darwin specifically identified the life habits of ichneumonid wasps as evidence that the world was not being run by a beneficent God. This group of wasps and many others are parasitic. The female wasp lays her eggs inside the larvae of butterflies or beetles. The wasp eggs hatch inside the host and proceed to systematically devour the living host in a way that keeps the poor creature alive for as long as possible while it is being eaten. By the time the wasp is ready to emerge, the parasitized larva has been reduced to an empty shell. This arrangement is discussed with wonder by Kirby and Spence:

> In this strange and apparently cruel operation one circumstance is truly remarkable. The larva of the Ichneumon, though every day, perhaps for months, it gnaws the inside of the caterpillar, and though at last it has devoured almost every part of it except the skin and intestines, carefully all this time *avoids injuring the vital organs*, as if aware that its own existence depended on that of the insect on which it preys!
>
> (Kirby and Spence 1815: 345; emphasis added)

Not only is the wasp cruel, but it is also self-aware in Kirby and Spence's rendering. Darwin, however, had quite a different take on the arrangement. In his letter to Gray, Darwin stated:

> But I own that I cannot see, as plainly as others do, & as I should wish to do, evidence of design & beneficence on all sides of us. There seems to me

> too much misery in the world. I cannot persuade myself that a beneficent & omnipotent God would have designedly created the Ichneumonidae with the express intention of their feeding within the living bodies of caterpillars, or that a cat should play with mice.
>
> (DCP-LETT-2814)

And this is not the only instance of Darwin finding insect habits difficult to reconcile with a loving God. In *Descent of Man* (1871), Darwin noted structures in male insects that they used to fight other males or subdue females during mating (Figure 5.4). While noting that parental and filial affections are

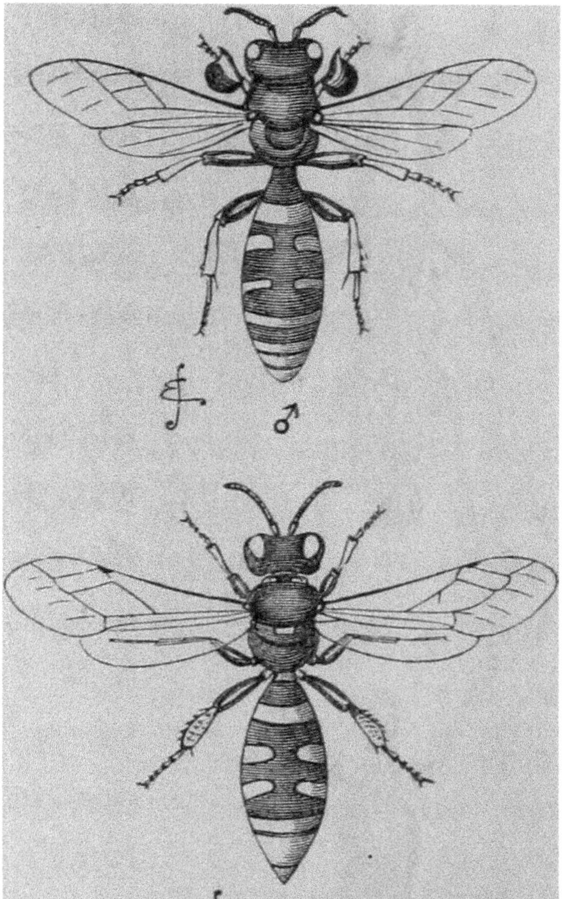

FIGURE 5.4 Differences in appearance between male (top) and female (bottom) slender-bodied digger wasps (*Crabro cribrarius*) from Darwin's *Descent of Man* (1871, vol. I: 333). Illustration by Edward A. Smith.

at the center of social relationships, some insects exhibit savage behaviors. "So it has almost certainly been with the unusual and opposite feeling of hatred between the nearest relations, as with the worker-bees which kill their brother-drones, and with the queen-bees which kill their daughter-queens; the desire to destroy, instead of loving, their nearest relations have been here of service to the community" (Darwin 1871: 823). In the previous sentence, we see the reason for Darwin's conversion from one who believed wholeheartedly in natural theology to a scientist with a materialistic view on the origins of life on this planet. The force of natural selection had supplanted God's will in Darwin's understanding of the world. He is explicit about his conversion: "The old argument of design in nature, as given by Paley, which formerly seemed to me conclusive, fails, now that the law of natural selection has been discovered" (Barlow 1958: 73). Indeed, the savage behavior in bees arose not because the benevolent Creator made it so, but because it redounded to the benefit of the bees. Thus, insects were critical in facilitating Darwin's conversion from a natural theologist to a materialistic scientist.

As Tennyson (1850: Canto LVI) noted, "Nature [is] red in tooth and claw." Darwin clearly rejected the idea that the inexhaustible cruelty seen in nature could be the work of a loving God. The origins of Darwin's conversion were at once professional and personal. Darwin stated at the close of the *Beagle* voyage that he had begun to distrust the historical evidence supporting the Bible (Browne 2006). Darwin also took offense at the notion that close relatives such as his father and grandfather, whom he knew to be upstanding men of good character, would endure everlasting punishment because of their disbelief. Darwin called this a "damnable doctrine" and could not believe in a God of mercy who would prefer eternal punishment to reformation (Moore 1989). The death of Darwin's daughter, Annie, at the age of ten appears to be the final blow to his flagging belief in a deity. The road from belief to deism or agnosticism had been a long one for Darwin. He, himself, admitted that he was "very unwilling to give up my belief." However, the entries in Darwin's restored autobiography clearly show a decline in faith by the time of its writing (Barlow 1958: 86).

While Darwin's aversion to theodicy appears to have been the basis for his conversion to a nonbeliever in a beneficent Creator, he was not alone. Others in Victorian society had independently reached the same conclusion as Darwin: that a loving, omnipotent God was inconsistent with the abundant suffering in the world. An example is the Victorian poet, Matthew Arnold, whose 1867 poem, *Dover Beach*, includes a stanza which is a metaphorical description of the retreat of religion from the Victorian world:

The Sea of Faith
Was once, too, at the full, and round earth's shore

Lay like the folds of a bright girdle furled.
But now I only hear
Its melancholy, long, withdrawing roar,
Retreating, to the breath
Of the night-wind, down the vast edges drear
And naked shingles of the world.

REACTION TO DARWIN IN THE AGE OF INDUSTRY

Charles Darwin returned from the voyage of the *Beagle* in October 1836. However, his magnum opus, *On the Origin of Species*, was not published until 1859. Darwin spent much of the twenty-three years between his return and his book's publication extensively documenting what he could about variation in nature and building a legalistic case for natural selection as the mechanism by which species changed. Darwin suffered from two lacunae in his knowledge base: first, he didn't know where variation came from and, second, he didn't understand how traits were inherited. Given those gaps, he feared that his ideas about natural selection would be unfavorably reviewed (Browne 2006; Quammen 2006).

Darwin had reason to worry about the public reception of his ideas. His work would refute the conventional notion that a loving God created everything in nature, including man, according to His perfect plan. But even worse, Darwin's mechanistic view of how and why species change over time would separate mankind from its exalted status as God's chosen image-bearer with a soul. How could such a being arise from the random forces of nature? How could people live without the assurance of their special status? Denying human supremacy would be a very noticeable and immense problem for Darwin's theory (Quammen 2006).

And Darwin had a front row seat to another controversy generated by second-guessing religion. In 1844, the book entitled *Vestiges of the Natural History of Creation* was published anonymously. The book explained many kinds of evolution from planetary to geological and biological. It was initially well-received by Victorian society and was widely read, even by Queen Victoria, because it was written in a conversational style that did not require years of scientific study, and created a buzz among staid Victorians who felt that they were engaged in discussions of great import. However, because some of the verbiage in *Vestiges* contradicted accepted tenets of natural theology, Victorian scientists who were also clergy came out against both the book and the author—whoever he might be (Bowen 1845; Sedwick 1845). The author, subsequently revealed posthumously to be Robert Chambers, earned the enmity both of scientists and the clergy by 1850 (Secord 2000). The naturally timid Darwin witnessed this controversy and appears to have had two reactions. The first was to take some encouragement that *Vestiges*' author had publicly discussed evolution, thereby partially paving the way for Darwin. But, because Darwin saw the rampant and vicious criticism endured by the book's author, he was even more determined to buttress his arguments with data. Ever-cautious, Darwin went back to his study, and was only pushed to share his ideas publicly when he received a letter from Alfred Russel Wallace in which Wallace enumerated nearly the same theory Darwin had been working on for decades. But Darwin's manuscript stood at more than 250,000 words and was more than halfway complete while Wallace only had a short letter (Stauffer 1975). This imbalance was resolved when a joint paper by the pair was presented to the Linnaean Society of London in 1858 (Darwin and Wallace 1858) and Darwin released *Origin of the Species* the next year (Darwin 1859).

Soon after publication of *The Origin of Species*, the reviews started coming in. Entomologists were a source of some initial support for Darwin's thesis (Kritsky 2008). The British entomologist Henry Walter Bates, discoverer of the eponymous batesian mimicry whereby a nontoxic species (e.g., viceroy butterfly) mimics a toxic species (e.g., monarch butterfly), announced his support for Darwin publicly at the Entomological Society of London meetings. Other British scientists were often vociferously negative, even among those who knew Darwin well and generally respected him. This included Adam Sedgwick, a geologist and minister at Cambridge whom Darwin met when he attended the university. Sedgwick once stated that the goal of science was "to teach us to see the finger of God in all things animate and inanimate" (Hull 1975). Sedgwick was, thus, a devoted natural theologist and conservative in his scientific thinking. Sedgwick rejected Lyell's ideas about the age of the earth and referred to Darwin's theory of evolution as "a dish of rank materialism cleverly cooked and served up" (Quammen 2006: 209). Anatomist Richard Owen panned Darwin's book as ridiculous for many reasons, some less than professional, but particularly for suggesting that mankind was some sort of mutated ape (Hull 1975).

In the US, a British transplant, Benjamin D. Walsh, not only wrote to Darwin to pledge his support, but also to suggest that he would actively tamp down any anti-evolution sentiment that arose in the States (Walsh 1864; Sheppard 2004). The Darwin Correspondence Project reveals active communication between Darwin and Walsh. An 1864 letter showed the strength of Walsh's commitment to Darwinian ideals by taking on America's primary anti-Darwin academic: Louis Agassiz, an influential Swiss geologist who was transplanted to Harvard. Agassiz, a devoted natural theologist, argued until his death that God was the active source and attentive steward of all creation. In one memorable letter to Darwin, Walsh said, "I do not know what the European Scientific World thinks of Agassiz, but here he is popularly considered as the incarnation of Science & as infallible as the Bible. He strikes me as a very much overrated man, who perpetually allows his imagination to get the better of his judgement [*sic*], & who not unfrequently argues for victory instead of for truth" (DCP-LETT 4663).

Walsh also played the important role of providing scientific evidence of evolution among insects. Walsh (1864a) used dozens of insect examples to show that insects change over time according to Darwinian principles, and attacked Agassiz for not carefully reading and understanding Darwin's book before refuting it. In a second paper, Walsh analyzed the feeding preferences of plant-eating (phytophagous) insects and described how insects of the same species that feed on different plants have different morphologies (Walsh 1864b). Walsh then advanced a Darwinian argument, recognizing the paucity of genetic information available, that explained how this might lead to the development of new species:

> Even with the little we know of the Law of Inheritance, we might infer *a priori*, that when from particular circumstances a Phytophagic Variety, including both the sexes, has fed for a great many generations upon a particular plant ..., it would be likely to transmit to its descendants ... a tendency to select that particular plant upon which to deposit its eggs .... If this were to happen ... for an indefinitely long period of generations ... it would cease to be possible for that insect ... to feed upon any other plant than that to which it became habituated by the Laws of Inheritance.
>
> (Walsh 1864b: 405–6)

With extraordinary prescience, Walsh thus suggested in 1864 a concept reminiscent of sympatric speciation in the twentieth century (Berlocher and Feder 2002). T.H. Huxley is often referred to as "Darwin's bulldog." However, it can be convincingly argued that B.D. Walsh was Darwin's bulldog *with data*.

Perhaps surprisingly, in reaction to several dismal reviews, the reticent Darwin initiated what can only be termed a publicity campaign. Darwin's correspondence with Asa Gray, an American botanist at Harvard, revealed the efforts of Darwin and his supporters to shape public opinion (Dobbs 2011). For instance in 1860, Darwin encouraged Gray to publish a review of the *Origin of Species* in influential periodicals such as *The Atlantic Monthly*. Gray complied, writing a lengthy review detailing that Darwin's theory of natural selection could be compatible with natural theology (Gray 1860; Browne 2010). Gray even defended Darwin against the aforementioned Louis Agassiz, who was Gray's colleague at Harvard. Thus Darwin, a man who assiduously avoided controversy, spent much of the rest of his life beseeching his supporters to contend with the many tempests his ideas begat.

The controversy Darwin inspired endured even in death. Although Queen Victoria discussed knighting Darwin with several prime ministers (Quammen

2006), the honor was never conferred. The Queen was, after all, both head of state as well as head of the Church of England. Many Anglican clergy negatively viewed Darwin's work, and the Queen was loathe to offend them. But, owing to the lobbying effort by William Spottiswoode, President of the Royal Society at the time of Darwin's death, Darwin was not buried as expected at St. Mary's churchyard in his home village of Down, but in Westminster Abbey near the graves of John Herschel and Isaac Newton (Clark 1984).

STATUS OF NATURAL THEOLOGY AT THE END OF THE AGE OF INDUSTRY IN BRITAIN AND THE UNITED STATES

Darwin's theory certainly did not eradicate natural theology by the turn of the twentieth century in Britain and the US. Generally, older scientists who were natural theologists continued to believe that science and religion were natural partners while younger scientists, whose views were formed after Darwin's work was published, were more likely to adopt a materialistic view of science. Thus, natural theology lost currency among young recruits. However, a more nuanced view is needed to fully explicate the standing of natural theology in the latter part of the Age of Industry.

Natural theology was very much alive in the early twentieth century. In many respects, it played an accommodationist role, attempting to resolve conflicts between science and religion. Reconciling these two epistemologies was a goal for some progressive clergy who recognized that church attendance had declined precipitously in the latter part of the nineteenth century. Many clergy tried to liberalize religion so that it would not be viewed as an oppositional force to science, and some even stated that evolution was valid and operated on every species except humankind.

Ironically, natural theology with a scientific twist appeared to become something of a Trojan Horse which, having gotten partway into the religious tent, threatened to abolish it altogether (Brooke 1991). It didn't take much to stir up anxiety among clergy and philosophers who still argued that nature's laws were a testament of divine activity. But perhaps Samuel Coleridge put the case most succinctly when he argued that people should naturally want to be Christian; there was no way to beat them into compliance through application of reason. The era of accommodation, thus, reached an ignominious end by 1920 in Britain. However, the situation in the United States, while in some ways parallel, reached a different outcome.

Interest in both the natural history of insects and natural theology as a means to understand God's beneficence thrived in the United States just as it did in Europe. Asa Fitch (1809–1879) was among the most prominent and influential entomologists in the US (Barnes 1988). He was also among the first to be recognized with professional status. Fitch was a medical doctor by training, but had a considerable interest in insects which eventually became his vocation. He was named the first official entomologist of New York State in 1853. After this time, Fitch avidly studied insect life histories and how to remediate pests. At his death, Fitch left behind of 148 notebooks with full life histories of 55,000 species of insects in the state of New York.

Despite having dramatically advanced entomological science in the US, Fitch was also an ardent natural theologist. As reported in Howard (1930: 45): "Fitch and Harris belonged to the type exemplified by dear old Kirby, Rector of Barnham. All three of these men combined with their love of nature the love of nature's God." Howard recounted a specific story about Fitch that demonstrated his devotion to both God and entomology:

> It is related by Dr. Fitch's daughter that, in the course of evening prayers one night, the good old gentleman was reading to the assembled family his customary chapter from the Bible, when a moth attracted by the light, dropped on the open page of the Good Book. Here was a dilemma. The spirit of the naturalist, however, overcame the religious feeling, and in a somewhat shamefaced manner the eminent entomologist reached out for his butterfly net and captured and bottled the specimen before finishing the chapter. Was the act an irreverent one? And does the fact that the moth afterwards proved to belong to a species then new to science justify the interruption of the religious exercises?
>
> (Howard 1930: 45–6)

Among the scholarly class in the US, as previously noted, Louis Agassiz at Harvard was unyielding in his devotion to natural theology. However, voices of change were emerging from scientists with a materialistic perspective, for example Benjamin Walsh and Asa Gray. Later US scientists were also instrumental, making great strides in genetic research that would clearly reveal how heredity worked. The impressive contributions of T.H. Morgan, for instance, using what would become the "workhorse" of genetic research for decades, the fruit fly, demonstrated that mutations were the source of genetic variation (Morgan 1910). This combined with the rediscovery of Mendel's work were critical in elucidating a material basis for heredity, and closed both gaps in Darwin's theory of evolution.

As the US was to become a tour de force in biological research in the twentieth century, one might expect that Americans would become less enamored with natural theology. However, that was not the case. Although acceptance of Darwin increased over time in Europe and reached the point of consensus by scientists and the lay public alike, natural theology remained prevalent in the US while acceptance of Darwinian evolution waivered over time.

In the early years of the twentieth century in the US a very important phenomenon was underway: compulsory education. During the years 1900–1920, new high schools popped up all over the US prompting a surge of enrollment from two hundred thousand in 1900 to over two million in 1920. As E.J. Larson (1997) documented, this meant that for the first time, farm kids were learning about evolution and sharing it with their unschooled parents. While US scientists were increasingly convinced that natural selection was the mechanism by which organisms became suited to their environments, the lay public had a much different reaction. Combatting the acceptance of "godless" evolution with mandatory Bible reading in school became common, mostly in the South (Cox 2004). Religious fundamentalist rallies emerged in 1919 (Marsden 1980).

The battle between natural theology and natural selection would replay itself many times in the US during the twentieth century and would eventually emerge as "Intelligent Design." In a poll taken by the Pew Research Center in 2019, 98 percent of scientists agreed with the statement "humans evolved over time," while only 33 percent of nonscientists believed that humans evolved due to natural processes (Masci 2019). This 65 percent difference in acceptance of evolution between US scientists and the lay public is both astonishing and frustratingly persistent, but can be traced, in many ways, to public devotion to natural theology that took root in the Age of Industry.

CHAPTER SIX

Insects as Symbols

Making Meaning with Insects in the Nineteenth Century

ADAM DODD

> *The butterfly, the representative of the soul, is prepared in the larva for its future state of glory; and if it be not destroyed by the ichneumons and other enemies to which it is exposed, symbolical of the vices that destroy the spiritual life of the soul, it will come to its state of repose in the pupa, which is its Hades; and at length, when it assumes the imago, break forth with new powers and beauty to its final glory and reign of love.*
>
> (Kirby and Spence 1815: 77)

What was the insect symbol's cultural function and appeal in the Age of Industry? The idea that a meaningful correlation is to be found between the insect's cultural role and the rapid onset of industrial society during the nineteenth century may initially seem tenuous. Yet between 1800 and 1920, much of western society underwent a kind of strange metamorphosis of its own, transforming rapidly from something traditional and bucolic, to something distinctly modern and machine-driven. The echoes of this period continue to resonate throughout our contemporary, twenty-first-century society, and shape our increasingly complex relationships with insects. This chapter focuses on the co-operation of insect symbolism (an ancient cultural practice), and nineteenth-century industrialism (an unprecedented and revolutionary socioeconomic project). If we understand "culture" as the shared beliefs, values and practices of social groups, then to pursue the cultural history of nineteenth-century insect symbolism is to examine how the representation of insects exhibits the collective, often abstract concerns of a people in a period of unique and profound social change.

Symbols are cultural artefacts, used to stand in for, or represent, something else—an idea, a belief, a concept, a myth—something elusive and not easily representable as itself. Symbols are fundamental to human communication and culture, but as Hogue (1975) and Kritsky and Cherry (2000) have shown, history is replete with *insect* symbolism; indeed, insects have figured, to varying degrees, in the symbolic practices of most cultures worldwide. This is perhaps unsurprising, since humans have always cohabited in very close proximity with insects. But the diverse insect symbols of the nineteenth century in particular can tell us what the insect comes to *mean* for modern society itself. This was a society characterized not only by sprawling, sudden, and highly organized developments in agriculture, manufacturing, transportation, trade, science, and urbanization, but also by the disruption of cherished beliefs about human beings and our place in the natural world.

THE INSECT AND MODERNITY

By 1800, organized and systematic efforts to produce reliable, objective knowledge of insects were well underway in western Europe. These endeavors, which began in the second half of the sixteenth century, were direct expressions of modernism, which can be understood as "a set of discourses concerned with the possibilities of representing reality and defining eternal truth" (Walsh 2002: 7). Representations of insects became increasingly fundamental to knowledge of insects; prior to the early modern period, representations of insects were mostly restricted to the illustrated margins of illuminated manuscripts. These medieval portrayals were essentially decorative, and not intended to facilitate knowledge of insects in any substantial way. Indeed, prior to the late sixteenth century, most insects had been traditionally regarded by western societies as inconsequential (owing to their minuscule size) and detestable (owing to their confronting habits). Some were pleasing to the eye, but many were not. In the absence of both the methodology and the instruments necessary to produce accurate, reliable knowledge, insects were usually regarded aesthetically, when they were regarded at all.

Central to modernity is a collective belief in the efficacy of progress, understood as "rational advancement through increments of perpetual improvement" (Walsh 2002: 7), including the advancement and improvement of knowledge about the natural world. Yet, as modern society marched ever onwards, the past it would leave behind retained a powerful hold on the thoughts and feelings of those it swept up. As a kind of antidote to the rapid onset of an increasingly secular, industrialized society, the past itself became enchanted, romanticized, and idealized. The widespread, popular appeal of natural history during the nineteenth century, which was characterized by an often deeply quixotic view of the past and an intensely aestheticized outlook on the natural world, had a direct bearing on the development of insect symbolism during this period. Nature at large was routinely conceived and represented as at once both instructional and wondrous. It was in this sense that insect symbolism found utility in both the emerging entomology textbooks of the period, and those publications intended to arouse enthusiasm for insects in lay audiences; it was not always clear where the symbolic insect ended, and the literal insect began. The enchantment of nature and a collective fascination with the past emerged as responses to the widespread rootlessness arising from the temporal and spatial dislocation, and general disorientation, exacted by the onset of industrialism during the nineteenth century.

The late entomologist and pioneer of cultural entomology Charles Hogue has suggested that "the simplest explanation, and probably most often the correct, for adopting a species into symbolism is a recognition in it of some anthropomorphic quality, whether anatomical, behavioral or ecological. By association the quality gives the insect a special significance in the lives of the people" (1975: 3). Indeed, anthropomorphic representations of insects abound in the nineteenth century, across both scientific and popular texts. These symbolic depictions speak to diverse, often complex apprehensions about the role and place of the human in relation to society and nature; hence, some insect species have historically been seen as more significant of human concerns, and hence more interesting, than others. Insects exhibiting social behavior, such as ants, termites, and bees, emerge as conspicuous examples during the nineteenth century of a society symbolically working through issues of class hierarchy and social order. On the other hand, crickets, flies, and praying mantids, conveying an often charismatic individuality, have been especially prone to anthropomorphic portrayals that refer to the individual, rather than society. Tied up in the symbolic portrayal of many insects from the nineteenth century onward is an unsteady relation of beauty and violence. Fundamentally, however, the insect symbol has in modern western societies become increasingly shaped by, and continues to problematize, a social dichotomy of inclusion and exclusion embedded within a consistently transgressive natural world.

Entomology and nineteenth-century western society share a fundamental co-dependence, since it was this society's alleviation of its members from an historically entrenched proximity to nature—through urbanization—that created the conditions in which modern entomology was to develop, and in which modern symbolic meanings of particular insects were to take shape. The more urbanization elevated people from direct engagement with insects and the natural world, the more significant did symbolic representations of insects become. Insects were frequently read as delicate, miniature "traces" of nature in an increasingly industrialized, artificial environment, and hence as catalysts of escapism conducive to the most enchanting ambitions of contemporary natural history. Although the observation and collection of insects had been an "enlightened," sophisticated and pleasurable pursuit of the landed gentry since the eighteenth century, in the early nineteenth century it developed into a hobby of the emergent middle class, and became part of a wider trend in recreational "rambles" through the countryside, or through one's own garden. Crucial to the convergence of visual culture and popular science, the insect figured prominently in the new spectatorship provided by the popularization of the microscope.

The simultaneous scientific and cultural development of microscopy, and the inherent escapism of entomology—its ability to provide a sense of departure from normal parameters of time, place, and self—served to inspire symbolic representations of insects that crossed over into scientific romance and, at the turn of the twentieth century, shaped the earliest works of science fiction. The symbolic insect facilitated a growing desire to read (and an ability to purchase) books about insects that variously described the denizens of the natural world so many had left behind in their migration to the burgeoning cities. Childhood, nostalgia, and a deep longing for a return to the world of pastoral life shaped portrayals of insects in western science and popular culture throughout the nineteenth century. Indeed, many of these portrayals established representational conventions—both scientific and popular—that are visibly adhered to today.

As printing methods improved and production costs declined, the inclusion of illustrations in printed works became more viable. This development was essential to the cultural emergence of the symbolic insect in the nineteenth century. Although a number of authors had devised ingenious methods with which to verbally describe the minute details of unfamiliar specimens to their readers, pictures quickly became critical to the portrayal of insects, and by the end of the nineteenth century the number of plates a volume could boast became a valuable selling point. Thus, images of insects became increasingly fundamental to the ways in which insects were imagined. Importantly, the ways in which insects and their "world" were depicted in these early illustrations exhibit a debt to the literary tradition that preceded them, specifically, their formulation of the so-called "insect world" as a metaphorical, and often symbolic "otherworld" that resounds within the world of the human. Just as numerous popular entomology books had offered an escape into nature from the modern city, so too did many images of insects seek to emulate an enchanted, anthropomorphized otherworld, sometimes by attempting to mimic what the insect world might look like from the insect's point of view. As part of a wider theme in popular natural history that sought to engender an intimacy

Perhaps the most widespread symbolic connotation of the anthropomorphized insect, however, is the phenomenon of metamorphosis. Hogue suggests that "there is even reason to believe that the progenitor of all [bugfolk] was Psyche, Goddess of Greek Mythology who represented the soul and borrowed her wings from the butterfly, that insect which symbolized resurrection after death" (1979: 36). This is one of the more agreeable symbolic meanings given to insects; it visibly intersected with fairy lore during the nineteenth century (see Bown 2001), and with the emergence of psychiatry as a scientific discipline. Anthropomorphic insects are, by definition, creatures between fixed states; they are creatures-as-process, signifying a nature that is always "becoming." This intermediary quality becomes especially significant when the educational and instructional roles of such figures are foregrounded—the most well-known example being the Talking Cricket in Carlo Collodi's *The Adventures of Pinocchio* (1881–1883), later developed into "Jiminy Cricket" by Ward Kimball for the Walt Disney Company (Figure 6.1). In the symbolic representation of metamorphosis, the process is presented not merely as a biological phenomenon, but as a spiritual transformation, a transition to a perfect, "higher state of being." Many nineteenth-century entomology books see the biological process of metamorphosis as epitomized by the winged insect's final ascendancy to the sky; one of the most comprehensive examples of this idea can be found in what was the primary entomological text in English for much of the nineteenth century, Kirby and Spence's *Introduction to Entomology*:

> ... insects not only mimic, in a manner infinitely various, every thing in nature, they may also, with very little violence, be regarded as symbolical of beings out of and above nature. The butterfly, adorned with every beauty and every grace, borne by radiant wings through the fields of ether, and extracting nectar from every flower, gives us some idea of the blessed inhabitants of happier worlds, of the spirits of the just arrived at their state of perfection. Again, other insects seem emblematical of a different class of unearthly beings. When we behold some tremendous for the numerous horns and spines projecting in horrid array from their head or shoulders; – others for their threatening jaws of fearful length, and armed with cruel fangs; when we survey the dismal hue and demoniack air that distinguish others, the dens of darkness in which they live, the impurity of their food, their predatory habits and cruelty, the nets which they spread, and the pits which they sink to entrap the unwary, we can scarcely help regarding them as apt representations of evil demons, the enemies of man, or of impure spirits, for their vices and crimes driven from the regions of light into darkness and punishment.
>
> (Kirby and Spence 1815: 12)

FIGURE 6.1 *The Adventures of Pinnochio*, C. Collodi, trans. M.A. Murrat, London, 1892. Author's collection.

Kirby and Spence regarded sensitivity to the symbolical meanings of insects as "a more profitable mode of studying them, as furnishing moral and spiritual instruction" (1815: 24). Because the spiritualization of insect metamorphosis is an example of parallel insect mythology, it suggests that insect behavior has directly shaped the content of human narratives and beliefs about the soul, narratives, and beliefs that may not be entirely the invention of cultural processes, but rather, deeply shared, human responses to the natural world. As Hogue observes,

> it is logical that the changes attending developmental metamorphosis led unrelated cultures to a parallel adoption of winged adult insects as symbols of the soul. The quiescent and mummy-like pupa would appear to represent the death of the earthbound larva, while the drastic transfiguration to a resplendent, flying adult, surely signals a miraculous resurrection. Perhaps this also explains why angels have wings.
>
> (1979: 3)

INSECT SYMBOLISM AND THE VISUAL REPRESENTATION OF ANIMALS

Modern insect symbolism can be understood in terms of changing public attitudes towards animals that accompanied the emergence of industrialized, urban society in the nineteenth century. In his widely influential essay, "About Looking," Berger writes that "zoos, realistic animal toys and the widespread commercial diffusion of animal imagery, all began as animals started to be withdrawn from daily life" (1991: 26). Yet this was also a period in which humans were withdrawn from the daily life of animals, and in which categories of "wild" and "domesticated" animals were reconfigured. Recognized as an animal, the insect can be seen as subject to many of the same institutional and ideological practices and representations as animals in general. But the withdrawal of animals from daily life to which Berger refers does not refer to insects in quite the same way as it does to other animals. Unlike mammals and birds, insects seem impossible to domesticate. Yet they often inhabit our homes, usually in secret, where they may feed and reproduce, and some feed on our bodies, spreading disease. These qualities, which demonstrate a multifaceted autonomy from human control accompanied by an enduring intimacy with human spaces, make the insect a comparatively unique example of modernity's reconfiguration of the animal. If insects are somehow not animals to the same extent as horses, cows, birds, and so on (perhaps "not as conscious" or less individuated from their environment), then our sense of removal from animals is enhanced; we are not obliged to acknowledge our continual existence amongst the (wild) animals that are insects. Indeed, the presence of the insect within domestic space is usually obscured by lived fantasies of its absence, fantasies which make the insect's appearances within the home all the more uncanny when they do occur—and amplifying entomophobic responses of horror, disgust, and revulsion.

More broadly, this kind of paradox is inherent to what Lippit calls a "cliché of modernity," in which "human advancement always coincides with a recession of nature and its figures – wildlife, wilderness, human nature, and so on" (2000: 1). Following Berger, Lippit observes that

> animals never *entirely* vanish. Rather, they exist in a state of *perpetual vanishing*. Animals enter a new economy of being during the modern period, one that is no longer sacrificial in the traditional sense of the term but, considering modern technological media generally and the cinema more specifically, *spectral*. In supernatural terms, modernity finds animals lingering in the world *undead*.
>
> (2000: 1)

Lippit's description of the spectral animal (as figured by technological media) suggests at once a nostalgic, uncanny and unconscious place for the insect, one evoked by Berger's observation that

> children in the industrialised world are surrounded by animal imagery: toys, cartoons, pictures, decorations of every sort. No other source of imagery can begin to compete with that of animals. The apparently spontaneous interest that children have in animals might lead one to suppose that this has always been the case ... Yet it was not until the 19th century that reproductions of animals became a regular part of the decor of middle class childhoods – and then, in [the twentieth] century, with the advent of vast display and selling systems like Disney's – of all childhoods.
>
> (1991: 22)

In a broad, historical sense, the industrial production of insect imagery begins in a kind of collective, cultural childhood, one located in the nineteenth century. From this period onward, to be born was to be born into an environment of often symbolic insect imagery; imagery that coincided with (eventual) direct experience of real insects in daily life, something not necessarily the case for elephants, monkeys, cows and horses—animals increasingly encountered only in zoos or in remote rural locations.

The "image"—something that exists both mentally and visually, something "visualized" and beheld—is intimately tied to the insect itself. The Latin *imago*, as appropriated by entomology, directly associates the image, and hence the imagination, with perfection of physical form as expressed in the final stage of the insect's metamorphosis. As Burt has observed of animals in general, the insect is not outside the changing dynamics of technology and culture, but integrally related to their development and articulation within the public domain. Most saliently, "the changing configurations of visibility and invisibility ... are what determine both the nature and power of animal representation" (Burt 2001: 222). It is in this sense that the insect symbol, as figured by modern industrial societies, speaks most directly to the human beings who have created, and exist within, those societies.

INSECT SYMBOLS IN POPULAR ENTOMOLOGY BOOKS

The introduction of the European public to the nature and variety of insect life began in the first half of the nineteenth century. Prior to this period, books about insects were comparatively rare, expensive, and somewhat specialized, making them generally inaccessible to all but those with a special interest in the subject. But from the beginning of the nineteenth century, a number of intersecting social and cultural developments occurred that contributed to the growth of popular interest in insect life. As Lightman has observed, "the growth of an educated middle class, and therefore a large reading audience, and the invention of new printing technologies made possible the birth of a mass market, which in turn opened up new career opportunities in journalism and writing" (2000: 652). In addition to Victorian science's progression toward specialized knowledges, this period was also "an age of popularization; the second half of the century saw a veritable explosion of publications on popular science. As professional scientists began to pursue highly specialised research, a market emerged for popularizers who could act as mediators between the professionals and the reading public" (Lightman 2000: 653). Entomology, along with botany, was particularly compatible with this rise in popular science books, since most readers could, with very little effort, observe insects and plants for themselves in their garden or in the field.

As the book became an indispensable tool of entomology, so too did the most popular entomology books tend to be those that were educational, pleasurable, and edifying of nature, and this made them particularly conducive to the contemporary objectives of natural theology and the representational conventions of insect symbolism. Along with the improving public profile of entomology itself, natural history books for amusement and instruction became increasingly prominent throughout the nineteenth century as the printed word and its associated imagery increasingly reached an audience that was not only larger than ever before, but whose number included more than one social class. While this period effectively marks entomology's movement from the realm of the landed gentry and aristocracy to the general public at large, it also marks the movement of a kind of pastoral theology into the publications of the increasingly industrialized cities. Consumed by social classes for whom the surrounding environment seemed less and less like the wonderful nature described in popular texts, images came to stand for a theological, symbolic connection to the divine that would ideally result from the actual emersion in nature itself, an experience increasingly reserved for those still living in rural environs. If one could not commune with nature in person, one could do it "in spirit," with the assistance of (increasingly realistic) images and immersive narratives.

On the history of science popularization, Lightman suggests that instead of approaching visual images in popular science with the idea that popularizers differed radically from professionals in using illustrations as a means of simplifying science for a mass reading audience, scholars should now begin to analyze the rhetorical role of visual images in both popular and scientific works, whatever the intended audience.

(2000: 654)

The significance of nineteenth-century entomology books, from this perspective, is that most of them do not sit comfortably within a binary model of scientific and popular cultures, particularly in their employment of symbolic portrayals of insects. As well as helping nineteenth-century readers to imagine new and unfamiliar parts of nature, the visual style of popular entomology books often worked to acquaint readers with nature itself—as if the whole of nature was another world that had not been completely recognized or appreciated. In some cases, this view of nature seems clearly to have been a reaction to the felt artificiality and oppression of the Victorian city. Arabella Buckley's *Life and Her Children: Glimpses of Animal Life from the Amoeba to the Insects* (first published in 1880), having as its main object "to acquaint young people with the structure and habits of the lower forms of life" (Buckley 1917: v), provides one example (Figure 6.2). Although it is a book for children, in style and content it is not remarkably different from contemporary nature books for adults. Buckley opens the first chapter by wondering "whether it ever occurs to most people [not merely children] to consider how brimful our world is of life, and what a different place it would be if no living thing had ever been upon it?" (1). This is obviously a rhetorical question; the implication is that, in fact, most people have not considered this. From here the child reader is introduced not just to the natural world, but to life itself, as if they are a stranger to it, and the language used constantly encourages them to "look" and to "see"—not literally, but rather to visualize and imagine what is described:

> There is scarcely a street or alley where, if it be neglected for a time, some blade of grass or struggling weed does not make its appearance, pushing its way through chinks in the pavement or the mortar in the wall; no spot from which we cannot see some insect creeping, or flying, or spinning its web, so long as the hand of man does not destroy it.
>
> (2)

Having been reminded of the small traces of nature still visible in one's man-made environment, the reader is then taken out of the stifling conditions of the city to "see" nature in all its complex majesty:

> And when we go into the quiet country, leaving man and his works behind, how actively we find life employed! ... *Look* at the little water-beetles skimming on the surface of the shady wayside pool, *watch* the snails feeding

on the muddy bank, *notice* the newts putting their heads above water to take breath, and then remember that, besides these and innumerable other animals visible to the naked eye, the fairy-shrimp and the water-flea, and other minute creatures, are probably darting across the pond, or floating lazily near its surface; while the very scum which is blown in ridges towards one corner of the pool is made up of microscopic animals and plants.

(2; italics added)

FIGURE 6.2 *Life and Her Children: Glimpses of Animal Life*, Arabella Buckley, London, 1917. Author's collection.

Like Keats's "guessing" and Eliot's imagination, the language used "allows for the perception of objects in the absence of [the reader's] physical vision" (Krasner 1992: 15), in this case, the omnipresent microscopic animals and plants that we must "remember." It is as if the new knowledge of the profundity of extant invisible life revealed through the microscope invigorated not just a general appreciation or recognition of the profundity of life per se, but its remembrance. This kind of ephemeral description is effectively illustrated on the gilded cover of the 1917 edition, showing a young boy and girl—as cherub-fairies with insects' wings—immersed in an Art Nouveau-style scene surrounded by the dendritic forms of plants and underwater sea life.

The sheer unfamiliarity (or perhaps more specifically, the uncanniness) of many insects, their behavior, and their "world" made them particularly malleable subjects, and this was conducive to their symbolic configuration in a religious context, along teleological lines, according to contemporary cultural conventions. During the nineteenth century, the interpretation of insect activity as dramatic allegory was a common tenet of the period's natural theology that often presented even the most alien forms as evidence of God's multitudinous, creative power. The association of movement—or animation—with life meant that the appreciation of God's creatures was compatible with an appreciation of their actions, and as the microscope and the increasingly close observation its use inspired revealed more and more knowledge of insect physiology to the observer, so too did the known complexities of insect behavior escalate in kind. At this time ecological entomology did not exist as a systematic account of natural systems, and so the natural theological commitments of many authors seeking to explain the behavior of insects often resulted in the conversion of prosaic instances of insect activity into dramatic, episodic, and anthropomorphic narratives. This certainly made the books they were selling all the more compelling, but it also worked to overturn the common disregard for insects that natural theologians saw as an affront to the marvelous yet miniature creations of God. This approach is particularly pronounced in the case of fabular insect literature, in which reader resistance to the contrived imposition of human narrative structures onto insect behavior is effectively suspended. In drawing real moral lessons from a perception of insect activity as both narrative and allegorical, natural historians drew the relationship of book and nature even closer. Nature was not only in itself a symbolic text, indicative of the designs of the Creator, but contained meaningful narratives in which moral truths were encoded—Nature was, for all intents and purposes, a book. This conception afforded those nature books that adopted natural theology a degree of epistemic prestige enjoyed by few other genres of the period. Kirby and Spence had, in their *Introduction to Entomology*, helped to initiate this transformation of the popular natural history book early in the nineteenth century, writing that "whichever way we turn our eyes on the objects of creation, above – below – athwart, *analogies* meet us in every direction, and it appears clear, that the Book of Nature is a Book of Symbols, in which one thing represents another in endless alternation" (1826: 43). From this position, at which nature becomes symbolic text—representative of something other than itself—the myriad ways in which insects represented both human beings and divine will were free to develop, sometimes with fantastic consequences.

The stories of insect lives (or insect "tribes") as told by entomologists were often so enchanting that fantasies of immersion in the insects' world—of becoming part of their story—began to appear in the literature (Figure 6.3).

FIGURE 6.3 *The Population of an Old Pear-Tree, or Stories of Insect Life*, E. van Bruyssel, London, 1888. Author's collection.

Van Bruyssel opened his fictional, richly illustrated account *The Population of an Old Pear-Tree* (first English translation from the French, 1870), an *Alice's Adventures in Wonderland*-styled adventure that takes the author into the insect world, with the declaration that

> When certain blue spirits begin to flit about me, I depart from my study to go and read, in what I am allowed, even by my clerical uncle, to call my book of devotions. The devotions I mean are not in my book-case. No publisher, if he ever thought of such a thing, could bring them out. They are a page of the book of Nature, opened in the country, under blue sky, displayed at all seasons. Every one has looked at that book, but few of us understand it in any degree, and the man who could read it all would be the wisest in the world.
>
> (1888: 1–2)

Although van Bruyssel's narrative relates the protagonist's trip through an insect world of constantly fluctuating scale, in which the actual senses of the insects are imparted to him, it also seeks to provide some accurate descriptions of insect appearance and behavior. Despite the book's obvious fantasy, and many overtly anthropomorphic portrayals, its essential message is that true wisdom is to be derived from the knowledge of nature, knowledge achieved by reading nature as a book of symbols, and by reading books about nature.

The association of nature and the book was a notion fundamental to texts seeking to unite an adult and juvenile readership by introducing readers to the wonders of nature, of which the insect world was an increasingly celebrated example. The illustration to the frontispiece of *Pear-Tree* exemplifies this intention. The perspective of the reader is miniaturized and set at the scale of the insects, centipedes, and spiders, and within what may be the trunk of a tree or perhaps a burrow in the hillside. The darkness of this insect space contrasts sharply with the light of the outside world, where in the distance—and thus appearing smaller than most of the insects—we see the human protagonist reading a book while leaning against a tree in the field. The insects in front of us stand upright on two legs and raise their "arms" joyously as if to welcome the protagonist into their secret world. But the fact that he seems oblivious to their presence—and is instead deeply involved with his book—suggests that his entry to their world is conditional upon his absorption with the text. As we read the story, we discover that the protagonist does not actually have a book with him in the field; the book is included in the opening illustration as a metaphor of nature and of the insect world. Thus the division between book and nature, and more specifically, book and insect world, is all but collapsed—while the reader actually holds *The Population of an Old Pear-Tree* in their hands.

While texts such as this one, appearing after Darwin's *On the Origin of Species* (1859), can perhaps be read as a contemporary response to Darwin's removal of a divine guiding force from nature, natural theology was an established theme in insect books before Darwin's work appeared. In one example, *Insects and their Habitations: A Book for Children* (1834), a small book published under the direction of the Committee of General Literature and Education, appointed by the Society for Promoting Christian Knowledge, the author points out that:

> We may learn many useful lessons from the history of insects; and, by observing them, we shall find that they set us an example of various good habits. Though they are the smallest of God's works, [microbes, apparently, do not qualify] yet His power and wisdom visibly shine forth in them; and we shall see fresh cause as we proceed, to adore the great and wise Creator, who formed them out of nothing.
>
> (1834: 5)

Here, the study of nature and insects is clearly intended to be not an end in itself but merely a stepping stone to the appreciation and worship of the Creator. The "various good habits" exhibited by insects usually involved tireless industry that seemed suggestive of a strong Christian work ethic. Other insect habits, however, were more problematic, and suggestive of real inequalities that existed in human societies. *Insects and their Habitations* addressed the behavior of ants, for example, making an awkward attempt to resolve the brutality of contemporary European militarism, colonialism, and slavery through reference to the natural behavior of particular ant colonies.

> Ants not only fight pitched battles ... they also besiege neighbouring cities, and take them by storm. The object they have in view is a most extraordinary one, and could hardly have been believed, if it had not been observed, and related by so many people. The ants we are now going to speak of are warlike and powerful, but not industrious like most other kinds. In order that they may have slaves to do their work for them, they attack the nest of a dark-coloured kind called negro-ants, and carry off their young. To these, when they are grown up, all the labour is left; they build and repair the common dwelling, collect food, attend to the young, and even feed their masters, and carry them about the nest. Every year the light-coloured ants add to the number of their slaves. At the season when there is a proper supply of workers in the negro ant-hills, they send out spies into the neighbourhood, and prepare for marching ... Having been taken when they were quite young they do not feel the change, and are quite as happy as they would have been among their own nation, where they would have the same tasks to perform.
>
> (1834: 56–7)

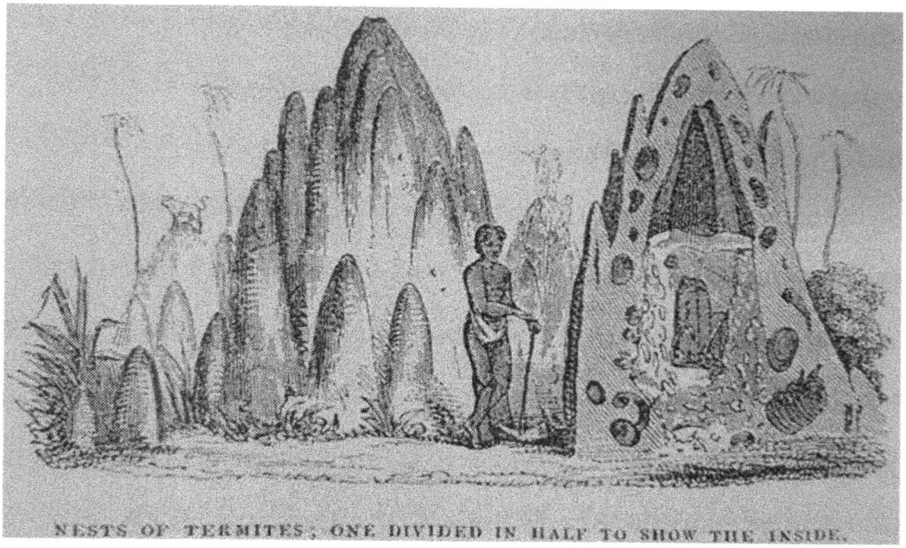

FIGURE 6.4 *Insects and their Habitations: A Book for Children*, London, 1834. Author's collection.

Hence, particular interpretations of ant behavior were used to symbolize, and implicitly naturalize, some of the more problematic practices of European society itself, particularly colonialism and slavery (Figure 6.4). *Insects and their Habitations* (which, it should be reiterated, was a "book for children") accompanied this description of ant behavior with a version of James Basire's engraving, based on Henry Smeathman's original drawing, of termite mounds Smeathman had examined while in Africa in the late eighteenth century. Of particular note in this image is the native positioned between the two termite "cities," and the obviously European explorers in the background, standing atop one of the mounds to see into the distance, as if exploring an alien environment. The world of the social insects—typified by efficient social order, hierarchies, and "architecture"—is presented as both strangely similar to the world of Europeans and markedly different from the "savage" world of the African native, whose culture is evidently incapable of constructing the very emblem of civilization: the city. What emerges from such a configuration is a view of the insects as "little people."

INSECTS AS "LITTLE PEOPLE"

The appearance of anthropomorphized insects in nineteenth-century popular culture is, to a significant degree, reflective of scientific and philosophical shifts of the period and their developing influence on the ways in which insects were understood. Anthropomorphic portrayals of insects were usefully employed by a number of authors because of their inherent appeal to the childlike imagination as a mode of awareness conducive to amusement and instruction. The images made possible for the eye by the microscope required not just new ways of looking at nature, but new ways of thinking about and representing nature. Popular illustrated entomology books of the nineteenth century are particularly vivid examples of this shift. While these texts do contain an (often impressive) amount of scientific information—in the form of facts about insects—they also illustrate a form of psychological adjustment to these facts themselves, a framing of the ontology of the microscopic within allegorical contexts. These kinds of representations allowed the Victorians to transpose a vast array of desires and anxieties about themselves, their society, and their future, onto insects and their "world." Popular insect books accompanied a period of environmental, as well as intellectual transition in Victorian culture in which definitions of the insect (and indeed of nature itself) were being negotiated and re-configured. In conjunction with a visual culture inspired by the influx of new visual media forms in the nineteenth century, this redefinition of the insect often involved anthropomorphic representations and interpretations that assisted a conception of insects as "little people." Despite their romantic leanings (and perhaps *because of* their allusions to the convergence of nineteenth-century entomology and fairy lore), such texts constitute not just vivid examples of how insects were incorporated into contemporary symbolism, but how such portrayals affected contemporary conceptions of human beings themselves.

The composite anatomy and behavior of the anthropomorphized insect, despite its strangeness, refers to real circumstances of the human world and to collective notions of human selfhood. The likening of insects to human beings is, fundamentally, an attempt to initiate, or draw attention to, a sense of communion between supposedly disparate beings. In this sense, the appearance of anthropomorphized insects in the nineteenth century can be seen as part of a wider trend in the symbolic representation of supernatural beings comprising various amalgamations of ontologically distinct entities. While the feeling of disconnection from natural environments that beings such as miniature, insectoid fairies represent can be seen as a result of nineteenth-century industrialism and a fading of, or a departure from, traditional rural life, values, and folklore, the bugfolk in particular operated to symbolically configure the many newly discovered differences and similarities—both anatomical and behavioral—between insects and human beings. Insect allegories became increasingly popular, and the bugfolk referred to an ongoing, ecological relationship that was not entirely within the realm of fantasy or myth, since human beings really are surrounded by insects almost all of the time. Despite the multitude of supernatural beings popularized in art and literature during the nineteenth century, such as fairies, gnomes, pixies, goblins, and so on, they were cohesive in their spatial origin, which was regarded as some kind of hidden domain of the natural world—an "underworld." A spatial metaphor, that is, for the actual realm of the insect as imagined by entomologists—the "insect world." The supernatural beings revived from folklore (and particularly the fairy) were often seen as personifications of "nature spirits," and in this sense were seen as fulfilling an important ecological role, which included the education of humans about nature and their own place within it. In reflection, the insect's animated capacity for education—its ability to teach us valuable lessons about the hidden realms of nature simply through its daily activities—was a quality appreciated by many entomologists from the nineteenth century onwards.

Fantasies of immersion in the insect world and of some kind of communion with its inhabitants may have helped to relieve the dissatisfaction (of some) with the reality that insects and their world can, primarily, only be known and experienced visually. In a childlike reaction to the violence and death that pervades the insect world, many writers anthropomorphized their insects simply to make the unfamiliar, familiar. As Bown writes, the difficulty faced in the representation of insects also involved attempts to reconcile two kinds of gaze: the "knowing" gaze, that could see nature as it really was, and the "loving" gaze, that could maintain a romantic vision of nature. Consequently, for many nineteenth-century entomologists, "the study of insects becomes a way of enchanting the natural world, of covering over 'the bare, the barren, and the death-like' and turning it into fairyland. The 'loving' gaze of Entomology imagines the natural world as a quasi-human, fairy-like domestic life taking place among the flowers or under the stone" (Bown 2001: 127).

For Bown, the imagination of the loving gaze "transforms nature by three devices: first, the insects are imagined as humans (anthropomorphised); second, they are invested with attributes of the past (made nostalgic); third, the world is seen at the insect's scale (miniaturised)" (127). A clear inspiration for the invention of such fantastic creatures is a desire to escape the increasingly industrial reality of Victorian society and to reconcile its removal of the folk from traditional, rural environments. The fairy was the ultimate manifestation of this yearning, since "by looking at and reading about fairies, the Victorians could imagine themselves as being still in the world they had given up for modernity. They shaped fairyland into the negative image of their own disenchanted world, and saw their own disappointed forms transfigured in the shape of the fairy" (1). The primary corporeal connection of insects and fairies is of course their wings, and "it is then a short step to turning insects into fairies and using this to cover over insects' alien qualities" (125). As Bown shows, the first published images of a fairy with insect wings were Thomas Stothard's illustrations for a 1798 edition of Alexander Pope's *The Rape of the Lock*, and these established the convention for popular nineteenth-century portrayals, which presented fairies as miniature, fair-skinned folk with insect wings, immersed in natural surroundings, devoid of technology.

A direct convergence of insects and fairies can be found in Charlotte Maria Tucker's *Fairy Frisket; or, Peeps at Insect Life*, first published in 1874 and reprinted several times after. It imparts contemporary entomological knowledge to children in the form of a fairy tale. Writing under the pseudonym A.L.O.E., Tucker tells the story of two sibling fairies, the female Fairy Frisket, her older brother, the wise, fatherly fairy, Know-a-bit, and their introduction of two young boys to the wonders of the insect world (Figure 6.5). Fairy Frisket, we are told,

> is scarcely more than five inches in height ... A wreath of tiniest heather-bells encircles her hair; if that can be called hair which is so fine it hangs over the fairy's shoulders like a golden mist, which the puff of a child's breath could set floating upon the air. [Her] dress is formed of petals of the blush-rose, fashioned by fairy fingers ... The wings that spread at the fairy's back are such as the dragon-fly wears: they are transparent as glass, and, as they quiver in the silvery rays, appear to be tinted with every hue in the rainbow.
>
> (Tucker 1893: 10–11)

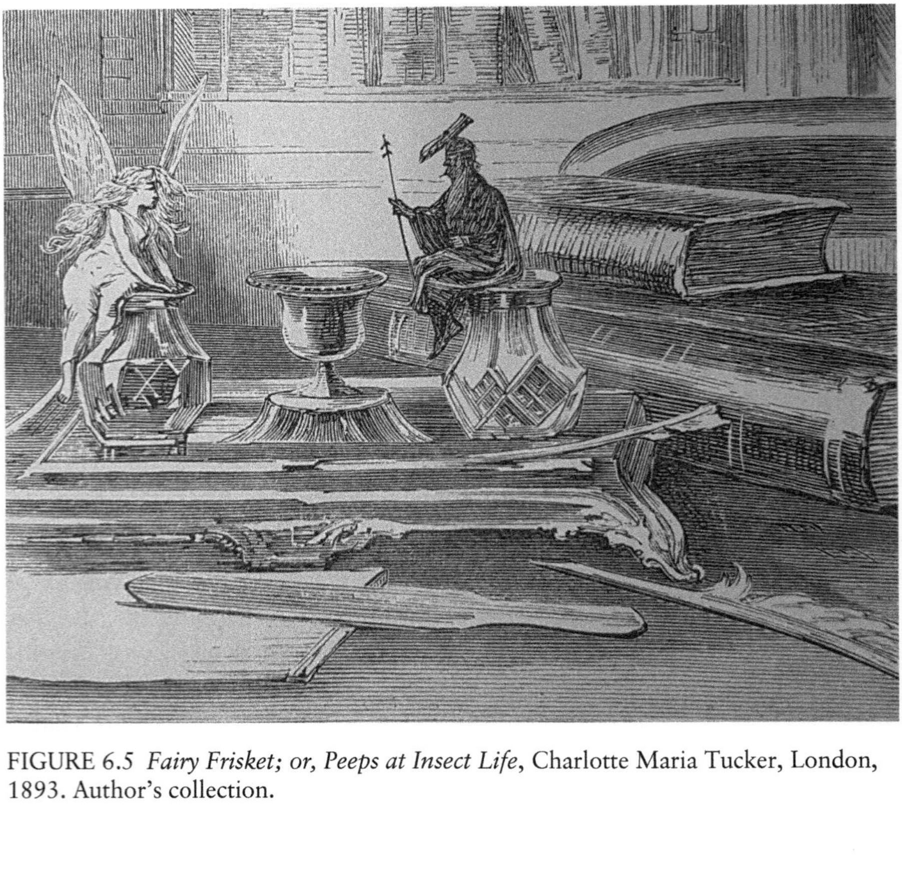

FIGURE 6.5 *Fairy Frisket; or, Peeps at Insect Life*, Charlotte Maria Tucker, London, 1893. Author's collection.

Fairy Frisket finds humans, with their lumbering, "meat-fed" bodies, offensive, brutish, and stupid. Know-a-bit, however, has spent the last four hundred years learning the ways of the human world, and reading its books, and thus has a more considered view. Know-a-bit has folded his wings under a robe and taken to wearing spectacles, much to Fairy Frisket's disdain. After a very long absence, the two meet one moonlit evening in Squire Philimore's study and beginning debating the pros and cons of human beings. Together they decide to "educate and amuse" the two boys of the house, Philibert and his guest, Sydney, by introducing them to the insect world. In this narrative the distinction between insect world and fairyland is completely dissolved, climaxing with Fairy Frisket's actual transformation of the two boys into ants (so that they may more thoroughly understand the insect world), and then her own transformation into an ant to serve as a guide through a tour of the ants' nest. The book contains many detailed illustrations of insects at various stages of development, as well as cross-sections of ants' nests and termites' mounds, all of which are wholly consistent with nineteenth-century entomological conventions. Indeed, there is nothing remarkably different about the visual representation of the insect world as it appears in *Fairy Frisket*, and its appearance in contemporary non-fiction insect books, such as John Wood's *Insects at Home* (1872).

Silver alludes to just how extensively the insect had captured the Victorian imagination and subsequently shaped interpretations of fairies into the twentieth century, quoting Edward Gardner, secretary of the Theosophical Society and, with Sir Arthur Conan Doyle, a supporter of the reality of fairies: "Allied to the lepidoptera, or butterfly genus ... rather than to the mammalian line ... they are as real as we are, and perform functions in connection with plant life of an important ... character" (1999: 54). For Gardner, fairies were a kind of insect with human attributes, rather than the opposite—true bugfolk (as defined by Hogue). However, this was not a widely shared interpretation. Despite their similarities, and their reflection of similar desires and anxieties, important distinctions can be drawn between the bugfolk and fairies in both form and function. Although the bugfolk do possess many fairy-like qualities, they are much closer in form to actual insects than are fairies (who often border on the angelic), and are typified by their attempts to impart both moral lessons and real entomological knowledge. By drawing parallels between insect and human societies, the bugfolk often engage in allegorical social commentary. Importantly, in the insect world, as in the fairy world, the benevolence of its inhabitants is unstable and regarded with skepticism. We can understand the bugfolk as "a composite that unites attributes held to be categorically distinct and/or at odds in the cultural scheme of things in *unambiguously* one, spatio-temporally discrete entity" (Carroll 1990: 43).

INSECTS AS "GERMS WITH LEGS"

The insect became permanently associated with narratives of bodily invasion and invisible threats in 1877, when Patrick Manson demonstrated that certain mosquitoes can act as carriers of a filarial parasite of man. Similarly, although flies had been associated with death and decay for millennia—owing to their habits—their role in disease transmission did not become apparent until the bacteriological studies of the late 1880s and 1890s (Tomes 1998: 99). These discoveries tied the insect to deeper, more mysterious processes of the microscopic, and thus it became even more representative of nebulous "unseen forces"—primarily, disease. The association of insects with disease transmission, once empirically verified, had a direct bearing on the increasing prevalence of entomophobia in industrialized societies. The pathology of entomophobia is closely linked to modern, urban life, primarily because most people living in hygienic, urbanized societies have little direct experience of insects and are conditioned to think of insects generally as invasive, biting, stinging "pests" that threaten the purity domestic spaces.

It is difficult to overstate the significance of Manson's discovery (generally considered to be the birth of medical entomology), but perhaps not so difficult to observe its effect on the symbolic representation of insects. While many were already renowned as serious threats to essential crops (rural space), and associated with death and the putrid (the fly), now insects were seen as vectors of lethal bacteria that could, potentially, spread very quickly through the densely populated cities. From the turn of the twentieth century, mass-produced insect imagery became a ubiquitous feature of urban environments in conjunction with the commercial industries of medical and agricultural entomology; arthropod encounters were thus increasingly mediated by portrayals that did not always serve primarily to engender logically informed, or ecologically sound, insect-human relations. Insects such as mosquitoes, flies, and cockroaches were portrayed in ways specifically engineered to promote fear, paranoia, and the sale of chemical insecticides. While flies and mosquitoes represented palpable threats to human health, the cockroach—largely sterile—represented a symbolic threat to urban hygiene. All three, however, came to represent a persistently transgressive natural world that seemed to constantly "push back" or "creep in" as modern industrial society continued to expand.

The natural theology that had shaped much of the insect symbolism of the nineteenth century somewhat abruptly gave way to the new findings of medical entomology at the turn of the twentieth century; the case of the "house fly" posed a particular challenge to those wishing to maintain a view of insects as collectively emblematic of providence. Indeed, the early decades of the twentieth century were marked by a kind of "war on flies" in which the insect came to

FIGURE 6.6 *The House-Fly at the Bar*, New York, 1909. Author's collection.

stand for "disease" in the broadest sense, far beyond the specific diseases it was capable of (unintentionally) distributing. In his *Book of the Fly* (1915), Hardy proclaimed confidently that

> the old fanciful dogma that everything existing was actually created "in the beginning," and "for a purpose," was once ardently championed as controverting aggressive Voltairean atheism, but it must be now recognised as an unwarranted assumption, deduced from an orthodox doctrine of "design," which in itself seems acceptably agreeable with the idea of unity, consistency, and perfection in Creation and The Creator. In fact the said "fanciful" dogma never really was an integral part of Christian Catholic doctrine. The house-fly, as we know it, is absolutely the developed product of human insanitation; scientifically and practically it is a new "species" of an old "genus" established by a long course of breeding in man-made environments.
>
> (Hardy 1915: 5)

The escalating secularism of early twentieth-century society reshaped—and in many ways, made redundant—the symbolic representation of insects that occurred through most of the nineteenth century. In conjunction with the emerging literary genre of science fiction, as a marked development from scientific romance, the insect came to stand for an alien threat to human concerns in an increasingly globalized world, framed by discourses of "biomilitarism." While occupying quite disparate historical roles in relation to human beings, however, it is the insect's ability to cooperate with contemporary, industrialized society that continues to determine the value of its life, its acceptable impact on human life, and hence its symbolic relevance.

CHAPTER SEVEN

Insects in Literature and Language

"Crawling at your feet, you may observe a Bread-And-Butterfly"

ROSALIND WHITE

Insects were everywhere in nineteenth-century culture: creeping into the margins of children's books and Christmas cards, manifest in the incipient genres of horror and science fiction, sparking epistolary friendships born from specialist periodicals, and, of course, featuring in countless narratives of natural history. The insect, in many ways, became the unwitting receptacle for the cultural anxieties of the age: operating as a rare point of reference that resided in the homes of both the wretched and the well-to-do. The sheer diversity of insect-life—their capacity for easy capture, collection, and comparison—as well as the evident affinity between metaphor and metamorphosis, in part accounts for their ubiquity. As the population swelled (city inhabitants increasing from 20 to 80 percent as the century progressed), townsfolk became a part of an ever-expanding web of simultaneity. The sudden proximity between neighbors, together with the arrival of the penny-post, railway, and printed press, meant that now—more than ever before—the budding hive-mind of humanity mirrored the collective-consciousness of insect societies. Naturally, there was a renewed interest in entomology. Insects and their cousins were anthropomorphized as engineers and architects of the microscopic realm, an ant nest was thought to rival the pyramids of Egypt, while spiders that could sail through "fields of ether in [these] air-light chariots" were likened to aeronauts aboard a hot air balloon (Kirby and Spence 1817: 336).

As the "effluvia of modernity" cast an encroaching shadow upon the countryside, many felt as though the landscape of the past had been lost to the industrial present (Bown 2001: 40). Insect collecting—in this context—became part of a nostalgic bid to capture lost nature in an increasingly urban Victorian Britain.[1] By the 1860s Charles Darwin, a self-proclaimed "decayed entomologist" (Darwin Correspondence Project 1854) had published *On the Origin of Species* which famously used the metaphor of an entangled bank, "with various insects flitting about, and worms crawling through the damp earth," to articulate how ecological interdependence and his theory of natural selection intertwined (1859: 60).

As we will explore, a complex web of ambivalent emotions and anxieties underpinned the metaphorical might of the insect. This chapter will pay particular attention to the manner in which questions of time, scale, and domesticity became embedded into the Victorian's understanding of the insect realm.

INSECT-MANIA

On a summer's day in 1827 a young Charles Darwin, tearing the bark from an old tree, uncovered two rare beetles. With a beetle bound in each hand, he glimpsed a third "new kind, which [he] could not bear to lose." The next logical step for an ardent beetle-hunter was to stow the specimen safely into his mouth, and attempt to withstand the "intensely acrid fluid, which burnt [his] tongue."[2] Darwin, much to his chagrin, was eventually compelled to spit the beetle out. It is hard to convey the dynamism, passion, and insatiable curiosity alive in the entomologist of the nineteenth century. (Though Darwin's tale, no doubt, affords us a modicum of his ardor.) By the 1860s beetle-mania was in its heyday. Children were screened for "the brow of an observer" (Darwin 2010: 6) by enthusiastic entomologists, and achieved teenage rebellion by assuming a best-loved insect that differed from their parent's favored choice. As Darwin remarked to Katherine Lyell—"my Boys are all butterfly-hunters, and [as] all young & ardent lepidopterists despise from the bottom of their souls coleopterists!"[3] So thorough were the Victorian entomologists that their twentieth-century counterparts struggled to find any new butterflies across Britain (Forbes 2009: 7). Darwin's avowal that worms will "revolutionize the world,"[4] like Roger Hamley's use of a wasp's nest as a clumsy token of courtship in Elizabeth Gaskell's *Wives and Daughters*, paints a picture of the entomologist as an eccentric specimen thoroughly determined to graft entomology onto his everyday life. In *The Romance of Natural History* Philip Henry Gosse casts his mind back to what it was like to be a "schoolboy, just infected with entomological mania." Gosse's "youthful savant" acknowledges the "bewilderment" of those who witness the peculiarities of his hobby, but any embarrassment he may feel is quite literally eclipsed by the promise of capturing a sought-after moth. Unabashed, he tails the object of his curiosity with as much abandon as an airborne moth chases the light.

> An hour before sunset he had been seen ... carrying a cup filled with a mixture of beer and treacle. With this he had bent his steps to the edge of a wood, and with a painter's brush had bedaubed the trunks of several large trees, much to the bewilderment of the woodman and his dog. Now the sun is going down like a glowing coal behind the hill, and the youthful savant again seeks the scene of his labours, armed with insect-net, pill-boxes, and a bulls-eye lantern.
>
> (1861: 24)

By 1855 the pursuit of entomology had, according to H.T. Stainton, doubled in popularity over the past twenty years. The first half of the century witnessed a steadily increasing swarm of men and women combing the countryside: armed with an assortment of nets, phials, collecting boxes, magnifiers and setting boards—fondly referred to as their "weapons of the chase" (Salmon 2000). Forgoing the railway for the stagecoach, entomologists were known to net airborne insects en route aboard the windswept coach-top, their specimens pressed beneath the warmth of their seats.[5]

TAXONOMY OF AN ENTOMOLOGIST

Across nineteenth-century literature the entomologist became a familiar archetype, a recognizable sub-species under the principle genus of the naturalist. As such, fictional entomologists can be grouped together according to a number of observable characteristics. Typically they are male,[6] in possession of an insatiable level of curiosity, and characterized by numerous eccentricities directly connected to their passion for bug hunting. Entomology-evoked emotion (in literature and correspondence) frames the pursuit as habitual, compulsive, and often bodily exhilarating. Alfred Russel Wallace recalled the "blood rush[ing] to [his] head," at the sight of new Ornithoptera, whilst Charles Darwin felt "like an old war-horse at the sound of a trumpet" whenever he heard "of the capture of rare beetles."[7]

In a letter to her publisher, Elizabeth Gaskell writes of "a made up story" (that would become *Wives and Daughters*) with a character who is much "like

Charles Darwin": a man who is "unpolished but works out for himself a certain name in Natural Sciences" (1996: 731). Throughout *Wives and Daughters,* Roger Hamley (the character Gaskell alludes to) much like Darwin or Wallace, is emotionally attuned to each and every aspect of entomological life. His "eyes are always wandering about" capable of "seeing twenty things" where most men would see one, and he is liable to "bolt into a copse" upon spying a specimen "fifteen yards off" (73). As his father, the Squire, tells us, Roger habitually observes the smallest of insect-homes, and is able to discern "what insect or spider made it, and if it lived in rotten fir wood, or in a cranny of good sound timber, or deep down in the ground, or up in the sky, or anywhere" (73). He "habitually" moves "with light and well-planted footsteps," taking care to avoid "treading unnecessarily on any plant" for the sake of an "insect [that] might develop itself in that which now appeared but insignificant" (73). As a fervent entomologist Roger experiences a sense of temporal thickening, allowing him to register every imperceptive movement and afford empathy to the smallest of creatures. The delirium of detail inherent in the insect realm calls for the entomologist to modulate his every minute. In an earlier novel, *Mary Barton* (1848), Gaskell articulates how an entomologist's inclination to "pore over every new specimen" might extend into other aspects of his or her life (39). Job Legh "directly put[s] down some insect, which he was impaling on a corking-pin" to listen to Mary in her time of need. He registers every detail of her change in demeanor, despite beginning "a leisurely inspection" of a candle:

> "Thou hast gotten a bit of pink in thy cheeks,—not much; but when last I saw thee, thy lips were as white as a sheet. Thy nose is sharpish at th' end; thou'rt more like thy father than ever thou wert before"
>
> (335)

Mary recalls how though "she spoke hastily, and (to her own idea) in an unnatural manner" Job and his granddaughter "do not seem to notice it," instead they provide Mary with their undivided attention (355). Job's identity as an entomologist is allied with his lack of societal judgment. Moreover, his hobby allows him to circumnavigate class boundaries and even act as a mediator between the classes:

> "What think yo of a lawyer? I know one, Mr Cheshire, who's rather given to th' insect line – and a good kind o' chap. He and I have swopped specimens many's the time, when either of us had a duplicate. He'll do me a kind turn I'm sure."
>
> (355)

Job is a part of a dedicated coterie of "Manchester workmen" whom Gaskell presents as a valued node of the wider network of natural historians.[8] Entomology acts as a cross-cultural shibboleth that remains in place even when the matter is not of an entomological nature. These artisans are treasured for their capacity to observe that which may have escaped the attention of other men—men who are unaccustomed to everyday labor:

> There are entomologists, who may be seen with a rude-looking net, ready to catch any winged insect, or a kind of dredge, with which they rake the green and slimy pools; practical, shrewd, hard-working men, who pore over every new specimen with real scientific delight. Nor is it the common and more obvious divisions of Entomology and Botany that alone attract these earnest seekers after knowledge. Perhaps it may be owing to the great annual town-holiday of Whitsun-week so often falling in May or June, that the two great beautiful families of Ephemeridae and Phryganidae have been so much and so closely studied by Manchester workmen, while they have in a great measure escaped general observation.
>
> (39)

As Anne Secord notes, it is likely that Gaskell adapted this passage from an 1843 article "Memoir of Mr Hobson" written in support of artisan naturalists by John Moore.

> Moore, a gentleman, had invited Edward Hobson, a handloom weaver turned warehouseman, to accompany him on an angling trip to Derbyshire, and he wrote of Hobson's delight in the variety of water-bred flies, especially "the two great families Ephemeridae" and "Phryganidae," which appear to have escaped the attention of our most careful entomologists.
>
> (2015: 135)

Both Gaskell and Moore choose to forgo common names, such as mayfly and caddis, for taxonomic nomenclature. This choice not only validates the work of the artisan but also foregrounds the fact that the entomologist, irrespective of his class or occupation, has the exactitude to differentiate between different families of a given species.

Like the artisan entomologist of the inner cities the English parson-entomologist was a familiar fixture of the English countryside. Reverend Gilbert White's *The Natural History and Antiquities of Selborne* (1789) is cited as formative inspiration for countless naturalists: Charles Darwin, Henry Walter Bates, and Alfred Russel Wallace, to name but a few. Contemporary critics saw in White's account of his vicarage "a place that lingers beyond the spatio-temporal horizon of modern life" (Menley 2004: 46). As Patrick Armstrong explains in his study of the parson-naturalist:

> There were times in the past when the vicar or rector of a rural parish had the cure (or care) of just a couple of hundred souls, and at a time when administrative duties were less onerous than today, time must sometimes have hung heavily ... these men used their long association with a single parish to make a detailed study of a particular locality.
>
> (2000: iv)

Mr Farebrother, the parson-naturalist of George Eliot's *Middlemarch* (1874), is presented as equally invested in the entomological and anthropological microcosms his parish provides: training his "quick gray eyes" indiscriminately on the goings on of man, woman, and beetle:

> [Farebrother] came like a pleasant change in the light, arresting little Louisa with fatherly nonsense as she was being led out of the room by Miss Morgan, greeting everybody with some special word, and seeming to condense more talk into ten minutes than had been held all through the evening.
>
> (118)

Intimately acquainted with his congregation, as one might be with the ecology of a beloved drawing room terrarium, Mr Farebrother—like Gaskell's Job Legh—converses with such careful consideration that those around him all but question the verisimilitude of time. Directly after this introduction, we learn that Farebrother has "made an exhaustive study of the entomology of this district" (118). He even speaks of his parochial duties as though they are entomological practices: "I prepared her for confirmation—she is a favorite of mine" (126).

Farebrother's role in the wider plot of *Middlemarch* is as a foil to the character of Tertius Lydgate—an idealistic young doctor that, fatally, thinks of others as satellites to his own reality. Lydgate, fascinated by the human anatomy, feels a "sense of secrecy and obscenity in connection with his internal structure" (144). His avid interest in the unseen complexity that lies beneath the veneer of the body belies his ignorance of the larger vision of relations—relations that would be patent to an entomologist such as Farebrother. Lydgate's propensity to dismiss "the hampering threadlike pressure of small social conditions" as "trivial Middlemarch business" is directly opposed with the watchful eye of the natural historian (130). It is this dismissal that determines Lydgate's obliviousness to both his wife's interiority and the "little woven future" she has in store for him (85). Farebrother and Lydgate first interact under the pretext of entomology—as Farebrother acknowledges, "we collectors feel an interest in every new man till he has seen all we have to show him" (118). Whilst Lydgate continues to display his disregard for provincial ecology, suggesting that "all country towns are pretty much alike," Farebrother repeatedly demonstrates his awareness of both societal and entomological intricacy. Urging Lydgate to "look at these delicate orthoptera!" Farebrother suggests Lydgate take up entomology to endear his neighbors to him—"that would be good discipline, you know, for a young doctor who has to please his patients in Middlemarch" (127). Ignoring Farebrother's suggestion, Lydgate insists his patients will have to "put up with him," as he looks "rather absently at the insects ranged in fine gradation." Lydgate's lack of interest in giving "scrutiny to each draw" of Farebrother's collection foreshadows his—ultimately ruinous—inability to comprehend the idiosyncrasies of provincial life (127).

As we have discussed, the study of insects was dynamic, the parameters of the field shifting with each fresh wave of entomological correspondence. Farebrother, like his real-life counterparts, benefits from the hive-mind of such a wide-reaching community:

> "I have not yet told you that I have the advantage of you, Mr Lydgate, and know you better than you know me. You remember Trawley who shared your apartment at Paris for some time? I was a correspondent of his, and he told me a good deal about you. I was not quite sure when you first came that you were the same man. I was very glad when I found that you were. Only I don't forget that you have not had the like prologue about me."
>
> (126)

Farebrother's entomological friendships have afforded him with the capacity to harness the observations of like-minded men who are further afield. Lydgate is spoken of as though he is a documented genus of beetle from a far-off region, first classified by a Parisian entomologist. Lydgate is, of course, quite unaware of the web of correspondence that has been documenting him—"having quite lost sight" of Trawley.

Farebrother's "den," jam-packed with specimens such as "pickled vermin, and drawers full of blue-bottles," is also home to "a pier glass" (125). In Eliot's parable of the pier glass (which opens chapter 27) the narrator calls to mind a pier glass "minutely and multitudinously scratched in all directions" (194). Her metaphor illustrates how one cannot truly perceive the web of society if they view it from an egoistical center, for when a lighted candle is placed upon the glass "the scratches will seem to arrange themselves in a fine series of concentric circles round that little sun," when—in actuality—"the scratches are going everywhere impartially and it is only your candle which produces the flattering illusion of a concentric arrangement" (194). By incorporating a pier glass into Farebrother's entomological den Eliot infers that the study of insects—in the context of her novel—denotes the displacement of mankind's egoistical center. This inference arises once more when, later in the novel, Eliot highlights Dorethea's self-effacing capacity for observation through her "frequent interchange of visits" with Mr Farebrother. On such visits Dorethea—motivated by her belief in "Mr Farebrother, like another White of Selborne, having continually something new" to contemplate—dresses "hastily" to arrive "at the parsonage rather earlier than was necessary" (582) so that after tea they might talk "more than usual [...] dilating [...] on the possible histories of creatures that converse compendiously with their antennae" (582).

HOW DOTH THE LITTLE BUSY BEE ...

Entomology was often framed as a precursory practice to other forms of natural history. Philip Henry Gosse, who would go on to popularize the aquarium and write an anticipatory rebuttal of Darwin's work, traces his love of natural history back to the day he won a copy of Kaumacher's edition of *Adam's Essays on the Microscope* (1798) which contained detailed instructions on the collection of insects. "I was delighted with my prize," he recalled, "it focused the wandering rays of science that were kindling in my mind and I resolved forthwith to collect Insects" (Croft 1993: 24). Likewise, Charles Lyell, credited with unearthing geologically deep time through his theory of uniformitarianism, recalls his adolescent self, witnessing a store of specimens swell in size: framing his feeling of gratification, as what "the phrenologist would term accumulative propensity." Love of such a "desultory occupation" led to intimations "that the pursuits of other boys [was] more manly." Ultimately, the disrepute by which "his hobby was held had a considerable effect on [his] character," his sensitivity to "the good opinion of others" eliciting a belief that entomology must be pursued "in stealth" (Murray 1881).

A sense of Lyell's adolescent anxiety can be traced through countless texts and articles trained on insect life. Leonard Jenyns in 1825 lamented the "contempt and distain which some of our moderns behold the study of insects" (Jenyns, Natural History Diary extract, qtd. in Armstrong 2000: 107). Kirby and Spence's *Introduction to Entomology*, the first port of call for the budding entomologist, depicts entomology as a "distressed damsel," abandoned and in need of rescue (1). Despite their impassioned entreaty, *The Edinburgh Review* describes them as pedlars of an idle pursuit with no "intellectual eminence," and refers to entomology as the embarrassing art of "impaling butterflies"—a practice wholly at odds with "a busy country" such as Great Britain (1822). In 1882 *Punch* magazine produced a scathing satire of John Lubbock, a member of parliament who produced popular work on the study of ants, bees, and wasps. Underneath a caricature of Lubbock, comically depicted as a bee hovering above a wealth of delicate flowers, they attached a parody of Isaac Watts' didactic verse "How doth the little busy bee" (Figure 7.1):

How doth the banking busy bee
Improve his shining hours
By studying on bank holidays
Strange insects and wild flowers!

FIGURE 7.1 Sir John Lubbock, by Edward Linley Sambourne, August 19, 1882, *Punch*'s Fancy Portraits, no. 97. Wikimedia Commons.

In 1871 Lubbock successfully passed the Bank Holidays Act, which created four Bank Holidays: Easter Monday, Whit Monday, the first Monday in August, and Boxing Day. Referencing his recent achievement, *Punch* confines the entomologist's study to "bank holidays," and in doing so explicitly houses the entomologist's work within the sphere of leisure. The poem contrasts his languid, pleasant hours spent observing "strange insects and wildflowers!" with the verse of the original poem that valorizes the busy bee's industrious aptitude for labor. The concluding exclamation mark further highlights the apparent disparity between the two exercises. Issac Watt's original poem directly spurns "idleness and mischief," the last verse framing the "healthful play" of a child's "first years" as the trifling by-products of "Satan's" influence. *Punch*'s choice of poetic contrast therefore indirectly suggests that entomology is nothing but the childish "work" of "idle hands."

ENTOMOLOGICAL ELOQUENCE

As Kirby and Spence explain in the preface to the first edition of *An Introduction to Entomology*, "in the minds of most men, the learned as well as the vulgar, the idea of the trifling nature of [entomology] is so strongly associated with that of the diminutive size of its objects, that an Entomologist is synonymous with everything futile and childish" (1815: ix). Literary scholars have built upon this association—Susan Stewart in *On Longing* (1984), a fascinating analysis of the miniature as metaphor, explains how toys, like the dollhouse, allied the miniature world with childhood and femininity. Thus, failure to outgrow a "childish" delight in insects was often frowned upon. One such literary child who is paradoxically outgrown by several insects on her Lilliputian odyssey is, of course, Alice.

Alice's first entomological encounter in Wonderland is with the Caterpillar (Figure 7.2). Tottering on tiptoe to see over the edge of a mushroom, she locks eyes with a "large caterpillar," "with its arms folded" (46). With scientific scrutiny, "The Caterpillar and Alice look at each other for some time in silence." Through his first words, "Who are *you*?," The Caterpillar attempts to taxonomically classify Alice. Her response, "I—I hardly know, sir, just at present—at least I know who I *was* when I got up this morning, but I think I must have been changed several times since then" is met with irritation (47). Perplexed by Alice's belief that her change in morphology could alter her nomenclature, the Caterpillar pushes further. Alice's observation that "being so many different sizes in a day is very confusing" is met with terse disagreement. The Caterpillar becomes affronted by Alice's contemplation that changing in size so often would feel "very queer," as, thin-skinned by nature, he has a rather personal perspective on metamorphosis and the constancy of change. *The*

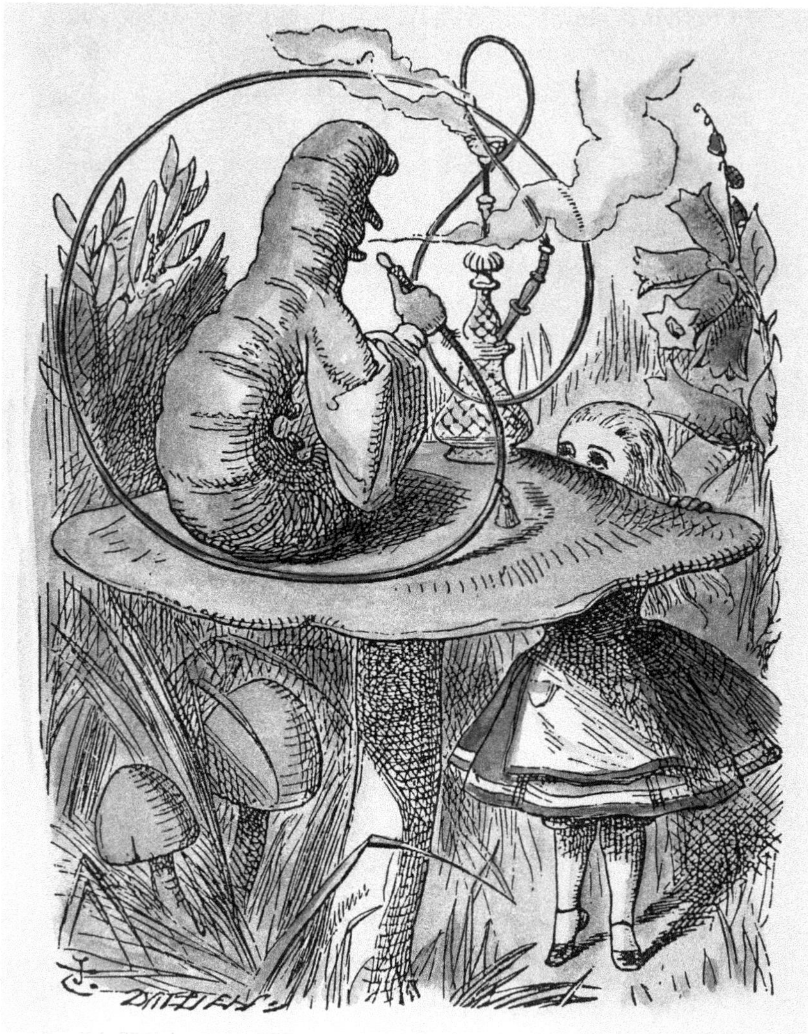

FIGURE 7.2 Alice Meets the Blue Caterpillar, by John Tenniel. The Granger Collection/Alamy Stock Photo.

Nursery Alice published in 1890 with amendments for children and enlarged illustrations, brings an increasing realism to the caterpillar's morphology by zooming in on the ingenuity of John Tenniel's illustration.

> "Do you see its long nose and chin? At least, they look exactly like a nose and chin, don't they? But they really are two of its legs. You know a Caterpillar has got quantities of legs: you can see some more of them, further down.

> What a bother it must be to a Caterpillar, counting over such a lot of legs, every night, to make sure it hasn't lost any of them!"
>
> (17)

Lewis Carroll's decision to blur the boundary between an anthropomorphized insect and an entomorphized human is indicative of an increasing flurry of work that sought to inhabit the mind of the insect and delve into its interior life. Darwin's *The Expression of Emotions in Man and Animals* considered the spectrum of entomological emotion including "anger, terror, jealousy and love" (1872: 350). As A.S. Packard, an American entomologist, expressed in 1873 "from what little we know of their psychical endowments, we see enough to convince us that ... [they possess] like passions with ourselves, so much so that we find our-selves unconsciously judging their actions by our own feelings" (13). Carroll took this idea a step further in his original draft for *Alice Through the Looking Glass* in a suppressed chapter dubbed "The Wasp in the Wig."[9] There is critical debate as to the veracity of this chapter, with some scholars believing that the extract is unrefined, or lacking in Carrollian humour. It is most likely that this episode was removed at the behest of illustrator John Tenniel, who had grumbled to Carroll in their correspondence that a wasp in a wig was "beyond the appliances of art" (Cohen 2003: 15). Like the Caterpillar whose legs are arranged "exactly like a nose and chin" the Wasp resembles an entomorphized creature that is first described as "something very much like an old man," his face tied up by "a yellow handkerchief" (295). In a letter to the illustrator of *Sylvie and Bruno*, Harry Furniss, Carroll contemplated the physiognomy of insect-kind:

> Some writers say that the full-face of a spider, as seen under a magnifying-glass, is very striking. Could you find one in some book of entomology, or look at a live one?
>
> (Cohen 2003: 117)

His request that Furniss consult "some book of entomology" or observe a magnified spider in order to become lost in the micro-details of its "face," suggests he may have carried out a comparable venture for his Wasp in the Wig. Countless natural history books of the period correspond with Carroll's observation of a wasp as a wigged creature with a yellow handkerchief tied around its face. Edward Latham Ormerod, for example, in *British Social Wasps* (1868) marvels at "the little yellow spot between the roots" of the antennae; a description that remarkably evokes an antennae bracketed toupee. Ormerod goes on to describe the clypeus of the wasp as a "yellow central patch, which constitutes so large a portion of a wasp's face" (35). Such a striking patch could be imagined as a taut handkerchief.

Alice pointedly approaches the wasp "rather unwillingly, for she was very anxious to be a Queen" (293). This passage, in a sense, entomologically establishes Alice's prospective royalty. Carroll's chessboard world, in which powerful queens govern pawns mirrors the colonial culture of bees, ants, and wasps. The male wasps, or drones, like the kings of the chessboard are impotent and inert, powerless without their Queen. The Wasp in the Wig echoes a number of fantastical natural history works that framed the peculiarity of Hymenoptera queen-making as a wondrous "monarchy of marvels and strange customs" (Budgen 1879: 233). Our titular Wasp in the Wig is emphatically male. Moreover, perhaps to reinforce his status as a "drone," Carroll intimates that this wasp is also lacking in social mobility. His working-class identity is evident from his language, his diction and use of slang suggesting humble beginnings: "Worrity, worrity! Can't you leave a body alone?" (294). As Morton Cohen and Martin Gardner point out, the Wasp's preference for brown sugar is also characteristic of the Victorian lower classes, who would have found brown sugar cheaper than white.[10]

Throughout the Wasp in the Wig, Alice is framed as both a prospective Queen Bee and a deficient insect: in need of compound eyes and set of strong mandibles. Adhering to what entomologists dubbed "the principle of similarity" the wasp systematically scans Alice's body, zeroing in on prospective parallels, drawn to her as a white butterfly might "suddenly change its course and fly down to a piece of white paper" (Packard 1881a: 372). Entomologists believed that such behavior encouraged insects to recognize others of their own species, and thus, ensured their survival.

> "Your jaws are too short," the Wasp went on ... [and] your eyes — they're too much in front, no doubt. One would have done as well as two, if you must have them so close."
>
> (297)

Despite his waspish personality Carroll's wasp—as a drone—has no sting: becoming more amiable as the chapter progresses. Alice first encounters the wasp immobilized: "shivering as if he were very cold" and grumbling about his "old bones" (293). (Wasps becoming increasingly motionless as the temperature drops.) In the spirit of Carrolian wordplay, by the end of the episode, Carroll's Wasp in a Wig no longer has a bee in his bonnet. Alice reads him the paper, bringing him news of an intrepid insect "Exploration Party" lost to a lake of treacle, and gently persuades him to move into the sun.

ALICE THROUGH THE MICROSCOPE

Chapter 3 of *Alice Through the Looking Glass* begins as though under the convex lens of a magnifying glass:

> "What *are* those creatures, making honey down there? They can't be bees—nobody ever saw bees a mile off, you know—" and for some time she stood silent, watching one of them that was bustling about among the flowers, poking its proboscis into them, "just as if it was a regular bee," thought Alice. However, this was anything but a regular bee: in fact, it was an elephant—as Alice soon found out, though the idea quite took her breath away at first.
>
> (169)

Lewis Carroll's phantasmagoric vision of elephantine bees pollinating enormous flowers that resemble "cottages with the roofs taken off, and stalks put to them" is a part of an enduring morbid fascination with the concept of gargantuan insects. Today, the giant-bug genre, following in the footsteps of H.G. Wells, is almost exclusive to horror or science fiction; in our everyday domestic encounters we see insects as virulent invaders of our home, and thus regard the idea of colossal creepy-crawlies as repulsive or horrifying rather than exciting or fascinating. Kirby and Spence in their *Introduction to Entomology* unpick the emotional predication behind such an attitude:

> [Insects] are usually pointed out to us by those about us as ugly, filthy, and noxious creatures; and the whole insect world, butterflies and perhaps some few others excepted, are devoted by one universal ban to proscription and execration, as fit only to be trodden under our feet and crushed: so that often, before we can persuade ourselves to study them, we have to remove from our minds prejudices deeply rooted and of long standing.
>
> (1856: 2)

Alice's first thoughts are not clouded with fear or disgust but prompted by buoyant curiosity, as she wonders "what quantities of honey [such a gargantuan bee] must make." Like the entomologists of her era Alice experiences time dilating, as she stands silently observing the bustling bees. During this moment of temporal thickening Alice maintains the clarity of mind to correctly identify the bee's elongated appendage as a "proboscis," its tubular mouthpart so large it resembles an elephant's trunk. One might imagine Charles Darwin to have dreamt of a similarly fearless encounter during his student days as an enthusiastic beetle hunter. His friend from university, Albert Way went so far as to produce a caricature entitled "Darwin and his Hobby" of young "Charley" astride a giant beetle, butterfly net in hand (Figure 7.3).

FIGURE 7.3 "Darwin and his Hobby," by Albert Way. Cambridge University Library. CUL DAR 204, f. 29. Reproduced by kind permission of the Syndics of Cambridge University Library.

Such entomological fervor continues to pervade this chapter, aptly entitled "Looking-Glass Insects." Aboard the train, scale recedes as Alice sits alongside a goat, a beetle, and a gnat "the size of a chicken" (175). The train guard looks at Alice "first through a telescope, then through a microscope, and then through an opera glass" (280). The guard's selection of various mechanical eyes, oscillating in aperture, draws attention to the incongruity of Alice's spatial similitude with insect-kind and illustrates the dizzying velocity with which optical instruments continued to evolve. By 1898 Fred Saxby would set out instructions on how to take a photograph through a fly's eye, attempting to capture an image of Queen Victoria with "the most infinitesimal lens known to science" (Saxby 1898). With their broad segmented eye structure made up of thousands of minuscule lenses, insects are able to see with panoramic clarity and an infinite depth of field—the swiftest of movements appearing as though in slow motion. The compound microscope, much like an insect's compound eye, utilized numerous lenses in order perceive the world in high resolution. Through the train-guard's series of optical instruments therefore, Carroll—like many of his contemporaries—coupled entomological vision with the mechanical eyes of modernity.

The microscope, Isobel Armstrong tells us, provoked a "thesaurus of transparency unique to lens culture"—a vocabulary ranging from the "semi-pellucid" to the "crystal" (2008: 329). This new lexicon went on to take root beneath the microscope's slide, as countless creatures were rendered comparably crystalline. Alongside sea-side specimens, the limpid clarity of insects frequently took center stage: granting men, women, and children the chance to see beyond entomology's exoskeleton and bear witness to the inner-workings of life itself. Mary Ward, author of *A World of Wonders Revealed by the Microscope* (1859), takes great pleasure unveiling the translucent frames of insect-kind:

> Their wings are beautifully transparent, consisting of a delicate membrane, stretched, as it were, to a sort of ornamental network covered on both sides with beautiful scales, laid in rows like the feathers on a bird ... when these scales are viewed as transparent objects, they no longer appear green, but the pattern on them, when viewed with a high magnifying power, assumes a strange and almost startling appearance.
>
> (10)

Likewise, in *Evenings at the Microscope* Philip Henry Gosse marvels at a slug's "glistening glass-like" tongue (49). According to Gosse, the "wonderworking instrument" of the microscope enchanted the "coverings" of various insects: giving their tissue a ghost-like transparency, which exposes the clockwork activity of their "viscera" (29). By 1859 Carroll had purchased the latest microscope of the day, manufactured by Smith & Beck in London. It was, according to curator Carolyn Vega, "something that he had for his whole life and took incredible care of" (Vega 2015). In a letter to his sister Elizabeth, Carroll recollects the wondrous experience of placing an insect beneath the microscope.

> This is a most interesting sight, as the creatures are most conveniently transparent, and you see all kinds of organs jumping about like a complicated piece of machinery ... everything goes on at railway speed, so I suppose they must be some of those insects that only live a day or two, and try to make the most of it.
>
> (July 5, 1851, qtd. in Cohen 1979: 19)

Scholars have recognized that this epistolary episode, is, in essence, chapter 3: Looking Glass Insects "in embryo" (Smith 2015: ch. 2). In his letter Carroll intimates that the unveiling of the sub-visible world, in turn, granted humanity with an entomological-eye's view of existence: allowing us, for the first time, to comprehend the spatio-temporal subjectivity of an insect specimen. In both this letter and chapter 3 of *Through the Looking-Glass*, insect-time is calibrated by the speed of a steam engine: the infinitesimal span of an insect's life registering, to us, as a flickering glimpse behind glass that vanishes in the blink of an eye. However, as Carroll acknowledges, the smaller the specimen, and the faster its metabolic rate, the slower it experiences time. Insects, equipped with lightning-quick brains, are capable of retrieving and processing great swathes of data from their panoramic eyes. Fine-tuned by the velocity of their metabolism, they move through life as though suspended in time, veering away from a butterfly net or fly swatter with relative ease. As A.S. Packard puts it "those who observe the ways of insects have noticed their extreme sensitiveness to external impressions, that their motions are ordinarily rapid and nervous" (1881a: 354).

Aboard the looking-glass train, the conversation in Alice's carriage continues to play with the elasticity of time and scale. As Susan Stewart deduces, "there are no miniatures in nature; the miniature is a cultural production, the product of an eye performing certain operations, manipulating and attending in certain ways to, the physical world" (1984: 55). Scale and time, then, are both determined according to a correspondence with that which is familiar. Our perception of size, likewise, is troubled by distance—that which appears small is far away, while that which appears outsized must be disturbingly proximate. This is perhaps why the looking-glass train guard, baffled by his carriage of unlikely passengers, looks at Alice through both telescope and microscope. Natural history books strove to articulate our spatial and temporal remoteness from insects. As the narrator of *Episodes of Insect Life* explains, "insects, their size considered, outstrip the speed of all other animals in traversing the ground ... Until [we are] able to jump three times as high as St. Paul's, or to vault a quarter of a mile, [we] must yield the palm of leaping to the flea and the locust" (Budgen 1879: 182). To Alice the gnat in the carriage is "the size of a chicken" as she holds her own body as the designated point of comparison. However, at this stage in the novel, Alice herself is the size of a pawn. The gnat therefore, when measured against the height of a chess piece, becomes quite the ordinary sized insect.

The gnat and beetle, holding their gargantuan tickets that "quite seemed to fill the carriage," intimate that Alice should not be regarded as a passenger, but as precious cargo that belongs in the "luggage." Like a curious specimen, stowed in a mason jar the gnat puns Alice "S'must be labelled '[G]lass with care,'" whilst another voice suggests "she must go by post, as she's got a head on her." Notwithstanding the use of "head" as Victorian slang for postage stamp, this assertion suggests that the gnat and beetle have met fellow insects capable of living, for a time, without their heads. Various naturalists of the age made use of entomological life to contemplate hierarchies of pain and paid special attention to cases of insect decapitation. Insects impaled on a pin and set in cork could wriggle for days before they finally died. One naturalist was shocked to see that a dragonfly he had just pinned through the thorax still clutched a struggling fly in its forelegs and proceeded to eat it (Conniff 2010: 354). As George Henry Lewes affirmed "an insect pinned to the table will continue to eat and a headless fly or worm will writhe and twist if touched" (1860: 333).

As the chapter continues, Carroll makes special use of the gnat for spatio-temporal speculation. The gnat is dwarfed by the presence of others at the level of the sentence, his dialogue appearing on the page in diminutive typeface to contrast with the default text.

> "I know you are a friend," the little voice went on: "a dear friend, and an old friend. And you wo'n't hurt me, though I am an insect."
>
> (172)

Throughout his short stay in *Through the Looking Glass* the gnat sighs "deeply ... two large tears [at one point] rolling down its cheeks." His sigh is so "wonderfully small" that Alice "wouldn't have heard it at all, if it hadn't come *quite* close to her ear" (176). Acutely aware of the ephemerality of his existence, the gnat appears hesitant to play a larger role in the conversation, repeatedly suggesting Alice "might make a joke" on this or that, and sighing mournfully when he has to do so himself. The gnat seems to believe that if only Alice were to make the jokes he might live longer—as though his life span is somehow relative to the size or duration of his role in the novel. One might imagine that, to the gnat, Alice's speech hurtles along at a breakneck pace—while he is forced to speak in slow motion in order to be intelligible to both Alice and to us. Sadly, the gnat appears to be correct in his deduction—his last "melancholy little sigh" predicating his departure:

> This time the poor Gnat really seemed to have sighed itself away, for, when Alice looked up, there was nothing whatever to be seen on the twig.
>
> (176)

JABBERWOCKY TAXONOMY

In the second half of this chapter Alice becomes a budding entomologist and encounters an assortment of curious specimens. Just as the Wasp and Caterpillar sought to classify Alice by determining her characteristics, taxon, and nomenclature, Alice learns to taxonomize various species of looking-glass insect. The rocking-horse fly swings itself from branch to branch and lives on sap and sawdust (Figure 7.4). The ephemeral Bread-and-Butterfly has wings of bread-and-butter, a breaded body and a sugared skull, whilst the snapdragon-fly—who makes his nest in a Christmas box—has a body of plum pudding, wings of holly-leaves, and a brandy-burning raisin for a head. As each specimen is cataloged Alice makes a series of acute observations. She notes the rocking-horse fly has been freshly re-painted, determines that the bread-and-butterfly will go the way of the dodo (due to its fatally restrictive diet of "weak tea with

FIGURE 7.4 The Rocking-Horse Fly, by John Tenniel, *Through the Looking Glass*, 1871. Wikimedia Commons.

cream in it"), and theorizes that the insects of our world, fond of flying into candles, may metamorphose into snapdragon-flies.

This chapter, teeming with taxonomic wordplay, parodies the nineteenth-century frenzy for collecting and classifying specimens. At a time when prominent entomologists declared that insect-collectors "who do not wish to be utterly isolated, must learn to call [insects] by names by which other people will know them," many were bewildered or resistant to such an alien shift in language (Stainton 1856–1859: iv). Carroll draws attention to the peculiarity of the scientific community, framing entomologists and their kind as men who thrived on attaching jabberwocky neologisms to familiar forms; or who had a proclivity for pigeonholing nature, despite insisting that one species could slide into another. Why then, according to Carroll, could butterflies and bread not share a common ancestor?

Carroll goes on to satirize the single-minded intensity with which many entomologists captured and classified their specimens. Charles Darwin, for example, describes reading "the impressive words, captured by C. Darwin" in a publication as "glory enough for any man" (Evans 2017: 17). The strong warning to never leave a specimen unlabeled accompanies countless natural history books; as one solemn guide advises "not even for an hour," less you "forget it, or die" (Coues 1890: 51). As the chapter advances, Alice, in want of a label, becomes increasingly anxious she may lose her own name. Having earlier answered the question "What's the use of [insects] having names, if they won't answer to them?" with the sharp observation "it's useful to the people that name them, I suppose," Alice enters the wood "where things have no name": under the impression that if she is to acquire a new name it will almost certainly "be an ugly [presumably Latinate] one" (176).

MICROCOSMS IN THE PARLOUR

As an increasing number of men, women, and children fell under the spell of entomology the presence of the insect in the home became a cause for celebration rather than concern. Opening with a cricket and teakettle, rivals for a place on the hearth trilling an unwitting duet, Charles Dicken's Christmas story *The Cricket on the Hearth: A Fairy Tale of Home* (1845) exemplifies this cultural shift. Often dismissed as sentimental anthropomorphism, this tale of a family watched over by a chirruping "little household god" outsold *A Christmas Carol* and was credited with "vindicating the fame of Mr Dickens" after the "failure" of 1844s Christmas tale *The Chimes*. The hearth cricket's song, as *Sharpe's London Magazine* tells us in 1846, "is connected with the sacred name of Home":

> It came from a mysterious being, always hiding from their sight, and eluding their strictest search. It seemed now here, now there, beneath their feet, close at the ear, or ringing from the very centre of the glowing hearth. It was at times a strange, almost supernatural sound; for, in the stillness of night, when the lone watcher sat waiting for the return of absent friends, or when the alarmed family were listening to the awful beatings of the tempest, then did it rise with unwonted energy, thrilling on the ear, and paining the sense by its monotonous acuteness. Yet with all its shrillness it was a sound of comfort; it relieved the feeling of loneliness, and breathed, like the tickings of the old clock, the notes of sympathy and companionship.
>
> (*Sharpe's London Magazine* 1846)

The resonant sound of cricket stridulation is figured as "a strange, almost supernatural sound" that is at once elusive and omniscient—the cricket's song carries with it an emotional resonance that seems to thrum in harmony with the household—rising with a shrill urgency in tempestuous times, and winding down in metronomic sympathy "with a lone watcher waiting for the return of absent friends."

By the 1860s the middle-class home frequently incorporated a shelf, or cabinet, of curated curiosities. These miniature museums, ornamented with various fossils, flowers, or feathers, often included a rare scarab beetle or butterfly as the crowning glory of their collection. An interesting example of domesticated insect-mania is described in an article entitled "A Young Lady Artist's Mania for Capturing Musical Insects" published in *The New York Times* in 1880. Guests were solicited with an invitation to "attend a cricket concert" and reminded not to ring the bell lest they disturb the crickets in "full song." The crickets performed in their hundreds behind a fern-filled wirework house in the young lady's at-home studio. This was a time when many a naturalist sought to merge the parlor and the laboratory. Entomologist John Lubbock, for example, assembled over thirty glass ant nests in his home at High Elms. Lubbock's "microcosms in the parlour" (Clark 2009: 97) were doted on equally by high-society guests and his family. Alice Lubbock, in 1886 noted their eighteen-month-old daughter, Ursula, "is very fond of Dada's ants, she recognized them out of doors one day." Likewise, a love of worms fortified Charles Darwin's household. Seeking to verify the kinship between lower organisms and mankind, Darwin coaxed thousands of worms to make a home of his billiard table. In order to observe their senses, emotions, and appetite, he encouraged his family to entertain them: his wife, Emma, playing the piano, his son, Frank, the bassoon and his maid, Bessy, serenading them with the demanding refrain "Worms! Worms!" (Johnson 2012: 116).

A renewed interest in the habits of the insect world sparked a flurry of homely stories that synthesized science and didactism. *Episodes of Insect Life* (1879) by one "Acheta Domestica," for example, associates the study of entomology with intimate enquiry in one's own home. *Episodes of Insect Life* goes on to explicitly house the discipline of entomology within the domestic sphere: attempting to associate insects "as much as possible with our domestic habits, the summer's stroll, the winter's walk, whilst exhibiting them in their numerous relations" (ix). L.M. Budgen's choice of pseudonym, *Acheta Domestica,* a feminized version of the binomial or scientific name for the common house cricket *(Acheta domesticus)* likewise binds entomology to the domestic realm. Budgen's choice also situates the narrator, in a uniquely empathetic position alongside her subjects of study.

In a similar vein, Beatrix Potter, at the close of the long nineteenth century, gave a myriad of children a window or magnifying glass into the interior lives of insect-kind, most notably in *The Tale of Mrs Tittlemouse* in which a wood-mouse's home is beset by insect intruders—a lost beetle and ladybird leave a trail of sticky prints, a spider (in search of Miss Muffet) trails cobwebs, and a family of bumblebees take up residence in the storeroom. However, after Mr Jackson, the toad—an insectivore, and the messiest guest of all—is expelled from Mrs Tittlemouse's home domesticity is restored:

> Next morning, she got up very early and began a spring-cleaning which lasted a fortnight. She swept, and scrubbed, and dusted; and she rubbed up the furniture with beeswax, and polished her little tin spoons.
>
> (2012: 84)

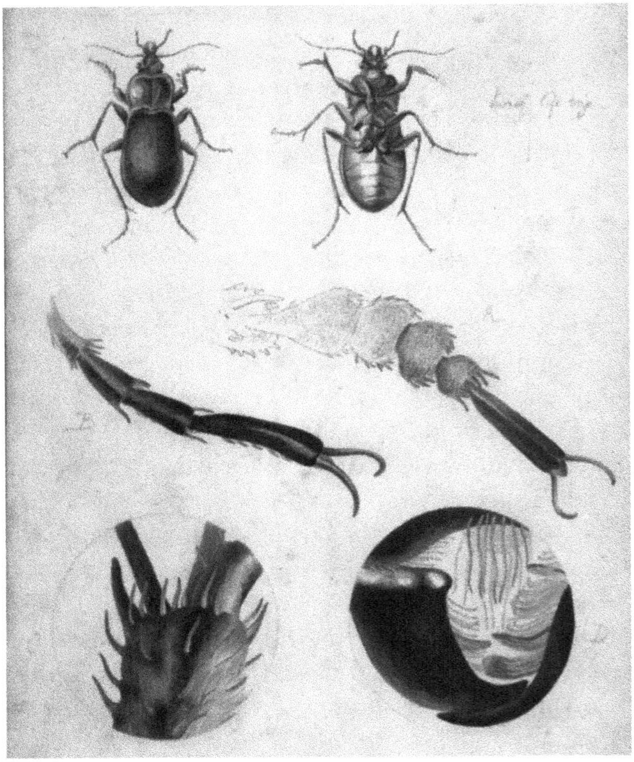

FIGURE 7.5 Magnified studies of a ground beetle (*Carabus nemoralis*), from *Studies of a Ground Beetle* by Beatrix Potter, Penguin Random House, 1887. Reproduced with permission of Frederick Warne & Co. Ltd.

Domesticity like entomology was, and is, often associated with childhood, minutiae, and triviality, or thought of as beneath the interest of the archetypal man. It is unsurprising, then, that the two became inextricably bound in both literary and scientific culture. As Maria Damkjaer has noted, domestic time was thought of "primarily in negative terms: not public, not historical, not productive, not measurable, [and thus] not narratable" (2016: 3). However, by the 1850s domesticity took on a "temporal turn": the intersection of time and place in the home eliciting "the realist drive to describe and define" (Damkjaer 2016: 3). The domesticated scale of these narratives sought to document the minutiae of everyday life. The home, like a spider's web, was often thought of as static or unchanging rather than as a work in progress—in need of ceaseless care and renewal. Writers such as George Eliot took great care to make the micro-mechanisms of daily life as textured and tangible as possible. Meanwhile, entomology was also associated with an illusion of inaction. Entomologists were thought of as motionless—capable of spending hour upon hour at home, their eye glued to their microscope. Equally, geological revelations charged the insect with a sense of a-temporality—as Samuel H. Scudder commented in 1890: "That creatures so minute and fragile as insects, [were] found fossil[ized], engraved, as it were, upon the rocks ... never cease[d] to be a surprise" (33).

In much of George Eliot's oeuvre the domesticated scale and the entomological scale become curiously concurrent. This pattern of association is outlined in *Silas Marner* (1861). Silas is freshly integrated into the hubbub of provincial life, after decades of isolation, now that he is accompanied by Eppie, "the little child [who] link[s] him once more with the whole world" (1994: 114). Everyday existence is experienced at a scale and speed that treasures domestic detail. Silas is transported right down to "the red lady-birds and the round pebbles" as he documents the children gazing at Eppie "like little dogs face to face with one of their own kind" or the servants carrying her out "to see if any cherries could be shaken down in the orchard" (114). Similarly, in *The Mill on the Floss* (1860) "the small pulse of the old English town," in all its "oppressive narrowness" resembles an ant-nest populated by the "emmet-like" (or ant-like) Dodsons and Tulivers (284). Eliot's protagonist, Maggie Tuliver, occupies "a little world apart" in the "great spaces of the Mill" admiring the "faery lace-work" of the spider webs: the spiders being a special "subject of speculation with her" (32). Like Lydgate's wife Rosamund, in *Middlemarch*, who dreams of their "little woven future," Maggie has a kinship with the spider and their place in the home.

In *Daniel Deronda* Eliot includes a poetic epigraph in chapter 51 that also allies domesticity with "insect-labor":

> She held the spindle as she sat,
> Errina with the thick-coiled mat
> Of raven hair and deepest agate eyes,
> Gazing with a sad surprise
> At surging visions of her destiny—
> To spin the byssus drearily
> In insect-labor, while the throng
> Of gods and men wrought deeds that poets wrought in song.
>
> (2014: 525)

In this chapter Deronda meets his mother who abandoned him to become a singer. The epigraph, therefore, articulates the perceived incompatibility between domestic motherhood and artistic ambition. Like a spider, Errina has "raven hair" and "agate eyes," which swirl with volcanic colors like a polished jewel. In her entomorphic vision she spins secreted filament or "byssus." As Wendy Williams notes, byssus is a type of silk used in ancient burial customs—her labor therefore, is tainted with the suggestion she is weaving her own shroud (2016: 32). Comparably, Miss Havisham, a figure of decayed domesticity in Charles Dicken's *Great Expectations*, blurs the boundaries between spider and spinster—a veil of cobwebs and spiders blanketing her wedding banquet.

Another recurring symbol of domestic imprisonment is, of course, the pinned insect under glass. For example, in an article from *Punch* magazine in 1870, the press toyed with the idea of whether a woman could be truly regarded as a "specimen of human reason" in the case of female suffrage (*Punch* 1870). In this caricature the Victorian female (hair pinned, thorax pinched, and waist bound in whalebone) is conceived as a mounted specimen herself. Similarly, in a *Punch* cartoon for 1881 the cage bustle of a woman's restrictive crinoline undergarments is likened to the exoskeleton of a "parasite" (Sambourne 1881). Symbolically associated since Ancient Greece with the trapped soul or psyche, the butterfly, was, unsurprisingly, often the insect of choice to express the trappings of domesticity. In Eliot's *Adam Bede* (1859) Hetty Sorrell is frequently described as having "a little butterfly soul" (115) or a disposition "as unsympathetic as butterflies sipping nectar" (89). The narrator, mindful of Hetty's future as a fallen woman, extends this metaphor:

> It is too painful to think that she is a woman, with a woman's destiny before her—a woman spinning in young ignorance a light web of folly and vain hopes which may one day close round her and press upon her, a rancorous poisoned garment, changing all at once her fluttering, trivial butterfly sensations into a life of deep human anguish.
>
> (215)

Likewise, in Gaskell's *Mary Barton* (1848) Mary's aunt Esther, a fallen woman, is known by the name of "Butterfly" (390). "Restless ... [at] her approaching death," and longing for "the open air," she is ultimately found dead in "a heap of white or light-coloured clothes" like a "poor crushed butterfly" (392).

Several writers conceive of a woman's marriage as an enclosed vivarium—she exists, under the whims of another as an adornment, her marital life a cruel parody of freedom behind glass. In Charlotte Brontë's *Shirley* (1849) marriage is envisioned as "inverting the natural order of insect existence"; Mrs Hellstone, the narrator comments, "would have fluttered through the honeymoon a bright, admired butterfly, and crawled the rest of her days a sordid, trampled worm" (89). In *The Tenant of Wildfell Hall* (1848) by Anne Brontë, Helen is "safely installed as the mistress of Grassdale Manor" by her husband who wants to have her "all to himself." In her naiveté, she imagines he thinks of her "as if [she were] some frail butterfly" and attributes his desire to confine her to his fear "of rubbing the silver off my wings by bringing me into contact with society" (159). When she does venture into society, she is forced to suppress her own "rooted principles in favour of a plain, dark, sober style of dress" and is adorned "for his gratification" in "costly jewels" like "a painted butterfly" (170). Comparably, in Oscar Wilde's *Dorian Gray* the Duchess of Monmouth confesses to Dorian that Lord Henry believes her husband married her on "purely scientific principles as the best specimen he could find of a modern butterfly." They joke that although the Duke does not stick pins in his wife, the Duchess' lady's maid, responsible for her mistress' hair, dress and jewellery, does so frequently (2012: 185).

A short story from *The Mask* (1866) entitled "The Painted Lady" burlesques the blurring of boundaries between men who chase beautiful butterflies and those who pursue beautiful women. In such a comparison, the female specimen is frequently framed as artificial or untrustworthy when viewed alongside her insect counterpart: "you know the butterflies are beautiful at a glance, because you know they are as nature painted them; but can you say the same of women?" (Lewis and Thompson 1868: 242–50). Towards the turn of the century those who sought to become social butterflies of high society adorned their bodies with mounted insects. As *Lipcott's Magazine* observed in 1876:

> the insect world has been ransacked of late ... bees, wasps, blue-bottles, dragonflies, butterflies, grasshoppers swarm around us ... bugs and beetles crawl around us, or, worse still, sit in the heart of coral and diamond flowers.
>
> (Lady Blanche Murphy, "The Age of Knick Knacks," in Lippincott 1876)

Most famously, the actress Ellen Terry, immortalized by artist John Singer Sargent, wore a gown laced with over a thousand iridescent beetle-wings at London's Lyceum Theatre in 1888 in her role as Lady Macbeth. In the throes of what Henry T. Fink called "Wasp-Waist Mania" (Fink [1887] 1902). *Punch* illustrator Edward Linley Sambourne satirized the extent to which the nineteenth-century woman became a walking entomological display. In his "Designs after Nature" series the exaggerated curve of a heavily crinolined bustle becomes the coiled shell of a snail, a corset mimics a beetle's segmented abdomen, and a woman's elaborately cobwebbed hairstyle is finished off with an antennae fascinator (Figure 7.6).

Any attempt to pick apart the fabric of nineteenth-century literature, across a given theme, is, of course, the study of a somewhat synthetic texture, fraught with gaps, biases, and roads not taken. My study—with a focus on how language interwove elaborate tokens of entomological fervor—has, thus, consciously chosen to follow threads such as scale, time, and domesticity. In an effort to chart how the human world responded to a glimpse into the world of insects, I have paid special attention to that which attempts to reconcile the two spheres. Efforts to interrogate the common ground we might share with insect-kind habitually addressed the question of an insect's subjectivity: how an insect

THE BUTTERFLY WAS A 'LADY.'

"At times methinks my soul hath wings,
At times it loves to crawl."
Old Ballad.

FIGURE 7.6 "The Butterfly was a 'Lady'," by Edward Linley Sambourne, *Punch*, June 17, 1871.

experiences time, scale, and emotion and at what point their experience might intersect with our own. Those with an interest in entomology became intimately acquainted with the value of observing that which lies beyond the looking glass of our own experience. Together, avid entomologists and readers of George Eliot or Lewis Carroll were taught to associate their own "glimpses of days that occur once, never to happen again" (Bellanca 2007: 77) with the emotions of innumerable insect ephemerae.

CHAPTER EIGHT

Insects in Art

The Age of Industry

DEIRDRE A. PRISCHMANN-VOLDSETH

People have a complex relationship with arthropods. These small, alien, creatures consume our food, drink our blood, feast on our flesh, and are responsible for immense human suffering and death. Arthropods are our superiors in geological history, abundance, and diversity, and are so interconnected to the planet's function that, ironically, *without* them people would suffer and die. It is no wonder then, that they provoke a wide range of emotional responses—fear, disgust, curiosity, joy. From the fatal kiss of a flea to the rapture of a butterfly's wing, arthropods bring people to their knees. It is a strange kind of power they wield over humanity.

The Age of Industry was a time of artistic expansion. The prevalence of Baroque and Rococo art in the seventeenth and eighteenth centuries gave way to Neoclassicism and Romanticism, and by the twentieth century there were multiple artistic movements, including realism, impressionism, neo-impressionism, and art nouveau, among others (Honour and Fleming 1995; Stokstad 1999). Subject material proliferated from the previously dominant themes of mythology and religion (Honour and Fleming 1995; Stokstad 1999). Representations of insects and the role they played in society also diversified, although a dichotomy remained between positively and negatively perceived arthropods. "We naturally divide insects into friends and foes" (Packard 1881: 67). Historically, insects and related arthropods like spiders, scorpions, flies, and grasshoppers were associated with evil, sin, malice, treachery, pestilence, death, etc. (Ferguson 1954; Impelluso 2004). Aesthetically pleasing or useful arthropods like butterflies and bees had positive connotations, including diligence, sweetness, eloquence, unity, piousness, resurrection, and the soul (Ferguson 1954; Impelluso 2004). The use of insects as symbols to represent human concepts and emotions dominated the visual arts for centuries (Frost 1937; Ferguson 1954) (also see Chapter 6 by Dodd in this volume). However, in the nineteenth century there was a transition in the use of insect imagery from metaphorical to fanciful and realistic. People were fascinated with living creatures and the natural world (see Chapter 5 by Meuti and Fisher). Artists captured the complex interplay between humans and insects, as well as interactions between insects and other species. There was a growing appreciation of insects, from celebrating their remarkable appearance, to documenting their diversity, to advancing scientific knowledge (see Chapter 1 by Johnson and Polaszek). Entomology blossomed in the Age of Industry, or perhaps what could arguably be called, the Age of Insects.

THE SCOURGE OF YELLOW FEVER

Insect-vectored diseases like yellow fever and plague changed the fabric of human history. Yellow fever is caused by an RNA virus (*Flavivirus*) that in urban disease cycles is transmitted by the mosquito *Aedes aegypti* (Barrett and Higgs 2007) (see Chapter 2 by Miller and Miller). Approximately 15 percent of people infected with yellow fever develop a severe form of the disease characterized by jaundice, bleeding, and organ failure, and of these, 20–50 percent will die (CDC 2015). Although the involvement of mosquitoes was suspected in 1848, their role as vectors was not confirmed until 1900 by the Yellow Fever Commission led by Major Walter Reed (Crosby 2006; CDC 2016). Thus, during the Age of Industry people were unaware what caused the disease and how it spread during early epidemics in the United States and Europe (Crosby 2006; Barrett and Higgs 2007; CDC 2016). In the mid-1800s, mosquitoes were portrayed as nuisance pests, as in Alfred Jacob Miller's drawing *Beach at Cape May* (see Figure 8.1), and the comedic illustration *The Mosquito War* (*Harper's New Monthly Magazine* 1854), where a boy's best efforts at battling a swarm of mosquitoes only results in a messy bedroom.

A series of colored plates by Charles-Philibert de Lasteyrie in 1820 illustrate the grim progression of yellow fever, the last image depicting a dying man with fresh blood and black stains on the bedsheets (Pariset and Mazet 1820). *Frank Leslie's Illustrated Newspaper* printed several articles and drawings on yellow fever epidemics in the United States:

> What is yellow fever? ... it is abrupt in its attack, commencing with a chill, not of a marked intensity, followed by pains in the head and loins, the skin

FIGURE 8.1 *Beach at Cape May*, Alfred Jacob Miller, 1840. Ink on paper. The Walters Art Museum, Baltimore, MD (37.2675). Gift of Decatur and Sally Miller, 1994. Wikimedia Commons.

becoming jaundiced or yellow, from whence the name; that then comes the vomiting of blood, called the black vomit, and then—the chances are dead against the sufferer ....

(Leslie 1878a: 43)

One observes distinct parallels between art on yellow fever and that of plague, but marked differences also exist. Victims of yellow fever were usually shown being transported to cemeteries for burial rather than being dumped en masse in pits:

Canal Street [in New Orleans] is the thoroughfare leading to the principal city cemeteries. From the turnings at the various cross streets the dread cavalcades of death are almost constantly filing in. They turn the corners

> from every quarter; they wind their long and sinuous way – the silent march of the dead – so many shadowy spectres, beckoning all in their train.
>
> (Leslie 1878b: 59)

Images of yellow fever sufferers were often set indoors, either in homes or hospitals, and convey a sense of order in the midst of tragedy and death. Some bystanders shrink back while others provide aid (Leslie 1878a, 1878b). Juan Manuel Blanes's oil painting *Episodio de la Fiebre Amarilla en Buenos Aires* (1871) exemplifies these trends. The painting features a baby sitting next to his mother's corpse while his father lays in the shadows, presumably dead or dying. An older child lets two gentlemen and a doctor inside while an onlooker shields his face with a cloth. The scene in Aparicio José's oil painting *Episode de la Fièvre Jaune à Valence* (1804) is more crowded and chaotic and takes place in a makeshift hospital where priests tend the sick and pray for the dead. Conversely, in a watercolor by an unknown artist (*Episode de la Fièvre Jaune*) priests abandon an ailing young woman underneath a tree. A cross rests in her lap and she grasps the hand of wrapped corpse who is likely her husband. Jacques Arago's lithograph of the 1821 yellow fever epidemic in Barcelona, Spain shows doctor André Mazet aiding the sick (*Mazet Dans Une Rue de Barcelone*, 1821). However, this image is strongly reminiscent of plague art, with dying people laying haphazardly in the streets, parents clutching children, husbands mourning wives, people covering their faces, bodies being carted away, and churches in the background (Hoffman 1964). The valiant doctor eventually succumbed to yellow fever (Hoffman 1964), as shown in *The Death of Mazet* by Henri Serrur (1861).

BLOODTHIRSTY FLEAS

Plague is perhaps the most notorious insect-vectored disease, but like yellow fever and mosquitoes, people didn't know fleas vectored the causative agent of plague (bacterium *Yersinia pestis*) from rodents to humans until the end of the century (Gage and Kosoy 2005). Therefore, it's unsurprising that artwork involving fleas focused on people grooming themselves to find and kill these blood-sucking "nuisance" pests. As in earlier centuries, many artworks were semi-erotic and showed partially naked women or men searching their body or clothes for fleas (see Figure 8.2), including *Seated Nude: The Flea* (1912) by Edvard Munch, who painted the iconic masterpiece *The Scream* (1893). Another famous artist, Henri de Toulouse-Lautrec, depicted a successful hunt in *La Toilette: Madame Fabre* (1891), as a woman pinches an offending insect between her fingernails. Other paintings focus on people picking fleas off animals.

FIGURE 8.2 *Un Pouilleux* (Flea Seeker), Mariano Fortuny y Carbó (1838–1874). Etching. National Gallery of Art, Washington, DC (2003.118.18). Gift of Frank Anderson Trapp. Wikimedia Commons.

William Blake was an English artist who believed in physiognomy, or judging someone's personality by their outward appearance, particularly their face (Butlin 2015). He painted in the dead of the night, supposedly conversed with angels and devils, and was deemed insane (Gilchrist 1907; Butlin 2015). One of his most famous works is *Ghost of a Flea* (1819–1820), which is an example of his "Visionary Heads" pieces that his associates John Varley and John Linnell imitated in their drawings (Gilchrist 1907; Butlin 2015). The painting shows a grotesque beast-like man standing on a starry stage, staring at its reflection in a golden cup. In Blake's "spiritual vision" that inspired this piece a flea told him:

> all fleas were inhabited by the souls of such men as were by nature bloodthirsty to excess, and were therefore providentially confined to the size and form of insects; otherwise, were he himself, for instance, the size of a horse, he would depopulate a great portion of the country.
>
> (Gilchrist 1907: 275)

Blake also created several plague-related drawings (detailed in Bennett 1984) that highlight many of the themes associated with this flea-vectored disease, such as corpses, caskets, mass graves, fire, survivors mourning the dead, and people praying.

THE SPECTER OF PLAGUE

Plague art in the nineteenth century was inspired by historical pandemics (i.e., Plague of Justinian 541 CE, Black Death 1347 CE, Bombay Plague 1894 CE), or more localized epidemics (Ligon 2006; Frith 2012). The manifestation of plague (bubonic, septicemic, pneumonic) depends on the route and location of infection, with grossly swollen lymph nodes (i.e., buboes) being a classic symptom of bubonic plague (Butler 1994). Ignorance about how plague was transmitted, along with recurring outbreaks, undoubtedly contributed to a continuing fear of the disease.

Superstition reigned in the absence of science, and people attacked religious groups and handicapped individuals, killed household pets, barricaded themselves indoors, and were advised to avoid dietary and emotional excess, have regular enemas, and abstain from sex (Ligon 2006). A colored etching by John Franklin elucidates concoctions thought to ward off the plague, e.g., balsam of sulphur, an electuary, plague water, vinegar, and an "Infallible Remedy by Constantine Rhodocanaceis," with the latter likely referring to the quack remedy "Constantine Rhodocanace's Grecian" (Leasor 2001). A more rational belief was that the contagion passed through infected air (Moote and Moote 2004), and so plague doctors wore masks with long beaks resembling a bird. Artists documented their imposing outfits, along with *monatti*, who disposed of corpses. Fear drove some individuals to extreme lengths to purify the air, as was the case with Solomon Eagle (a.k.a. Solomon Eccles), an eccentric who ran through the streets of London with burning coals strapped to his head, raving about God and the apocalypse (Bastian 1965). Multiple artists documented his antics and created various other plague-related illustrations, including death carts and burial pits. Edward Matthew Ward's series of chalk drawings (1848) captured the incredible devastation associated with this disease, including the horror of a positive diagnosis, parents cradling dead children, a cart stacked with bodies, and a woman staring helplessly into a pit filled with the dead.

These themes recur in many plague paintings, which feature a mix of dead and dying people being tended, mourned, or disposed of by the living. The bereaved maintain a physical connection to their loved ones, either clutching them in their arms or holding their hands. Mothers and fathers embrace stricken children, or the converse, while bystanders recoil from diseased individuals, hold their noses, or cover their faces. Corpses are scattered on the ground or slumped in the street. Carts and burial pits heaped with contorted bodies are more common than coffins and funerals and speak to how the epidemic disrupted traditional burial customs. Smoke, fire, and a sense of disorder and despair pervade much of the artwork. Less common indoor scenes feature sickbeds and hospitals, as in Francisco Goya's somber painting *Hospital de Apestados* (1808–1810).

Many people believed the plague was the wrath of God, and religious symbolism continued to be pervasive in plague art. A wood engraving by Edmund Evans (*A Street During the Plague with a Death Cart and Mourners*, 1864) shows the practice of marking plague-afflicted households with crosses and messages like "Lord have mercy on this house" (Doyle 1864). Artists personified death in various ways, including dark angels, grim reapers, or skeletons (see Figure 8.3). Jean Louis André Théodore Géricault, a French artist who painted victims of yellow fever, the insane, and severed heads and limbs, included a skeletal grim reaper near a cluster of bodies in *Drame de la*

FIGURE.8.3 *Die Pest*, Arnold Böcklin, 1898. Tempera on wood. Kunstmuseum Basel, Switzerland (Inv. 114). Deposit of the Gottfried Keller Foundation, Federal Office of Culture, Bern, 1902. Wikimedia Commons.

Peste (1808–1812). Brighter paintings with a more hopeful feel usually had fewer obvious, grotesque corpses and highlight a savior-like figure (e.g., spirit, priest, doctor) tending to the sick or saving people from the plague.

MYTHOLOGY AND ARTHROPODS

Myths are traditional stories used to explain natural events like disease epidemics or social phenomena. Several myths have inspired arthropod-related artwork, including the story of Arachne, which means "spider" in Greek. Arachne was a mortal who lost a weaving contest to the goddess Athena, after which Athena transformed her into a spider. Nineteenth-century images inspired by this myth show Arachne as a spider-woman hybrid, and often feature realistic spiders and webs.

The love story of Cupid and Psyche is even more renowned, and artists frequently used butterflies to symbolize the soul of Psyche, which means both "butterfly" and "soul" in Greek (Werness 2006; Nouvel-Kammerer 2007). Apuleius wrote the story in the second century CE, and according to a 1566 translation, Psyche is a beautiful young virgin who angers Venus (Roman goddess of love), unknowingly marries and then falls in love with Cupid (Amor, god of love and Venus's son), and eventually becomes a goddess (Adlington 1947). In artwork the couple are frequently shown embracing or in bed, but other scenes include Zephyrus (god of the west wind) or Cupid carrying Psyche, Psyche using a lamp to peek at her mysterious husband, Psyche attempting impossible tasks assigned by Venus, and Cupid waking Psyche from an enchanted sleep. The couple are usually shown as young adults, although sometimes one or both are children.

Artists used butterflies to symbolize the duo surrendering their souls to one another or to comment on the nature of love. White or yellow butterflies appear as decorations on their bed, or hover near Psyche and Cupid; occasionally the lovers are shown alone contemplating or touching single butterflies. Bertel Thorvaldsen created multiple two- and three-dimensional works featuring Psyche that are housed at the Thorvaldsen Museum in Copenhagen, Denmark. In his marble relief *Cupid and a Butterfly*, Cupid is shown poised to capture a butterfly with a net (see Figure 8.4). Antoine-Denis Chaudet focused on the torturous, binding aspect of love in his sculpture *Cupid Playing with a Butterfly* (1817), where Cupid and his minions hold a butterfly captive, although along the carved base, bees help free the butterfly. Goya's oil painting *Allegory of Love, Cupid and Psyche* (1798–1805) seems to align with Chaudet's sentiment, as it appears Psyche is teasing Cupid by holding a butterfly just out of reach.

Butterfly wings are frequently used in lieu of the actual insect, although their appearance and location varies. In most images Psyche has insect wings

FIGURE 8.4 *Cupid and a Butterfly*, Bertel Thorvaldsen, 1831. Marble. Thorvaldsens Museum, Copenhagen, Denmark (A397).

and Cupid has bird wings. However, Cupid occasionally has butterfly wings, along with Zephyrus and nude cherub-like children called zephyrs or putti (sing. putto). Statues of Psyche can be remarkably detailed, with two pairs of wings, some with venation and eyespots. In paintings the wings are variable in size, shape, and color. They're often transparent or light colored, but some are green or blue. Many have eyespots or other markings and look similar to real Lepidoptera wings, while others are less entomologically accurate. The amalgamation of the human body with insect wings was also a hallmark of fairyland paintings.

INSECTS IN FAIRYLAND

The public's fascination with fairies and wealth of art on the topic have been addressed by several authors (Schindler 1988; Maas 1997; Silver 1999; Wood 2000; Bown 2001; Mikl 2004; Schachat 2014; Talairach-Vielmas 2014; Forsburg 2015; Keene 2015) (see also Chapter 6 by Dodd). Fairy art melded romantic fantasy with the environment and stimulated people's interest in science and nature (Talairach-Vielmas 2014; Forsberg 2015; Keene 2015). The fanciful images may even have helped people overcome the "alien strangeness" of insects (Bown 2001: 127). Insect elements are omnipresent in fairy art, whether in the form of insects, fairies with butterfly wings, miniature people interacting with realistic arthropods, insect-human hybrids, or anthropomorphized insects.

Several paintings from this era refer to or feature characters from William Shakespeare's sixteenth-century play *A Midsummer Night's Dream*, including Hermia, Lysander, the mischievous fairy Puck, and the King and Queen of the Fairies—Oberon and Titania. Fairies resemble tiny people with wings, although their stature spans from human-sized to miniature. Other artists created different "classes" of insect-fairy hybrids, yielding a spectrum in the degree of amalgamation.

Virtually all the rendered fairies have insect wings on their backs, but like images of Psyche, the location, shape, size, texture, and color of the wings varies. The wings often stick out rather than lay flat lie against the back, and in rare instances sprout from the fairy's head. Transparent, long, and sometimes iridescent wings are reminiscent of damselflies and dragonflies, while shorter, wider wings resemble lacewings, although the venation is variable or absent. Representations of beetle wings are relatively rare. Most commonly, fairies have two pairs of colored, patterned butterfly or moth wings. Occasionally the wings resemble specific insect species, like monarch butterflies or tiger moths. Butterflies are also shown as fairy jewelry.

The portrayal of insect-fairy interactions is diverse. One example of insects posing a threat is seen in Robert Huskisson's *The Midsummer Night's Fairies* (oil on panel, *c.* 1847) where miniature soldiers protect Titania from a group of ants. Edward W. J. Hopley's painting *Puck and the Moth* (*c.* 1853) shows a cherubic Puck about to stab a moth with a spear-like stick, but based on Puck's gleeful expression and the prostrate moth's supplicating posture, it's unclear if this is for protection or sport. Some fairies and cherubs chase after insects with nets. Realistic-looking insects are present in many fairyland scenes, and the inhabitants appear to be coexisting peacefully. However, artists also show fairies using insects, with no distinction between willing servant and slave. Butterflies, bees, beetles, flies, grasshoppers, and damselflies, carry fairies on their backs, or are hooked to carts and chariots like aerial horses (see illustrations by Richard Doyle in Doyle and Allingham 1870 and Figure 8.5) and occasionally insects protect fairies, like an armed grasshopper flying to join the attack on a bat. Some insects are given human attributes, including a dragonfly playing a trumpet in Richard Dadd's *Fairy Feller's Master-Stroke* (1855–1856) (Fowle 2000a), although due to the long legs and wings that fold flat against the abdomen, the insect in question may actually be a crane fly.

Fanciful images of fairies and arthropods appeared on numerous postcards, trading cards, and other paper ephemera during this era, as well as children's books illustrated by E. Lejeune (*Les Petits Habitants des Fleurs*, Leneveux 1850), *The New Tom Thumb: with an Account of his Wonderful Exploits*, MacKenzie 1822), John Tenniel (*Alice's Adventures in Wonderland*, Carroll 1867), Fortuné Louis Méaulle (*Alphabet des Insects*, Becker 1883), Henry

FIGURE 8.5 *The Fairy Queen Takes an Airy Drive in a Light Carriage, a Twelve-in-Hand, Drawn by Thoroughbred Butterflies*, Richard Doyle, 1870. Color wood engraving and color lithography. Engraved and printed in color by Edmund Evans. The Metropolitan Museum of Art, New York. Gift of Lincoln Kirstein, 1970 (1970.565.74-13).

Justice Ford (Lang's series of Fairy Books from 1889–1913, e.g., Lang 1889), Arthur Rackham (e.g., *Peter Pan in Kensington Gardens*, Barrie 1906), Edith Brown (*The Cheerful Cricket and Others*, Marks 1907), Warwick Goble (*The Water-Babies: a Fairy Tale for a Land-Baby*, Kingsley 1909), and M.T. Ross (*The Butterfly Babies' Book*, Gordon 1914). Illustrations of anthropomorphized insects or human-insect hybrids also enlivened adult publications (Schachat 2014), including *Episodes of Insect Life* (Budgen 1849–1867), and work by famed French caricaturist Jean Ignace Isidore Gérard (J.J.) Grandville (Grandville 1844; Grandville et al. 1867; Stahl 1867).

A KALEIDOSCOPE OF BUTTERFLIES

If any insect has paved a path between Phylum Arthropoda and *Homo sapiens*, it is surely the butterfly. Known by many names, psyche (Greek), schmetterling (German), mariposa (Spanish), vlinder (Dutch), farfalla (Italian), papillon (French), these beautiful, complex creatures of the sky have occupied a unique place in human culture. For centuries artists have used them to symbolize transformation, resurrection, and the soul

(Ferguson 1954; Werness 2006; Nouvel-Kammerer 2007), as well as desire and sensuality (Kačāne 2012) (see also Chapter 6 by Dodd). Lepidoptera are undoubtedly the most frequently encountered insect in nineteenth-century art (Dicke 2000, 2004).

Butterflies transcended the confines of artistic movements and genres, and while many artists used them symbolically, others focused on their aesthetic appeal. The Pre-Raphaelite Brotherhood had a lasting impact on the visual and literary arts, and was formed in 1848 by a group of English artists who rejected the conventions of the Royal Academy in favor of a more natural style reminiscent of late medieval or early Renaissance art (Stokstad 1999; Meagher 2004; Prettejohn 2012). Members of this short-lived group, and those who adopted their style, used butterflies and moths in a symbolic context, but their realistic portrayal of insects likely contributed to the aesthetic use of insects in fine art (Meagher 2004; Kačāne 2012).

In Henry Alexander Bowler's oil painting *The Doubt: Can these Dry Bones Live?* (1855), a butterfly rests on a skull while two others flit around a graveyard, and in John Maler Collier's *The Butterfly (Portrait of Mabel Hamilton)* (date unknown) a girl on the riverbank solemnly watches a lone white butterfly. The colorful tortoise-shell butterfly resting on a blind girl's cloak is also symbolic (Millais 1899). William Holman Hunt uses a death's-head hawkmoth as a bad omen in his oil painting *The Hireling Shepherd* (1851) (Macmillan 1972). He used a live insect as a model for *Our English Coasts (Strayed Sheep)* (1852), where butterflies feed on flowers overlooking the English Channel (Fowle 2000b). Dante Gabriel Rossetti used butterflies to symbolize the soul in his oil painting *Sibylla Palmifera* (1866–1870) although they may signify desire and sensuality in *Venus Verticordia* (1864–1868) (Kačāne 2012).

The Aesthetic Movement (mid-1870–1880s) emphasized self-expression, beauty, and the art itself rather than narrative (Frelinghuysen 1999; Souter and McCain 2018). Artists used butterflies as stylistic elements, e.g., Albert Joseph Moore's *The Green Butterfly* (*c.* 1879–1881) and James McNeill Whistler's *Harmony in Grey and Green: Miss Cicely Alexander* (1872–1874). The latter artist even used a butterfly motif to sign his paintings, and because it changed over time it's been used to date his work and letters (see Figure 8.6; MacDonald et al. 2003; WHMA 2016). His signature is reminiscent of Aubrey Beardsley's stylized butterflies in illustrations for the book *Salome* by Oscar Wilde (Wilde 1904; Harris 1967). A fairy-butterfly hybrid poses alongside more ornate butterflies in Beardsley's book *Under the Hill*, where he writes:

> Huge moths, so richly winged they must have banqueted upon tapestries and royal stuffs, slept on the pillars that flanked either side of the gateway, and the eyes of all the moths remained open and were burning and bursting with a mesh of veins.
>
> (Beardsley 1904: 7–8)

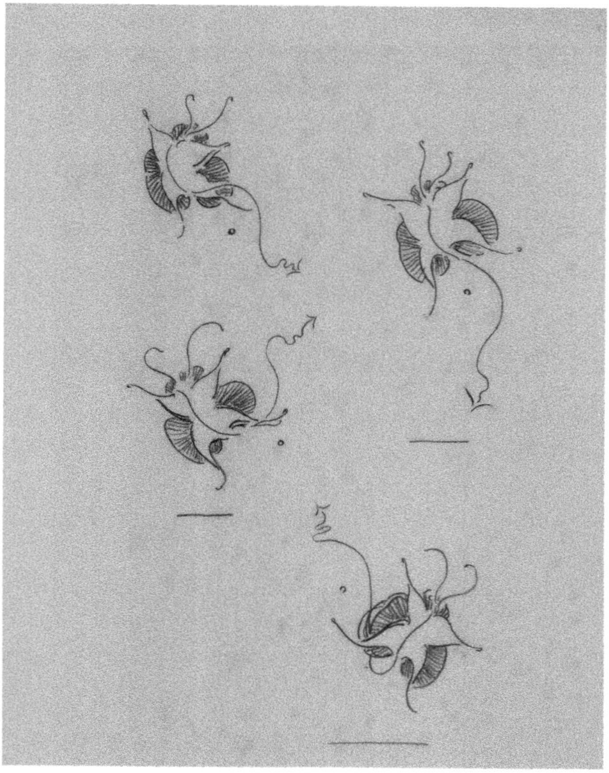

FIGURE 8.6 *Butterflies: Design for the Gentle Art of Making Enemies*, James Abbott McNeill Whistler (1834–1903). Drawing, crayon. Los Angeles County Museum of Art, CA, www.lacma.org (M.2015.73.48). Gift of Anita and Julius L. Zelman in honor of the museum's fiftieth anniversary.

The symbolist Odilon Redon drew three insect-spider hybrids in the early part of his career, which focused on black and white pieces, or "noirs" (Hauptman 2005). Later he shifted to colorful compositions, and several of these oil paintings feature butterflies, either as focal points, in floral still-lifes, or as elements within more fantastical, pre-surrealist pieces (Hauptman 2005). William Baxter Closson's grayscale engravings of butterflies and moths contrast with the colorful insects created by Redon and other artists, such as John Carter, who began his artistic career after an accident paralyzed his limbs (Mills 1868). He painted by holding a brush in his mouth, and his first work was a vividly rendered red, yellow, and black butterfly captured in his room (Mills 1868). The American artist Albert Bierstadt, although primarily known for epic landscapes, did several paintings featuring lone butterflies on a blank page. He apparently created these colorful "Bierstadt Butterflies" using a folded paper technique to create a symmetrical image and gave them to ladies who visited his studio as souvenirs (Hendricks 1974).

Henry Dalton used butterflies as both subject and artistic instrument by placing individual butterfly scales and diatoms on microscope slides to create elaborate "micromosaics" (Beughhold 2013). His miniature still-lifes featured flowers, birds, and fittingly, butterflies. Traditional micromosaics were made of small glass rods called tesserae and used to decorate objects or create jewelry (Gabriel 2000; Soros and Walker 2004), including pieces with Psyche and Cupid, butterflies, floral still-lifes with butterflies, and scarab beetles.

THE ALLURE OF INSECT ORNAMENTATION

People have decorated themselves or their homes with insects for centuries (Frost 1937; Kuehn 2003). In this era, it was fashionable to have insect-inspired items in one's home, and many ladies wore jewelry or clothing with insect motifs spanning multiple hexapod orders, with butterflies and moths being extremely popular (Akre et al. 1991; Liu 1998; Tolini 2002; McTier 2013). Due to Napoleon's Syrian campaign and the subsequent fascination with Egyptian culture dubbed "Egyptomania," scarab beetles were also a desirable design on jewelry and other decorative objects (Liu 1998; Soros and Walker 2004; Burleigh 2007; Brier 2013). But even flies were a source of inspiration according to author Jules Michelet:

> The goldsmith and the lapidary would do well to seek in them models and instruction. The soft insects, the flies, specially possess in their eyes truly magical irises, with which no casket of gems can bear comparison. ... Take the dull, gray, dusty, odious horse-fly, which lives on warm blood; its eye, to the magnifying-glass, offers the strange faëry spectacle of a mosaic of jewels, such as all the art of Froment-Meurice has scarcely invented.
>
> (Michelet 1875: 201)

François-Désiré Froment-Meurice was a celebrated French metalsmith and jeweler, and one of several artists who made jewelry with insect designs in this era, including René Lalique (a.k.a. Monsieur Butterfly; Heuzé 2003), Louis Comfort Tiffany (Munn 1987; Frelinghuysen 1999; Geoffroy-Schneiter 2007), the firm of Jaques & Marcus (see Figure 8.7), Lucien Gaillard, and Maurice Pierre Andre Daurat. Beatrice Whistler, the wife of James McNeill Whistler, created jewelry designs featuring fairies, as well as butterflies reminiscent of her husband's butterfly signature.

Using insects for personal ornamentation extended beyond traditional jewelry. "*En tremblant*" pieces vibrated on springs, thus creating the illusion that the insects were alive (McTier 2013), which were occasionally used in

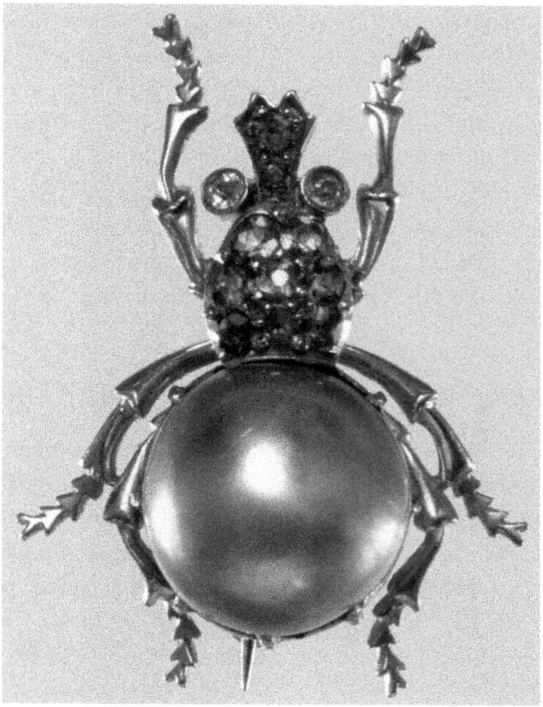

FIGURE 8.7 Brooch, *c.* 1882–1892, Jaques & Marcus. Gold, pearl, demantoid garnets, and diamonds. Length 1 1/8 in. (2.9 cm). Brooch marked on hook: J & M. Box marked inside lid: JAQUES & MARCUS / BROADWAY / CORNER 17th ST. / NEW YORK. The Metropolitan Museum of Art, New York (2001.246). Purchase, Susan and Jon Rotenstreich Gift, 2001.

elaborate coiffures. Women wore jewelry and clothes decorated with dead insects or body parts like elytra (hardened forewing of beetles) (Tolini 2002). People even secured live animals to their garments via chains or nets as personal "pets" (Long 2009). Beetles and butterflies were among the most popular insect pets, and luminescent fireflies were highly desirable (Cowan 1865; Long 2009); Kirby and Spence refer to them as "artificial diamonds" (1817: 413). In her book *Insect Ways on Summer Days*, Jennett Humphreys speaks as a firefly:

> A great many ladies in my own country wear me, and wear dozens and dozens of my sisters and my cousins at the same time. They put us in little pieces of gauze; they have the gauze fastened on to pins, and then stick the pins in their hair, and everybody cries, "How beautiful!"
>
> (1887: 111)

Wearing fashions inspired by nature was repeatedly mocked by Edward Linley Sambourne via his illustrations for the satirical magazine *Punch*, including ladies wearing outlandish garments and hats made from butterflies, beetles, bees, wasps, and spiders (Brewer 2004). The caption for his illustration "The Butterfly was a 'Lady'," which shows a woman sporting an outfit abounding with swallowtail butterflies, includes a quote from an old ballad, "At times methinks my soul hath wings, At times it loves to crawl" (*Punch* 1871: 245). Regardless, insect motifs remained en vogue, and collecting butterflies became a widespread phenomenon that transformed the masses into entomophiles.

ACQUIRING AN EYE FOR NATURE

Collecting natural objects is an activity interwoven with human history, and has emotional, social, economic, and scientific facets (Simmons 2016). Collections were housed in precursors to museums called cabinets of curiosity, which were known by a variety of names, occurred in an array of physical forms, and were important status symbols (Simmons 2016). The Age of Enlightenment and beginning of the scientific revolution marked a time of dramatic imperial expansion (Wallbank et al. 1971; Simmons 2016). The subsequent influx of collectibles from abroad not only furthered people's interest in biodiversity, but also inspired professional and amateur artists. Clubs dedicated to natural history formed an important link between private and institutional collections, in addition to furthering scientific disciplines (Osborn 1937; Sorensen 1995; Simmons 2016). In the 1700s, enthusiasm for natural history collections expanded to all of "polite society," and by the 1800s had spread to the general public and working class (Salmon 2000; MacGregor 2014). This was in part due to advances in transportation and more affordable collecting equipment (Wallbank et al. 1971; Philonoë 1998; Parks 2015). Leisure time was a hallmark of higher social standing and having the means to acquire and display a natural history collection in your home was a sign of financial success (Simmons 2016). Eventually, interest in the natural world transcended socio-economic classes. People believed anyone could access the wonders of nature in their own backyard (Leach 2014; Parks 2015).

Interests and motivations of insect collectors ranged widely, from amateur hobbyists, to those primarily interested in aesthetics, to competitive elitists who sought rare or exotic specimens, to people interested in gaining knowledge of God's creations (see Chapter 5 by Meuti and Fisher), to more serious collectors who hoped to advance science (Salmon 2000) (see Chapter 1 by Johnson and Polaszek). The public's passion for natural history and insect collecting began to decline in the twentieth century due to concerns related to conservation.

THE PASSIONATE PURSUIT OF LEPIDOPTERA

Engaging with nature and collecting butterflies were acceptable pastimes for ladies and children. Frederik Hendrik Kaemmerer's *The Scarab's Serenade* (oil on panel, date unknown) shows a girl sitting indoors watching a large beetle walk across a tambourine, and it is interesting to note she holds a thin branch, no doubt as a proxy with which to prod the beetle. This contrasts with paintings where women or children coexist with insects or gaze curiously at insects on plants, or butterflies on their bodies. Many paintings show women and youngsters with nets or chasing after butterflies, including the fauvist work by Henri Matisse *Boy with Butterfly Net* (1907). True to the focus on the outdoors that was a hallmark of the impressionists (Honour and Fleming 1995), the French artist Berthe Morisot painted her sister and children in grassy gardens with gauzy green butterfly nets in *Chasse aux Papillons* (1874) and *Dans le Parc* (1874) (Stuckley et al. 1987).

Paintings in outdoor settings often provide insight into the nuances of observing insects or collecting butterflies. Sir Lawrence Alma-Tadema's oil painting *Ninety-Four Degrees in the Shade* (1876) shows a youthful man lying beside his net reading an illustrated butterfly book. Jeanne Bôle depicts a girl holding a net, paintbrush, and sketchbook in *A Young Girl with a Butterfly Net* (1877), suggesting she is going to catch and document real butterflies. In *Junge Dame Beim Schmetterlingfangen* (date unknown) by Frederik Hendrik Kaemmerer, a young woman with a net on her lap appears to be holding a pinned butterfly between her fingers, and Émile Eisman-Semenowsky shows a woman placing specimens in a decorative container on her lap while a gauzy net lays at her feet in *Catching Butterflies* (date unknown).

Many clergymen were serious butterfly collectors, including William Kirby, Rector of Barham, England and co-author of *An Introduction to Entomology* (Kirby and Spence 1846; Salmon 2000) (see Chapter 5 by Meuti and Fisher). Collecting was considered a "highmindedly Christian pursuit" that people did on Sunday after church (Philonoë 1998: 8). Contemplating nature was encouraged, and outdoor recreation, especially in the country, was considered a healthful activity (Casto 2013). Thus, butterfly collecting simultaneously addressed multiple goals. Carl Spitzweg's painting *Der Sonntagsspaziergang* (1841) depicts one such excursion, with a well-dressed family walking through a grass field and a boy lagging behind wielding a butterfly net. *La Chasse aux Papillons* by Albert-Émile Artigue (date unknown) also depicts what is likely a Sunday walk, but has a slightly satirical element, as a gentleman on all fours attempts to catch butterflies with his hat.

MOCKING INSECT COLLECTING

Insect collecting was a widespread hobby, but anything that deviated from acceptable social norms was a target of mockery, whether it was the inept, buffoonish amateur or the zealous, eccentric professional. Early in the nineteenth century, entomology was in its infancy. Unlike other natural sciences, the study of insects was considered childish, and people viewed those who seriously pursued the subject odd at best, and often ridiculed them or treated them with pity and scorn (Kirby and Spence 1846). People who collected insects were even branded as insane (Kirby and Spence 1846). Kirby and Spence addressed objections against entomology, namely allegations that it is "trifling and unimportant, and deemed an egregious waste of time and talents," and "[Entomologists] are reprobated as unfeeling and cruel, and as tending to harden the heart" (1846: 47).

Satirical artworks about insect collecting primarily featured men. In some images insects comprised part of the clothes or body, and in George Spratt's lithograph *The Entomologist* (1831) the man's body is made almost entirely of insects. Many artists satirized the collector's intense passion for capturing insects regardless of circumstance. One example is a pair of paintings (date unknown) by the English artist Edward Hopley. In *The Chase* a man fixated on catching a butterfly runs into a stream, and in *Bagged* the man gazes happily at the butterfly in his net, oblivious to the fact he's knee-deep in water and his supplies are floating away. Thomas Rowlandson's etching *Butterfly Hunting* (1806) shows two foolish-looking gentlemen trampling garden plants as they run after butterflies with hinged racket-like nets (see Figure 8.8). Rowlandson also mocks insect collecting in *Summer Amusement, Bug Hunting* (1811) where a grotesque-looking couple gather what are likely bed bugs from their bed. The "oblivious enthusiast" was even mocked in a two-minute silent film from 1906, *Chasse au Papillon*, which was black and white except for yellow hand-colored butterflies (Clark et al. 2012). The film shows a man bumbling around irritating people as he attempts to catch a butterfly for his collection. *Un Renaud: Pris au Piège* (Nodier 1867) is a story in a satirical book illustrated by J.J. Grandville (Stahl 1867) featuring primates as academic scholars (perhaps in reference to Charles Darwin), a man-bird hybrid with bugs pinned to his hat brandishing a net, and tongue-in-cheek mockery of insect collectors:

> Do you know a person who is happier and at the same time more commendable to his friends and fellow-citizens than the one we meet in early morning, breathless and joyful, scouring the tall grass with his net, wearing in his buttonhole a ball armed with long pins to jab skillfully, and without causing it the slightest pain (because it never complained), the winged insect that the breeze carries?
>
> (Nodier 1867: 252)

FIGURE 8.8 *Butterfly Hunting*, Thomas Rowlandson, 1806. Print, hand-colored etching. National Gallery of Art, Washington, DC (1945.5.890). Rosenwald Collection.

Of special note is *The Entomologist's Dream* (1909) by Edmund Dulac, which appeared in a story about an entomologist who, in his quest to find a blood red butterfly for his lady friend, becomes insane, destroys his entire collection, and subsequently dies (d'Houville 1909).

THE ADVANCEMENT OF ENTOMOLOGY

Natural history collections played an important role in transforming people's understanding of the world, and exotic specimens fueled interest in entomology. Naturalists described the insect fauna at home and abroad and amassed sizeable collections of flora and fauna (see Chapter 1 by Johnson and Polaszek). Many scientifically minded individuals published books with detailed colored plates of visually striking butterflies or romanticized images of naturalists collecting in the field (Twidle 1920; Smart 1975). This contrasted with satirical images of bourgeois snobs who adopted the mantle of adventurers and scientists because they had the means to travel abroad (Cagidemetrio 2010). One example is Carl

Spitzweg's oil painting *Der Schmetterlingsfänger* (*The Butterfly Hunter*, 1840), where an odd, almost insectile-looking man grips his inadequately small butterfly net and goggles at large exotic butterflies flying through the jungle. As Thomas Henry Huxley said in a letter to Darwin's friend, Joseph Dalton Hooker, on July 29, 1859, "For the word 'Naturalist' unfortunately includes a far lower order of men than chemist, physicist, or mathematician. ... every fool who can make bad species and worse genera is a 'Naturalist'!" (Huxley 1901: 177).

Many naturalists collected and studied insects, but few self-identified as entomologists. People usually studied insects as an unpaid hobby and the few people who earned a living as entomologists were stigmatized (Kirby and Spence 1846; Osborn 1937; Elias 2014). However, entomologists persisted and formed scientific societies, field clubs, and journals dedicated to the study of insects. The Society of Aurelians formed between 1720 and 1742 in London, so named because butterfly collectors were known as Aurelians, after *aureolus*, meaning golden in Latin (Salmon 2000). *The Aurelians* (1909) by Cooke is a dignified portrait of two society members and hangs in the Royal Entomological Society of London (Salmon 2000). Over time entomologists gained more respect in society, the professional world, and academia, and there were fewer satirical images of insect collectors. The growing acceptance of entomology is highlighted in L.M. Budgen's richly illustrated book *Episodes of Insect Life*, who says via her cricket pseudonym *Acheta domestica* (shown on the back of a mayfly with butterfly net in hand):

> a hobby, we confess it, whereat even under her imposing name, Entomology, all her sister-ologies were once in the habit of kicking up their heels, as they left her disdainfully in the rear. Of late, however, she has been gaining ground, and now promises to keep up with the proudest jade among her fellows.
>
> (Budgen 1849: 13)

The number of books dedicated to entomology skyrocketed, with many intended for children or a general audience. Scholarly references for professionals and scientists usually focused on insect biology and taxonomy, as well as illustrations of tools used to capture and preserve insect specimens. Eighteenth-century books discuss clap-nets (also known as bat-folders, or bat-fowling nets) while later books feature scissor nets, racket nets, modifications of ring nets, and even aquatic collecting methods (Harris 1766; Ingpen 1839; Religious Tract Society 1839).

Many entomology books had beautiful color illustrations, and much of the artwork was done by women, since it was a socially acceptable way they could participate in scientific endeavors (Osborn 1937; Sorensen 1995; Le-May

Sheffield 2001; Gates 2002). Young ladies of the upper and middle classes were expected to have genteel hobbies such as drawing and painting, and plants, flowers, and insects were ideal subjects (Parks 2015). Several women were accomplished natural science artists (e.g., Anna Botsford Comstock, Louisa Anne Meredith, Mary Peart, Lucy Say, Harriet and Helena Scott) who created extraordinary detailed images of insects that were integral components of early entomological texts (Edwards 1868; Osborn 1937; Sorensen 1995; Ball 2005; Braby and Olsen 2011; Leach 2014). Hand-drawn or painted illustrations were gradually replaced by engravings, lithographs, etchings, and photographs.

AMONG THE PLANTS

Still-life painting flourished throughout Europe in the 1600–1800s, particularly in the Netherlands (Liedtke 2003; Meagher 2008). The arranged objects, including flowers, food, and animals, were highly symbolic, and insects appeared in many works (Dicke 2000; Liedtke 2003; Murphy 2017). The prevalence of still-life painting waned as time passed, but some artists, such as the French artist Jean Georges Hirn, continued to paint in a style reminiscent of the Dutch masters. Other still-life artists appeared to use insects as aesthetic elements. Pierre-Joseph Redouté specialized in painting flowers and was the court artist to Queen Marie-Antoinette and Napoleon's first wife, the Empress Josephine (Van Druten 2013). He published his art in several books, and *Choix des Plus Belles Fleurs et des Plus Beaux Fruits* (1827–1833) has multiple plates with butterflies resting on flowers. Paul de Longpré was a French artist who primarily worked in America, and whose use of insects was similar to Redouté, as his floral watercolors frequently featured bees in flight or on flowers (see Figure 8.9). The academic Arthur Harry Church generated a wealth of botanical illustrations and information on insect-plant associations (Mabberley 2000):

> In gardens the flowers [*Ribes sanguineum*] are visited by large numbers of bees, including the Hive-bee ... and Humble-bees (*Bombus terrestris* ...), and the Hive-bee is the most industrious and possibly most important pollinating agent. *Anthophora pilipes* ... is by far the most elegant worker .... *B. terrestris* is an extremely clumsy visitor, and probably does as much harm as good.
>
> (Church 1908: 157)

Artists used bees symbolically to represent various human concepts and emotions, including abundance, love, industriousness, eloquence, and the monarchy (Ferguson 1954; Lawrence 1993; Impelluso 2004; Werness 2006), and these insects are found in many art works. Napoleon, who crowned himself

FIGURE 8.9 *Poppies and Bees*, United States, 1906. Paul de Longpré (France, Lyons, active United States, 1855–1911). Watercolor on paper, mounted on cardboard. Sheet: 25 3/4 × 16 in. (65.41 × 40.64 cm), image: 23 × 13 3/8 in. (58.42 × 33.97 cm). Los Angeles County Museum of Art, CA, www.lacma.org. Charles H. Quinn Bequest (75.4.9).

emperor of France in 1804, actively promoted his power and regime through the use of symbols, including the eagle and bee. Bees adorned textiles, furniture, decorative objects, his sword, and throne, and featured in multiple paintings related to Napoleon (Le Bourhis 1989; Nouvel-Kammerer 2007). Napoleon was said to favor the bee because decorative gold bees were one of the treasures recovered from the tomb of Childeric I, Merovingian king of the Salian Franks in the fifth century (426–481), and thus these insects represented a connection to the origins of France (Jones 1898; Le Bourhis 1989; Wood 1994; Library of Congress 2023). When asked if he would like eagles on the mantle of his coronation outfit Napoleon apparently replied (although he may have been referring to Childeric I, not Chilperic):

> I would have gold stars, or rather gold bees. The latter would be a national emblem, for bees were found in the tomb of Chilperic. That insect is the symbol of industry. The stars will be for me, and the bees for the people.
>
> (Langon 1837: 116)

Oscar Wilde's fictional character Dorian Gray referenced bees in association with Chilperic I, another Merovingian king (539–584) (Wood 1994) as he ruminated about luxurious textiles, including "the mortuary cloth of King Chilperic, with its three hundred golden bees" (Wilde [1890] 2012: 72).

CARING FOR INSECTS

Beloved butterflies enchant and delight us, but people go to elaborate lengths to help bees survive. Bees are critically important plant pollinators and turn nectar into honey, a desirable natural sweetener and foodstuff. The intimate relationship between our two species has been recognized for centuries, and apiaries, beehives, and beekeepers feature in several nineteenth-century drawings and paintings, with artists documenting variations in hive design and beekeeping practices from different parts of the world (see Figure 8.10).

Art on this topic was particularly prevalent in Russia and eastern Europe, with artists usually depicting tall cylindrical hives, whether made from clay, mud or logs. Many of these hives have slanted wooden covers weighted down with stones, although some are topped with grass or straw, and others have ramps to the hive entrance. Skeps, or upside-down basket hives, are found in paintings from western Europe and America, sometimes with roof-like straw coverings, and occasionally stored within a larger shelter. Artists also painted rectangular box hives of various shapes and sizes. In Elena Samokish-Sudkovskaya's painting *At the Bee Yard* (1890), a box beehive opens from the side like a cabinet and has a complex interior design with multiple compartments. Beehives are usually clustered together outside, although hives are occasionally shown within larger open-sided structures, and in some images people work in apiaries or with open hives, thus providing insight into beekeeping practices.

BEYOND BEES AND BUTTERFLIES

Globetrotting artists like Martin Johnson Heade and Marianne North blurred the boundaries between art and science. The former was an American artist who painted landscapes, floral still-lifes, tropical plants like orchids, hummingbirds, butterflies, and beetles (Stebbins et al. 2000). The latter was a wealthy woman from England who traveled the world by herself, an occurrence virtually unheard of during this time period (Guelke and Morin 2001; Le-May Sheffield 2001; Payne 2016). Although primarily known for her botanical renderings (the gallery named for her at the Royal Botanical Gardens, Kew in Richmond, England displays more than eight hundred of her works), several of her oil paintings feature arthropods (e.g., butterflies, moths,

FIGURE 8.10 *The Honey-Bee*, Edward Bevan, 1827. Illustration in *The Honey-Bee; Its Natural History, Physiology and Management*, frontispiece. CC BY 4.0 (.b12817946). Wellcome Collection.

caterpillars, beetles, ants, and spiders). Her autobiography provides insight into her entomological experiences beyond the canvas, including from Brazil:

> ... the abundant vegetation here was covered with Garapatas [ticks], the most intolerable of insect-plagues, and at this season in their infantine and most venomous stage. One blade of grass might shake a whole nest on to the passing victim ... if left, the hundreds of tiny grains would diverge in every direction till they found places they fancied screwing their proboscis into, when they would suck and suck till they became as big as peas, and dropped off from over-repletion.
>
> (North and Symonds 1894: 159)

> Another unpleasant creature was a stinging caterpillar, whose hairs were as dangerous as a scorpion's tail; a rub against them might cause the hand or arm to inflame so that amputation was necessary to save life.
>
> (North and Symonds 1894: 150)

The devastation associated with contemporary or biblical plagues of insect herbivores was captured by several artists, including one featuring an army of locusts marching across the landscape with the pyramids in the background. One author says:

> in Africa and Arabia, he saw a flight of them extending above a mile in length, and half as much in breadth .... As they approached from the east, the density of the host obscured the solar rays, cast an awful gloom, like that of an eclipse ..., and caused a noise like a rushing of a torrent. ...
>
> (Carpenter 1833: 217)

Overall, compared to nectar- and pollen-feeding insects, there was much less artistic notice paid to herbivorous insects like locusts, Colorado potato beetles, and cockchafers (i.e., May bugs, June bugs, hannetons, maikäfer), although the latter and other types of beetles graced various paper ephemera such as tobacco cards (see Figure 8.11; Clark 2009), and were objects of curiosity in paintings featuring cats and dogs.

THE ANIMAL KINGDOM

Fine art involving ecological relationships beyond insects and plants was relatively rare in the nineteenth century, with one exception being household pets and birds. Upwards of twenty artists created paintings featuring cats and

FIGURE 8.11 *Child Chasing Beetle with Butterfly Net*, Allen & Ginter Company, 1887. Color lithograph, tobacco card from the Girls and Children series (N58) promoting Our Little Beauties Cigarettes. The Metropolitan Museum of Art, New York (63.350.202.58.8). The Jefferson R. Burdick Collection, Gift of Jefferson R. Burdick.

kittens, less commonly dogs and puppies, interacting with insects. Curiosity, playfulness, and mischievousness were dominant themes and the pets may have functioned as surrogate humans, reflecting society's fascination with the natural world. Animals engaged with a myriad of non-living toys as well as frogs, crayfish, snails, pet monkeys, and insects. Felines and canines stare intently at common insects like butterflies, bees or wasps, flies, and beetles, and in one painting two cats and a dog surround a large exotic weevil, apparently captivated by the strange insect flexing its elytra. Cats examining insects having a snack or resting on dishes or food was also common. Some artists allude to predator-prey relationships, with pets stalking or chasing insects. American artist Mary Russell Smith's paintings of domestic fowl highlights the importance of insects in the food chain, an element seen in other artwork featuring insects and amphibians.

HEART OF AN ARTIST

Vincent van Gogh (1853–1890, see Figure 8.12) was a famous Dutch artist known for his vivid paintings and struggles with mental illness (Department of European Paintings 2004). This artist connects many of the diverse entomological threads in this chapter via his iconic paintings and eloquent correspondence (all letters from Jansen et al. 2009). According to his sister Elisabeth, Van Gogh collected insects in his youth:

> how clever he was at catching the water insects; he would show his trophies to the children on his return. Such jolly little and big beasties! There were broad beetles, with their glossy backs; others with great round eyes, and crooked legs, that nervously wriggled the minute they left the water. ... And then, after he had prepared them, he would carefully pin them in a little box,

FIGURE 8.12 *Self-Portrait with a Straw Hat*, Vincent van Gogh, 1887. Oil on canvas. The Metropolitan Museum of Art, New York. (67.187.70a). Bequest of Miss Adelaide Milton de Groot (1876–1967), 1967.

> which first had been beautifully lined with snow-white paper, and neatly labeled with the names pasted above each insect – even the Latin names!
>
> (du Quesne van Gogh 1913: 5)

Van Gogh lived in Paris for two years prior to moving to Arles, France in February 1888, and continued to paint while at an asylum in Saint-Rémy, including several canvases that featured butterflies (Department of European Paintings 2004). He often wrote to his brother Theo and discussed the inclusion of a large yellow butterfly in *Small Pear Tree in Blossom* (1888) (letter 595, April 13, 1888), which is similar to the yellow butterfly seen in *Peach Tree in Blossom* (1888). He also talked about the insect that inspired the well-known painting *Giant Peacock Moth* (1889):

> Yesterday I drew a very large, rather rare night moth there which is called the death's head, its coloration astonishingly distinguished: black, grey, white, shaded, and with glints of carmine or vaguely tending towards olive green; it's very big. To paint it I would have had to kill it, and that would have been a shame since the animal was so beautiful.
>
> (letter 776, May 23, 1889)

The letter included a sketch of the moth alongside a scarab beetle, with the latter appearing in the painting *Roses* (1889) and later identified as the rose chafer beetle *Cetonia aurata* (Van Gogh Museum 2018). Van Gogh did several drawings of the moth prior to completing the painting *Giant Peacock Moth*, which has been identified as an emperor moth, *Saturnia pyri* (Schiffermüller) (Dicke 2000).

Cicadas feature in several of Van Gogh's letters from July, and he sketched them twice (letter 638, letter 790). He referred to their striking vocalization, saying they "sing at least as loudly as a frog" (letter 638, July 9 or 10, 1888), and "Outside the cicadas are singing fit to burst, a strident cry ten times louder than that of the crickets" (letter 787, July 6, 1888). He indicated, "Their song in times of great heat holds the same charm for me as the cricket in the peasant's hearth at home" (letter 790, July 15, 1889), and repeatedly mentions hot weather when talking of them, "I even work in the wheatfields at midday, in the full heat of the sun, without any shade whatever, and there you are, I revel in it like a cicada" (letter 628, June 19, 1888). "The olive trees are more in character … and I've tried to express the time of day when one sees the green beetles and the cicadas flying in the heat" (letter 805, September 20, 1889). The green beetles may refer to blister beetles (Family: Meloidae), as previously he mentions, "golden and green *Cantharides* swarming on the olive trees" (letter 638, July 9 or 10, 1888).

Van Gogh talks of cockroaches in his food (letter 787, July 6, 1888), and because he painted outdoors, he dealt with biting flies (letter 638,

July 9–10, 1888), was "eaten up by the mosquitoes" (letter 639, July 13, 1888), and had insects stick to his artwork. He writes to his brother, "I must have picked a good hundred flies and more off the 4 canvases that you'll be getting" (letter 515, July 14, 1885). Remnants of a grasshopper were found entombed in the oil painting *Olive Trees* (1889) (Dwyer 2017).

Van Gogh frequently used arthropods as metaphorical devices, including when ruminating about fame:

> You know the fireflies in Brazil that are so luminous that in the evening ladies stick them into their hair with pins. It's very fine, fame, but see, it is to the artist what the hairpin is to those insects. You wish to succeed and shine; do you really know what you want? Now I have a horror of success ....
>
> (letter 660, August 13, 1888)

He used metamorphosis to explore questions of human transformation, mortality, and life after death. In a letter to his sister Willemien he compares painters with herbivorous beetle larvae:

> Now I know that it's fairly impossible for the white potato or salad grubs that turn into May bugs later to be capable of forming credible ideas about their future overground existence. And that it would be rash of them to undertake overground studies to throw light on this question, since the gardener or others interested in salad and vegetables would immediately trample them underfoot as being harmful insects. But for parallel reasons I have little faith in the rightness of our human ideas concerning our future life. We can no more judge our own metamorphoses impartially and sagely than the white salad grubs can theirs. For the same reason that a salad grub has to eat salad roots for its higher development — so I believe that a painter has to make paintings — perhaps that there's something else after that.
>
> (letter 579, February 24, 1888)

He writes to fellow painter Emile Bernard:

> The butterfly, where does the butterfly emerge, from that sated caterpillar — the cockchafer from that white grub? ... I'd also very much like to know roughly what I'm the larva of myself, perhaps.
>
> (letter 633, June 27, 1888)

> ... the possibilities of doing painting in better and changed conditions of existence — an existence changed by a phenomenon perhaps no cleverer and no more surprising than the transformation of the caterpillar into a butterfly, of the white grub into a cockchafer. That existence of painter as butterfly

> would have for its field of action one of the innumerable stars, which, after death, would perhaps be no more unapproachable, inaccessible to us than the black dots that symbolize towns and villages on the map in our earthly life. Science — scientific reasoning — seems to me to be an instrument that will go a very long way in the future.
>
> (letter 632, June 26, 1888)

Vincent van Gogh died on July 29, 1890 from a self-inflicted gunshot wound to the chest (Department of European Paintings 2004).

US AND THEM

Insects are among some of the most diminutive animals, veritable specks in comparison to humans. Yet their impact is large, and our interconnections perhaps too complex to fully comprehend. In the Age of Industry, art featuring insects was incredibly diverse, not only in form, but also in how artists portrayed relationships between humans and insects. Throughout the era, artists defined insects within the context of human culture, whether to convey human concepts and emotions, or for self-expression. However, as the century progressed and insects became more embedded in human consciousness, people's portrayal of insects shifted more toward an appreciation and acknowledgment of these remarkable creatures and what they can teach humanity.

Author A.S. Packard muses about the relationship between arthropods and people in his book *Half Hours with Insects*:

> Insects are related to us in a thousand ways, and somehow, either by themselves or through their products, they are, more than we should at first imagine, constantly in our thoughts. ... How life to many would be glorified, and this world seem a brighter one, if the flea, the louse and the bed-bug were removed from its surface? How much difference would it make to the world of insects were man to be blotted out of existence? ... Imagine the eons and eons of measured eternity through which by day and night the busy hum of insect life has risen and fallen with no human ear to listen. But on the other hand blot out those busy servants of ours, ... and the world would be a poor home .... [People] would be forced then, if never before, to appreciate the place of man in creation, and would be taught that as an animal his life touches at many points the lives and interests of the humble creatures about him.
>
> (Packard 1881: 66–7)

NOTES

Introduction

1. The US held the greatest number of black slaves in the New World at the start of its Civil War but had imported less than 4 percent of the total number of imported slaves (Eltis and Richardson 2010: xix). President Lincoln's Emancipation Proclamation (1863) did not abolish slavery; that occurred in 1865 with ratification of the thirteenth amendment to the US Constitution.
2. Farming in Britain had traditionally been "open field," with villagers cooperatively using unfenced areas of common land to graze animals and gather fuel. But beginning in the seventeenth century and continuing into the nineteenth century, Parliament authorized a series of enclosure acts that reallocated and privatized these areas (particularly in the Midlands). Implementation of these acts changed the countryside dramatically, as landscapes that had once resembled a crazy quilt of open areas became more regularly shaped, with fewer woods and footpaths. The government argued that enclosure of the common fields would improve agricultural efficiency and productivity. The wealthy could take advantage of this system, but tenant farmers and others of little means could ill afford to purchase land, enclose it with fences or hedges, and pay tithes on it, impelling many to move to towns and cities for employment (Mokyr 2009). Mount (2013) provides a Baedeker-worthy account of the enclosure acts and their lasting impact on England's countryside.
3. National Army Museum. Available online: https://www.nam.ac.uk/explore/armies-east-india-company (accessed October 15, 2022).
4. The Company also traded illegally in opium, grown in India and smuggled into China, a practice that culminated in the mid-nineteenth-century Opium Wars and impacted the global silk trade (see Chapter 3) (Cannadine 2017; Dalrymple 2019).
5. Today the Royal Botanic Gardens at Kew possess the largest and most diverse plant and fungi collections in the world. Among its approximately seven million

herbarium specimens are a fern collected by Charles Darwin from the Galápagos and palms collected by Alfred Russel Wallace in the Amazon (Leydon 2013).

6. Two points bear mentioning. First, life expectancy data are affected by infant and child mortality rates (both of which were relatively high at the time) and life expectancies in Europe also approximated forty years. Second, between 1830 and 1860, life expectancy differences in the US also were greater for rural versus urban residents. Thus in Boston, considered among the healthiest of US cities, life expectancies for both sexes were in the upper thirties, compared with the mid-forties for residents of rural Massachusetts (Klein 2012).
7. Until the late nineteenth century, all university fellows and many school masters were required to take holy orders.
8. "Rotten boroughs" were districts with very few voters but the same voting power as more populous districts.
9. As the group spread across England to disparate areas, the purpose of the movement expanded to include opposition to corruption, wage cuts, and unemployment and grievances against the Church of England that led to demands for separation of Church and State (Charlton 1997).
10. When the movement failed, Andrew Carnegie's father and maternal grandfather, who were Chartists, moved the family from Scotland to the US, where Andrew became a steel industrialist and eminent philanthropist. Available online: https://www.carnegie.org/interactives/foundersstory/#!/ (accessed December 6, 2021).
11. In 1928, all women over age twenty-one were permitted to vote, thereby achieving parity with men at the ballot box.
12. Peel resigned as Prime Minister in the aftermath and fallout of the vote. It is not entirely clear that repeal of the Corn Laws had the desired impact on Ireland or that the Corn Laws were causative of the famine.
13. Van Gogh's letters from his London days frequently reference these British women authors and express his admiration for their literary depictions of everyday life.
14. Reverend William Kirby, Tory High Church clergyman, authored the seventh *Bridgewater Treatise* (1835). Prone to moralizing, Kirby averred that human biting and sucking insects were God's way of punishing "Fallen" man (Topham 1992; Clark 2009).
15. On display from Spitalfields weavers at London's East End was an array of woven silk (e.g., velvet, satin, *gros de Naples*, brocade, damask, and moiré). From the shop of Josiah Wedgwood, Darwin's maternal grandfather, one could admire fine pottery. Shockingly to modern readers, from Birmingham manufacturers for export to the southern US one could view leg-irons, shackles, and manacles used to restrain slaves (Hobhouse 1950).
16. Building the Museum. Available online: https://www.vam.ac.uk/articles/building-the-museum#slideshow=31131014&slide=0 (accessed December 7, 2021).
17. *A Brief History of the Science Museum*. Available online: https://www.sciencemuseum.org.uk/sites/default/files/2017-10/science-museum-history.pdf (accessed December 7, 2021).

18. Nathaniel (Charles) Rothschild (1877–1923). Available online: https://family.rothschildarchive.org/people/104-nathaniel-charles-rothschild-1877-1923 (accessed December 7, 2021).
19. Miriam Louisa Rothschild (1908–2005). Available online: https://family.rothschildarchive.org/people/122-miriam-louisa-rothschild-1908-2005 (accessed December 7, 2021).
20. An abbreviated list includes the Geological Society of London (1804), Royal Astronomical Society (1820), Royal Meteorological Society (1850), Royal Society for Horticulture (1861), Linnaean Society of London (1870), Royal Anthropological Institute (1871).
21. Scott's *Terra Nova* expedition aimed to advance scientific knowledge across multiple disciplines in contrast to Amundsen's expedition, whose goal was to reach the South Pole first (Martin 2012).
22. *British Antarctic ("Terra Nova") Expedition (1910–1913)*. Available online: https://onlinebooks.library.upenn.edu/webbin/book/lookupname?&key=British%20Antarctic%20%28%22Terra%20Nova%22%29%20Expedition%20%281910%2D1913%29&c=x (accessed January 14, 2022).
23. Both Darwin and Wallace's exotic voyages were inspired by Alexander von Humboldt's highly popular *Personal Narrative of Travels in South America*, a travelogue of his explorations in the Canary Islands and equatorial South America (Barlow 1958; Fagan 2008).
24. Chloroform, when available, was widely used as a surgical anesthetic during the US Civil War.
25. The unwillingness to admit women to the practice of medicine extended to the push to establish a legislative mandate for midwives in 1870. Those advancing the cause were vilified as "absurd," "time-wasters," and an "objectionable body of busy-bodies" (Worth 2002). The physicians objected not only to the professionalization of women practitioners but the potential loss of revenue they might suffer.
26. The Louisiana Territory, 827,000 square miles of land west of the Mississippi River, today comprises fifteen US states and two Canadian provinces.
27. The Homestead Act of 1862 was an articulation of the "free soil" policy or the idea that farms should be operated by individual farmers rather than by Southern slave-owners who bought large tracts and land and appropriated the labor of Africans to run them. The Homestead Act of 1873, also known as the Timber Culture Act, required land recipients of 160 acres to grow trees for timber. The National Stock Raising Homestead Act of 1916 required land recipients of 640 acres to raise livestock for consumption.
28. In 1846 Philadelphia was the only US city with a library, university, museum, and academy of natural sciences (Barber 1980).
29. Say and his future wife, Lucy, sailed down the Ohio River on the "Boatload of Knowledge" in 1826 to New Harmony, Indiana, where they founded a utopian society with Robert Owen and others. Owen, sometimes called the "father of British socialism," had emigrated from Wales, where he operated the largest cotton-spinning mill in England and established many worker reforms. Owing

to his activism, more humane regulations for mill workers were included in the British Factory Act of 1819, but decades would pass before his reforms would be instituted. Available online: https://www.bl.uk/people/robert-owen (accessed June 5, 2023).

30. John Eatton Le Conte, father of famed nineteenth-century Coleopterist John L. Le Conte, raised Say's hackles by shipping specimens off to Europe for identification and, at times, scientific description.

31. Kirby and Spence's *Introduction to Entomology* went through several editions and enjoyed wide usage throughout the nineteenth century (Allen 1976).

32. In 1860, 350,000 farmers in the Northeast subscribed to at least one agricultural journal, although elsewhere, access to the information could be limited by location, illiteracy, and lack of money (Hurt 2002). Periodicals with entomology as one focus included *American Farmer*, first published in 1819, followed in the 1830s and 1840s by *American Agriculturist*, *Ohio Cultivator*, and *Genesse Farmer*, to name a few. The *Practical Entomologist* and *American Entomologist* (later, *American Entomologist and Botanist*) appeared in the 1860s–1870s and were largely focused on insect pest problems and recommended solutions.

33. In 1876 Congress appropriated $18,000 to the Commission, which included A.S. Packard, Jr. and C. Thomas. The next year another $20,000 were appropriated. The Commission's reports included recipes to prepare grasshoppers as human food (see Chapter 3).

34. E.B. Lewis, 1998, "Thomas Hunt Morgan and His Legacy." Available online: www.nobelprize.org/prizes/medicine/1933/morgan/article (accessed December 7, 2021).

35. Lucy Say, first woman elected to the Academy of Natural Sciences, was both artist and scientific illustrator of her husband's works.

36. W.B. Tegetmeier, naturalist and editor of the British publication the *Field*, was a pigeon-fancier who also corresponded with Darwin.

37. UDHS Citizenship and Immigration Services. Available online: https://www.uscis.gov/about-us/our-history/overview-of-ins-history/early-american-immigration-policies (accessed December 7, 2021).

38. US Citizenship and Immigration Services. Available online: https://www.uscis.gov/about-us/our-history/mass-immigration-and-wwi (accessed December 7, 2021).

39. Carnegie stated that the man who dies rich dies disgraced. He founded the renowned scientific research institution that bears his name, 2,500 public libraries worldwide, and established the famed eponymous New York City concert hall.

40. An incomplete list includes C.V. Riley, B.D. Walsh, H.A. Hagen, O. Lugger, T. Pergande, J.R. de la Torre-Bueno, E.A. Schwarz, H. Ulke, H. Edwards, A.R. Grote, A. Koebele, T.D.A. Cockerell, and C.R. Osten Sacken (Mallis 1971).

41. US Centers for Disease Control and Prevention, 2018. Available online: https://youtu.be/U6Ccdk5wPvk (accessed December 7, 2021).

Chapter 2

1. Glover was originally hired in 1854 as the first Federal entomologist in the US Patent Office.

Chapter 3

1. The main insect harvested around the Great Salt Lake was a katydid, the Mormon cricket (*Anabrus simplex*). Individuals harvested food insects in early morning by picking insects from grass before flight muscles operate (Downs 1966). The main industrial scale harvesting of these food insects in 1800s included digging a hole 4–5 feet deep, 10–12 feet diameter and beating the grass with sagebrush (*Artemisia tridentata*) branches. People stood twenty feet apart walking toward the hole (Chittenden and Richardson 1905). With fire, two methods typically were used for orthopterans. A shallow pit was dug in a meadow with fire set on both sides of the pit. Wings burned off, making insect collection easy (Dixon 1903). Great Basin Native Americans built a large fire, made a bed of coals, and walked in circle toward the coals (Aginsky 1943).
2. There was a fifth component as well, the maize mushroom that also supplied nutrients. The Native Americans called this fungus an attractive name, huitlacoche, but the European immigrants not knowing its health-giving properties called it by a derogatory name, corn smut (Aydoğdu and Gölükçü 2017).

Chapter 4

1. See Berenbaum (1995), Waldbauer (2009), and Demillo (2020) for excellent historical overviews of sericulture and the silk industry.
2. To obtain 1 pound of raw silk, about 150–225 pounds of fresh leaves are needed to feed two thousand to three thousand silkworms.
3. Likely half or more of China's total production was consumed in-country.
4. Japan's silkworm eggs were free of *pebrine*, and during the mid- to late 1860s, Japan became the sole supplier of eggs for Mediterranean sericulture. In the late 1880s, Japan began producing improved hybrid varieties of silkworm eggs (Federico 1997).
5. Vanterpool (2010: 66), quote from *Syracuse New York Herald,* October 24, 1928.
6. With technological advances in transoceanic steamers and rail transport, total transit time for raw silk from Asia to New York was less than two weeks (Field et al. 2007).
7. At the time, British capital also traveled westward in pursuit of profitable investment abroad (Margrave 1975).
8. Population Reference Bureau, "Trends in Migration to the U.S." Available online: https://www.prb.org/resources/trends-in-migration-to-the-u-s/ (accessed May 22, 2023).
9. Congressional Research Service Report, "The Fair Labor Standards Act (FLSA) Child Labor Provisions," June 29, 2016. Available online: https://fas.org/sgp/crs/misc/R44548.pdf (accessed May 22, 2023).
10. Images of Union Cases containing photographs, Science Museum Group, London. Available online: https://collection.sciencemuseumgroup.org.uk/objects/co10499/octagonal-shellac-union-case-1851-union-cases-shellac-ambrotypes (accessed May 22, 2023).

Chapter 7

1. See Bown (2001) and Clark (2009).
2. See Darwin (2010: 6).
3. Charles Darwin to Katherine Murray Lyell. See Evans (2017: 112).
4. Recalled by M.C. Derby in a letter to Darwin. See Evans (2017: 11).
5. See Allen (1976: 123).
6. Though literary entomologists were predominately male there were a number of female Victorian entomologists, including: Alice Blanche Balfour (1850–1936), Mary Ball (1812–1898), Emma Hutchinson (1820–1909), Eleanor Ormerod (1828–1901), and Emily Shore (1819–1839). See my forthcoming thesis *"In my Mind's Cabinet": Epistemology, Emotion and the Impact of Natural History on Victorian Literature and Culture 1820–1880* for a further exploration of gender and natural history.
7. See Wallace (1869) and Darwin Correspondence Project (1854) (n.d.b).
8. See Anne Secord's extensive scholarship (1994, 2015) on working-class 'artisan naturalists' in the early nineteenth century.
9. Pages for "The Wasp in the Wig" refer to Carroll (2000).
10. See Carroll (2000: 294).

BIBLIOGRAPHY

Adlington, W. [1566] 1947. *Apuleius The Golden Ass: Being the Metamorphosis of Lucius Apuleius*. London: William Heinemann Ltd.

Agassiz, L. 1848. *Nomenclatoris zoologici, index universalis*. Soloduri: Jent et Gassmann.

Aginsky, B. 1943. "Culture Element Distributions: XXIV-Central Sierra." *Anthropological Records*, 8(4): 393–468. Berkeley: University of California Press.

Akre, R.D., Hansen, L.D., and Zack, R.S. 1991. "Insect Jewelry." *American Entomologist* 37(2): 90–5.

Alexander, A. and Dunkel, F.V. 2017. "Local Malaria Elimination: A Historical Perspective from Palestine 100 Years Ago Informs the Current Way Forward in sub-Saharan Africa." *American Entomologist* 63(4): E1–E14.

Allen, D.E. 1976. *The Naturalist in Britain: A Social History*. Princeton: Princeton University Press.

Allen, F. 1902. "The Silk Industry in America, June 1st, 1900." In United States Census Office, 12th census. Available online: https://www.google.com/books/edition/The_Silk_Industry_of_America_June_1st_19/1FznAAAAMAAJ?hl=en&gbpv=1&dq=inauthor:%22United+States.+Census+Office.+12th+census,+1900%22&printsec=frontcover (accessed December 2, 2021).

A.L.O.E. [C.M. Tucker]. 1893. *Fairy Frisket; or, Peeps at Insect Life*. London: Thomas Nelson and Son.

Anderson, B.C. 2015. "Evidence of Cochineal's Use in Painting." *The Journal of Interdisciplinary History* 45(3): 337–66.

Anelli, C.M. 2006. "Darwinian Theory in Historical Context and its Defense by BD Walsh: What is Past is Prologue." *American Entomologist* 52(1): 11–19.

Anelli, C.M. and Prischmann-Voldseth, D.A. 2009. "Using Beeswax and Cochineal Dye: An Interdisciplinary Approach to Teaching Entomology." *American Entomologist* 55(2): 95–105.

Anonymous. 1842. "Kirby and Spence's Entomology." *Gardeners Chronicle* 2: 808.

Anonymous. [1851] 1970. *The Crystal Palace Exhibition. Illustrated Catalogue, London 1851*. New York: Dover Publications.

Anonymous. 1859. "Italian Bees." *American Agriculturalist* 18: 346.

Appleman, P. 2001. "Darwin on Changing the Mind." In P. Appleman (ed.), *Darwin. A Norton Critical Edition*, 3rd edn., 3–20 New York: W.W. Norton.

Armstrong, D. 2012. *Science and Religion: Close Partners or Mortal Enemies?* Morrisville, NC: Lulu Press.

Armstrong, I. 2008. *Victorian Glassworlds: Glass Culture and the Imagination 1830–1880*. Oxford: Oxford University Press.

Armstrong, P. 2000. *The English Parson-Naturalist A Companionship Between Science and Religion*. Leominster: Gracewing.

Arnold, M. [1867] 1994. "Dover Beach." In C. Ward (ed.), *Dover Beach and Other Poems*. Mineola: Dover Publications.

Auerbach, J.A. 1999. *The Great Exhibition of 1851: A Nation on Display*. New Haven: Yale University Press.

Axtell, J. 1979. *White Indians of Colonial America*. Fairfield: Ye Galleon Press.

Axtell, J. 1981. *The European and the Indian: Essays in the Ethnohistory of Colonial America*. New York: Oxford University Press.

Aydoğdu, M. and Gölükçü, M. 2017. "Nutritional Value of Huitlacoche, Maize Mushroom Caused by *Ustilago maydis*." *Food Science and Technology* 37(4): 531–5. Available online: https://doi.org/10.1590/1678-457X.19416 (accessed October 12, 2022).

Baer, W.S. 2011. "The Classic: The Treatment of Chronic Osteomyelitis with the Maggot (Larva of the Blow Fly)." *Clinical Orthopaedics and Related Research* 469: 920–44. Available online: https://doi.org/10.1007/s11999-010-1416-3 (accessed February 20, 2019).

Ball, G.E. 2005. *The Art of Insect Illustration and Threads of Entomological History*. Edmonton: The University of Alberta Press.

Bancroft, H.H. 1889. *The Works of Hubert Howe Bancroft*, vol. XXVI, *History of Utah, 1540–1886*. San Francisco: The History Company.

Barber, L. 1980. *The Heyday of Natural History, 1820–1870*. Garden City: Doubleday & Company, Inc.

Barlow, N., ed. 1958. *The Autobiography of Charles Darwin, 1809–1882*. London: Collins.

Barnes, J.K. 1985. "Insects in the New Nation: A Cultural Context for the Emergence of American Entomology." *Bulletin of the Entomological Society of America* 31(1): 21–30.

Barnes, J.K. 1988. *Asa Fitch and the Emergence of American Entomology: With an Entomological Bibliography and a Catalog of Taxonomic Names and Type Specimens*. Albany: University of the State of New York, State Education Department, Biological Survey.

Barr, J. 1985. "Why the World was Created in 4004 BC: Archbishop Ussher and Biblical Chronology." *Bulletin of the John Rylands University Library of Manchester Manchester* 67(2): 575–608.

Barraclough, G., ed. 1982. *The Times Concise Atlas of World History*. Maplewood: Hammond, Inc.

Barrett, A.D.T. and Higgs, S. 2007. "Yellow Fever: A Disease That Has Yet to be Conquered." *Annual Review of Entomology* 52: 209–29.

Barrie, J.M. 1906. *Peter Pan in Kensington Gardens*. London: Hodder & Stoughton.

Bastian, F. 1965. "Defoe's Journal of the Plague Year Reconsidered." *The Review of English Studies* 16(62): 151–73.

Bates, H.W. 1863. *The Naturalist on the Amazons*. London: George Routledge & Sons, Limited.
Beardsley, A. 1904. *Under the Hill*. London: John Lane.
Becker, L. 1883. *Alphabet des Insectes*. Paris: J. Hetzel & Cie.
Bellanca, M.E. 2007. *Daybooks of Discovery: Nature Diaries in Britain, 1770–1870*. Charlottesville: University of Virginia Press.
Bennett, S.M. 1984–1985. "A Newly Discovered Blake at the Huntington." *Blake/An Illustrated Quarterly* 18(3): 132–9.
Berenbaum, M.R. 1995. *Bugs in the System: Insects and Their Impact on Human Affairs*. Reading: Addison-Wesley Publishing Company.
Berg, P.K. 2016. *Nineteenth Century Photographic Cases and Wall Frames*, 3rd edn. Leavenworth: Paul K. Berg.
Berger, J. [1980] 1991. *About Looking*. New York: Vintage.
Berlocher, S.H. and Feder, J.L. 2002. "Sympatric Speciation in Phytophagous Insects: Moving Beyond Controversy?" *Annual Review of Entomology* 47(1): 773–815.
Bethune, B. 2006. "The Back Pages." *Maclean's* (October 16).
Bettcher, C.W. 1918. "C.A.C.'s or Coast Artillery Cooties." *Liason* 1(23): 223.
Beughhold, A. 2013. "'Tis Not Unlikely Fancy Will Betray You Into Error': Das Museum of Jurassic Technology und der Verfremdungseffekt." *Andererseits: Yearbook of Transatlantic German Studies* 3: 27–38.
Bevan, E. 1827. *The Honey-Bee; Its Natural History, Physiology and Management*. London: Baldwin, Cradock and Joy.
Billings, J.D. 1888. *Hardtack and Coffee*. Boston: George M. Smith & Co.
Blanchard, R. 1889. "De la nomenclature de êtres organis's." *Compte-Rendu des Séances du Congrès international de Zoologie, Paris, 1889*: 333–404.
Bloom, H., ed. 1986. *Thomas Carlyle: Modern Critical Views*. New York: Chelsea House Publishers.
Bodenheimer, F.S. 1951. *Insects as Human Food*. The Hague: W. Junk.
Boisduval, J.B.A. 1832. *Voyage de découvertes de l'Astrolabe. Exécuté par ordre du Roi, pendant les années 1826–1827–1828–1829, sous le commendement de M. J. Dumont D'Urville. Faune entomologique de l'Océan Pacifique* .... Part 1. *Lépidoptères*. Paris: J. Tastu.
Boisduval, J.B.A. 1835. *Voyage de découvertes de l'Astrolabe. Exécuté par ordre du Roi, pendant les années 1826–1827–1828–1829, sous le commendement de M. J. Dumont D'Urville. Faune entomologique de l'Océan Pacifique* .... Part 2. *Coléoptères et autres ordres*. Paris: J. Tastu.
Bousquet, Y. and Bouchard, P. 2013a. "The Genera in the Second Catalogue (1833–1836) of Dejean's Coleoptera Collection." *Zookeys* 282: 1–219.
Bousquet, Y. and Bouchard, P. 2013b. "The Genera in the Third Catalogue (1836–1837) of Dejean's Coleoptera Collection." *Zookeys* 282: 221–39.
Bowen, F. 1845. *A Theory of Creation: A Review of Vestiges of the Natural History of Creation*. Boston: Otis, Broaders and Company.
Bown, N. 2001. *Fairies in Nineteenth-Century Art and Literature*. Cambridge: Cambridge University Press.
Brabin, B.J. 2014. "Malaria's Contribution to World War One – the Unexpected Adversary." *Malaria Journal* 13(497): 1–22.
Braby, M. and Olsen, P. 2011. *A Flutter of Butterflies*. Canberra: National Library of Australia.

Brewer, L.A. 2004. "Natural Fashions: An Introduction to the Animal Products Industry in London via the Caricatures of Edward Linley Sambourne for *Punch*, 1867–1874." MA thesis, S.U.N.Y. Fashion Institute of Technology, New York.
Brier, B. 2013. *Egyptomania: Our Three Thousand Year Obsession with the Land of the Pharaohs*. New York: Palgrave Macmillian.
Brock, W.H. 1966. "The Selection of the Authors of the Bridgewater Treatises." *Notes and Records of the Royal Society of London* 21(2): 162–79.
Brontë, A. [1848] 1996. *The Tenant of Wildfell Hall*. London: Wordsworth Editions.
Brontë, C. [1849] 1993. *Shirley*. London: Wordsworth Editions.
Brooke, J.H. 1991. *Science and Religion: Some Historical Perspectives*. Cambridge: Cambridge University Press.
Brooks, S. 1966. *Civil War Medicine*. Springfield: Charles C. Thomas.
Brown, W.H. 1919. *Stories of the Great War for Public Speakers*. Cincinnati: Standard Publishing, Company.
Browne, J. 2006. *Darwin's Origin of Species: A Biography*. New York: Grove Press.
Browne, J. 2010. "Asa Gray and Charles Darwin: Corresponding Naturalists." *Harvard Papers in Botany* 15(2): 209–20.
Browne, J.H. 1866. *Four Years in Secessia*. Detroit: W.H. Davis.
Brullé, A. 1832. *Expédition scientifique de Morée. Section des Sciences Physiques. Tome III, 1.re partie. Zoologie. Deuxième section – Des animaux articulés*. Paris: F.G. Levrault.
Buckley, A. [1880] 1917. *Life and Her Children: Glimpses of Animal Life from the Amoeba to the Insects*. London: Macmillan.
Budgen, L.M. [writing as Acheta domestica]. 1849. *Episodes of Insect Life*. London: Reeve, Benham, and Reeve.
Budgen, L.M. [writing as Acheta domestica]. 1851a. *Episodes of Insect Life*, Second Series. New York: J.S. Redfield, Clinton-Hall.
Budgen, L.M. [writing as Acheta domestica]. 1851b. *Episodes of Insect Life*, Third Series. London: Reeve and Benham.
Budgen, L.M. [writing as Acheta domestica] and Wood, J.G., ed. 1867. *Episodes of Insect Life*. London: Bell and Daldy.
Budgen, L.M. 1879. *Episodes of Insect Life*. London: G. Bell and Sons.
Burkhardt, G.S., ed. 2009. *Double Duty in the Civil War: The Letters of Sailor and Soldier Edward W. Bacon*. Carbondale: Southern Illinois University Press.
Burleigh, N. 2007. *Mirage: Napoleon's Scientists and the Unveiling of Egypt*. New York: Harper Collins.
Burt, J. 2001. "The Illumination of the Animal Kingdom: The Role of Light and Electricity in Animal Representation." *Society and Animals* 9(3): 203–28.
Butler, T. 1994. "*Yersinia* infections: Centennial of the Discovery of the Plague Bacillus." *Clinical Infectious Diseases* 19(4): 655–61.
Butlin, M. 2015. "Blake, Linnell and Varley and *A Treatise on Zodiacal Physiognomy*." In D. Dethloff, T. Murdoch, K. Sloan, and C. Elam (eds.), *Burning Bright: Essays in Honour of David Bindman,* 126–35. London: UCL Press.
Cagidemetrio, A. 2010. "Henry James and the American Evolution of the Snob." In S. Beckert and J.B. Rosenbaum (eds.), *The American Bourgeoisie: Distinction and Identity in the Nineteenth Century*, 45–61. New York: Palgrave Macmillan.
Cannadine, D. 2017. *Victorious Century: The United Kingdom, 1800–1906*. New York: Viking.
Cardon, D. 2007. *Natural Dyes: Sources, Tradition, Technology and Science*. London: Archetype Publications.

Carpenter, W.M. 1833. *Scripture Natural History: Containing a Descriptive Account of the Quadrupeds, Birds, Fishes, Insects, Reptiles, Serpents, Plants, Trees, Minerals, Gems, and Precious Stones, Mentioned in the Bible*. Boston: Lincoln, Edmands & Co.

Carroll, L. 1867. *Alice's Adventures in Wonderland*. London: Macmillan and Co.

Carroll, L. [1890] 2015. *The Nursery Alice*. London: Pan MacMillan.

Carroll, L. 2000. *The Annotated Alice: The Definitive Edition*. Ed. M. Gardner. New York: Norton.

Carroll, N. 1990. *The Philosophy of Horror, or, Paradoxes of the Heart*. New York: Routledge.

Casto, M. 2013. "Nineteenth-Century Nature Crafts." *Southeastern College Art Conference Review* 16(3): 273–92.

Cavendish, M. 2003. *Insects and Spiders of the World*. Singapore: Marshall Cavendish Corporation.

CDC (Centers for Disease Control and Prevention). 2015. "Symptoms, Diagnosis, & Treatment." Available online: https://www.cdc.gov/yellowfever/symptoms/index.html (accessed March 7, 2018).

CDC (Centers for Disease Control and Prevention). 2016. "Yellow Fever Virus." Available online: https://www.cdc.gov/yellowfever/index.html (accessed March 7, 2018).

Chalfant, W. 1933. *The Story of Inyo*, 2nd edn. Bishop: Chalfant Press.

Chapple, J.A.V. and Pollard, A., eds. 1997. "Gaskell to George Smith, 3 May 1864." In *The Letters of Mrs. Gaskell*. Manchester: Manchester University Press.

Charlton, J. 1997. *The Chartists: The First National Workers Movement*. London: Pluto Press.

Chauzat, M.-P, Cauquil L., Roy L., Franco S., Hendrikx P., and Ribière-Chabert, M. 2013. "Demographics of the European Apicultural Industry." *PLoS ONE* 8(11): e79018. Available online: https://doi.org/10.1371/journal.pone.0079018 (accessed October 14, 2022).

Chittenden, H.M. and Richardson, A.T. 1905. *Life, Letters and Travels of Father Pierre-Jean De Smet, SJ., 1801–1873*. New York: Francis P. Harper.

Christopher, C.J. and Gill, G.C. 2005. "Nightingale in Scutari: Her Legacy Reexamined." *Clinical Infectious Diseases* 40(12): 1799–805.

Church, A.H. 1908. *Types of Floral Mechanism: A Selection of Diagrams and Descriptions of Common Flowers*. Oxford: Clarendon Press.

Clark, J.F.M. 2006. "History from the Ground Up: Bugs, Political Economy and God in Kirby and Spence's Introduction to Entomology (1815–1856)." *Isis* 97(1): 28–55.

Clark, J.F.M. 2009. *Bugs and the Victorians*. New Haven: Yale University Press.

Clark, R.W. 1984. *The Survival of Charles Darwin: A Biography of a Man and an Idea*. New York: Random House.

Clark, S., Carter, T., Cowell, B., and Carter, P., eds. 2012. "Dying Light: Restoring Tinted Films. Part One: Toning and Tinting Film, October 2012." National Film and Sound Archive of Australia. Available online: https://www.nfsa.gov.au/latest/dying-light-1 (accessed April 14, 2018).

Clodfelter, M. 2017. *Warfare and Armed Conflicts: A Statistical Encyclopedia of Casualty and Other Figures, 1492–2015*, 4th edn. Jefferson: McFarland & Co., Inc.

Cloudsley-Thompson, J.L. 1976. *Insects and History*. London: Weidenfeld and Nicholson.

Coates, P. 2011. "Xenophobia." In D. Simberloff and M. Rejmanek (eds.), *Encyclopedia of Biological Invasions*, 705–8. Berkeley: University of California Press.
Cohen, M. 1979. *The Letters of Lewis Carroll: ca. 1837–1885*. Oxford: Oxford University Press.
Cohen, M.N. and Wakeling, E., eds. 2003. *Lewis Carroll & His Illustrators: Collaborations and Correspondence, 1865–1898*. Ithaca: Cornell University Press.
Colebrook, L. 1948. "Almroth Edward Wright. 1861–1947." *Biographical Memoirs of Fellows of the Royal Society* 6(17): 297–314.
Committee of General Literature and Education. 1834. *Insects and their Habitations: A Book for Children*. London: John W. Parker.
Comstock, A.B. 1905. *How to Keep Bees*. New York: Doubleday, Page and Company.
Comuni italiani. 2011. Cacie' punt. Available online: www.comuni-italiani.it. Archived from the original 2021, http://cheeseymarie.com/2014/07/ (acccessed September 5, 2021).
Conan Doyle, A. 1900. *The Great Boer War*. London: Smith, Elder & Co.
Conniff, R. 2010. *The Species Seekers: Heroes, Fools, and the Mad Pursuit of Life on Earth*. New York: W.W. Norton & Company.
Cook, A.J. 1894. *The Bee-Keepers Guide or Manual of the Apiary*, 15th edn. Chicago: G.W. York and Co.
Cook, L.M., Grant, B.S., Saccheri, I.J., and Mallet, J. 2012. "Selective Bird Predation of the Peppered Moth: The Last Experiment of Michael Majerus." *Biology Letters* 8: 609–12.
Cook, T., and Son. 1876. *Cook's Tourists' Handbook for Palestine and Syria*. London: Simpkin, Marshall & Co.
Costa Neto, E. and Dunkel, F. 2016. "Insects as food: History, Culture, and Modern Use Around the World." In A. Dossey, J. Morales-Ramos, and G. Rojas (eds.), *Insects as Sustainable Food Ingredients: Production, Processing, and Applications*, ch. 2. San Diego: Elsevier Publ. Co. Academic Press.
Coues, E. 1890. *Handbook of Field and General Ornithology: A Manual of the Structure and Classification of Birds*. London: Macmillan.
Cowan, F. 1865. *Curious Facts in the History of Insects; Including Spiders and Scorpions*. Philadelphia: J.B. Lippencott & Co.
Cox, W.F., Jr. 2004. *Tyranny Through Public Education*. Jackson, WY: Allegiance Press.
Coyne, J.A. 2010. *Why Evolution is True*. New York: Penguin Books.
Crafts, N. 2011. "Explaining the First Industrial Revolution: Two Views." *European Review of Economic History* 15(1): 153–68.
Crane, E. 1976. "History of Honey." In E. Crane (ed.), *Honey, A Comprehensive Survey*, 439–488. London: Heinemann.
Creese, M.R.S. 1998. *Ladies in the Laboratory? American and British Women in Science, 1800–1900: A Survey of their Contributions to Research*. Lanham: The Scarecrow Press.
Croft, L.R. 1993. "P.H. Gosse in Newfoundland and Lower Canada, 1827–1838." *Archives of Natural History* 20(1): 1–29.
Crosby, M.C. 2006. *The American Plague: The Untold Story of Yellow Fever, the Epidemic that Shaped Our History*. New York: Berkeley Books.
Crouzet, F. 1990. *Britain Ascendant: Comparative Studies in Franco-British Economic History*. Cambridge: Cambridge University Press.

Dadant, C.P. 1915. *Langstroth on the Hive & Honey Bee*. Hamilton: Dadant & sons. Available online: https://hdl.handle.net/2027/uiug.30112074927671 (accessed October 14, 2022).
Dalrymple, W. 2019. *The Anarchy: The Relentless Rise of the East India Company*. New York: Bloomsbury Publishing.
Damayanti, D., Jaceldo-Siegl, K., Beeson, W.L., Fraser, G., Oda, K., and Haddah, E.H. 2018. "Foods and Supplements Associated with Vitamin B12 Biomarkers Among Vegetarian and Non-Vegetarian Participants of the Adventist Health Study (AHS-2) Calibration Study." *Nutrients* 10(6): 722.
Damkjær, M. 2016. *Domesticity and Print Culture in Nineteenth-Century Britain*. London: Palsgrave Macmillian. Available online: https://link.springer.com/book/10.1057/9781137542885 (accessed October 14, 2022).
Dapson, R.W. 2009. "The History, Chemistry and Modes of Action of Carmine and Related Dyes." *Biotechnic & Histochemistry* 82: 173–87.
Darwin, C. 1839. *Narrative of the surveying voyages of His Majesty's Ships Adventure and Beagle between the years 1826 and 1836, describing their examination of the southern shores of South America and the Beagle's circumnavigation of the globe. Journal and Remarks. 1832–1836*. London: Henry Colburn. Available online: https://www.biodiversitylibrary.org/page/33517833 (accessed October 14, 2022).
Darwin, C. [1859] 2008. *On the Origin of Species*. Oxford: Oxford University Press.
Darwin, C. [1871] 2006. "The Descent of Man and Selection in Relation to Sex." In E.O. Wilson (ed.), *From So Simple a Beginning: The Four Great Books of Charles Darwin*, 767–1248. New York: W.W. Norton and Company.
Darwin, C. 1872. *The Expression of the Emotions in Man and Animals*. London: John Murray.
Darwin, C. 2010. *The Works of Charles Darwin*, vol. 29. Ed. Nora Barlow. New York: New York University Press.
Darwin, C. and Wallace, A. 1858. "On the Tendency of Species to Form Varieties; and on the Perpetuation of Varieties and Species by Natural Means of Selection." *Journal of the Proceedings of the Linnean Society of London. Zoology* 3(9): 5–62.
Darwin Correspondence Project (DCP). 1839. Letter no. 480, January 9. Available online: https://www.darwinproject.ac.uk/letter/?docId=letters/DCP-LETT-480.xml (accessed October 14, 2022).
Darwin Correspondence Project (DCP). 1854. Letter no. 1585, September. Available online: http://www.darwinproject.ac.uk/DCP-LETT-1585 (accessed January 31, 2018).
Darwin Correspondence Project (DCP). 1859. Letter no. 2532, November 22. Available online: https://www.darwinproject.ac.uk/letter/?docId=letters/DCP-LETT-2532.xml (accessed December 8, 2021).
Darwin Correspondence Project (DCP). 1860. Letter no. 2814, May 22. Available online: https://www.darwinproject.ac.uk/letter/?docId=letters/DCP-LETT-2814.xml (accessed October 14, 2022).
Darwin Correspondence Project (DCP). 1864. Letter no. 4663, November 7. Available online: https://www.darwinproject.ac.uk/letter/?docId=letters/DCP-LETT-4663.xml (accessed October 14, 2022).
Datta, R.K. and Nanavaty, M. 2007. *Global Silk Industry: A Complete Source Book*. Delhi: APH Publishing.
Daunton, M.J. 1995. *Progress and Poverty: An Economic and Social History of Britain, 1700–1850*. Oxford: Oxford University Press.

Davey, W. 1854. *The Illustrated Practical Mesmerist, Curative and Scientific*. Edinburgh: W.H. Lizars.

Dayrat, B. 2010. "Celebrating 250 Dynamic Years of Nomenclatural Debates." In A. Polaszek (ed.), *Systema Naturae 250, The Linnaean Ark*, 185–239. London: CRC Press.

D'Azevedo, W., ed. 1986. "Great Basin, Volume 11." In W. Sturtevant (ed.), *Handbook of North American Indians*. Washington, DC: Smithsonian Institution.

Debré, Pe. 1998. *Louis Pasteur*. Trans. Elborg Forster. Baltimore: The Johns Hopkins University Press.

Defoe, D. 1835. *A Journal of the Plague Year, or, Memorials of the Great Pestilence in London, in 1665*. London: Thomas Tegg and Son.

DeFoliart, G.R. 2002. "Insect Foods of North American North of Mexico Indigenous Populations." In G. DeFoliart, *The Human Use of Insects as a Food Resource*. Available online: https://insectsasfood.russell.wisc.edu/the-human-use-of-insects-as-a-food-resource/ (accessed November 13, 2021).

DeFoliart, G., Dunkel, F., and Gracer, D., eds. 2009. *Chronicle of a Changing Culture: The Food Insects Newsletter*. Salt Lake City: Aardvark Global Publishing.

Delaney, D.A., Meixner, M.D., Schiff, N.M., and Sheppard, W.S. 2009. "Genetic Characterization of Commercial Honey Bee (Hymenoptera: Apidae) Populations in the United States by Using Mitochondrial and Microsatellite Markers." *Annals of the Entomological Society of America* 102: 666–73.

Demillo, E.D. 2020. *The Butterfly Effect: Insects and the Making of the Modern World*. New York: Alfred A. Knopf.

Department of European Paintings. 2004. "Vincent van Gogh (1853–1890)." In *Heilbrunn Timeline of Art History*. New York: The Metropolitan Museum of Art. Available online: http://www.metmuseum.org/toah/hd/gogh/hd_gogh.htm (accessed May 22, 2023).

Depew, A.N. 1918. *Gunner Depew*. Chicago: Reilly & Britton Co.

Desmond, A. 2015. "Huxley, Thomas Henry." *Oxford Dictionary of National Biography*. Oxford: Oxford University Press.

Dessi, M.A. 2008. *Casu Marzu: la rivincita di un grande formaggio. Edizioni pubblicità Italiá. Numero 1*. Available online: https://www.pubblicitaitalia.com/carne/prodotti/eurocarni/2008/3/8001 (accessed July 4, 2023).

d'Houville, G. 1909. "Le Papillon Rouge." *L'Illustration: Journal Universel*, Numéro de Noel, v. 134, December 4.

Dick, E.N. 1970. *The Lure of Land: A Social History of the Public Lands from the Articles of Confederation to the New Deal*. Lincoln: University of Nebraska Press.

Dicke, M. 2000. "Insects in Western Art." *American Entomologist* 46(4): 228–37.

Dicke, M. 2004. "From Venice to Fabre: Insects in Western Art." *Proceedings of the Netherlands Entomological Society* 15: 9–14.

Dickens, C. 1845. *The Cricket on the Hearth: A Fairy Tale of Home*. London: Bradbury & Evans.

Dixon, R.B. 1903. "The Huntington California Expedition. The Northern Maidu." *Bulletin of the American Museum of Natural History* 17(part III): 183–4, 190.

Dobbs, D. 2011. "How Charles Darwin Seduced Asa Gray." *Wired Magazine*, April 28. Available online: https://www.wired.com/2011/04/how-charles-darwin-seduced-asa-gray/ (accessed May 18, 2020).

Dodd, A.P. 1927. "Notes on Parasitic Hymenoptera from Australia, with Descriptions of New Species." *Memoirs of the Queensland Museum* 9: 63–75.

Donkin, R.A. 1977. "Spanish Red: An Ethnogeographical Study of Cochineal and the *Opuntia* Cactus." *Transactions of the American Philosophical Society* 67(5): 1–84.

Donovan, E. 1842. *Natural History of the Insects of China*. London: Henry G. Bohn and John O. Westwood.

Doolittle, G.M. 1888. *Scientific Queen Rearing*. Hamilton: American Bee Journal.

Downs, J. 1966. *The Two Worlds of the Washo. An Indian Tribe of California and Nevada*. New York: Holt, Rinehart and Winston Inc.

Doyle, J.W.E. 1864. *A Chronicle of England, B.C. 55–A.D. 1485*. London: Longman.

Doyle, R. and Allingham, W. 1870. *In Fairyland: A Series of Pictures from the Elf-World*. London: Longman, Green, Reader and Dyer.

Dreyer, J.J. and Wehmeyer, A.S. 1982. "On the Nutritive Value of Mopanie Worms." *South African Journal of Science* 78: 33–5.

Dunkel, F. 2010. Personal communication, Montana State University student researching journals of his family during their arrival in the Great Salt Lake Valley.

Dunkel, F. 2017. *Incorporating Cultures' Role in the Food and Agricultural Sciences*. San Diego: Elsevier. Companion website available online: https://www.elsevier.com/books-and-journals/book-companion/9780128039557 (accessed May 22, 2023).

Dunning, J.W. 1879. "Frederick Smith." *The Entomologist* 12: 89–92.

Dupuis, C. 1974. "Pierre Andre Latreille (1762–1833): The Foremost Entomologist of his Time." *Annual Review of Entomology* 19: 1–14.

Du Quesne van Gogh, E. 1913. *Personal Recollections of Vincent van Gogh*. New York: Houghton Mifflin Company.

Dwyer, C. 2017. "The Grasshopper in the Van Gogh." NPR, November 8. Available online: https://www.npr.org/sections/thetwo-way/2017/11/08/562852372/the-grasshopper-in-the-van-gogh (accessed February 23, 2018).

Ealand, C.A. 1915. *Insects and Man*. New York: The Century Co.

Earp, C.A., ed. 2002. *Yellow Flag: The Civil War Journal of Surgeon's Steward C. Marion Dodson*. Baltimore: Maryland Historical Society.

Echard, J.-P., Bertrand, L., and von Bohlen, A. et al. 2010. "The Nature of the Extraordinary Finish of Stradivari's Instruments." *Angewandt Chemie International Edition* 49: 197–201.

Edinburgh Review, The. 1822. "History of Insects." June 31: 122–36.

Edwards, W.H. 1868–1872. *The Butterflies of North America*, series 1–3. Philadelphia: The American Entomological Society.

Eggleston, M.A. 2016. *The 5th Marine Regiment Devil Dogs in World War I: A History and Roster*. Jefferson: McFarland and Co., Inc.

Elias, S.A. 2014. "A Brief History of the Changing Occupations and Demographics of Coleopterists From the 18th Through the 20th Century." *Journal of the History of Biology* 47: 213–42.

Eliot, G. [1859] 1997. *Adam Bede*. London: Wordsworth Editions.

Eliot, G. [1860] 1994. *The Mill on the Floss*. Ed. C.T. Christ. New York and London: W.W. Norton & Co.

Eliot, G. [1861] 1994. *Silas Marner*. London: Wordsworth Editions.

Eliot, G. 1871. *Middlemarch: A Study of Provincial Life*. London: Blackwood and Sons.

Eliot, G. [1876] 2014. *Daniel Deronda*. Ed. G. Handley. Oxford: Oxford University Press.

Elliott, I.H. 1902. *History of the Thirty-Third Regiment Illinois Veteran Volunteer Infantry in the Civil War*. Gibson City: Regimental Association.

Eltis, D. and Richardson, D. 2010. *Atlas of the Transatlantic Slave Trade*. New Haven: Yale University Press. Introductory maps available online: https://www.slavevoyages.org/voyage/maps#introductory- (accessed December 6, 2021).

Empey, A.G. 1917. *Over the Top*. New York: G.P. Putnam's Sons.

Endersby, J. 2010. "Review of *Bugs and the Victorians*." *Reviews in History*. Available online: https://reviews.history.ac.uk/review/924 (accessed May 18, 2020).

Eng, R.Y.-Y. 1986. *Economic Imperialism in China: Silk Production and Exports 1861–1932*. Berkeley: Institute of East Asian Studies, University of California.

Essig, E.O. 1931. *A History of Entomology*. New York: The Macmillan Company.

Evans, S., ed. 2017. *Darwin and Women: A Selection of Letters*. Cambridge: Cambridge University Press.

Evenhuis, N.L. 1994. "The Publication and Dating of P.A. Wytsman's *Genera Insectorum*." *Archives of Natural History* 21: 49–66.

Evenhuis, N.L. 2018. "The Life and Work of Francis Walker (1809–1874)." *Fly Times*, Supplement 2: 1–101.

Fagan, M.B. 2008. "Theory and Practice in the Field: Wallace's Work in Natural History (1844–1858)." In C.H. Smith and G. Beccaloni (eds.), *Natural Selection and Beyond: The Intellectual Legacy of Alfred Russel Wallace*, 66–90. Oxford: Oxford University Press.

Fagan, M.M. 1918. "The Uses of Insect Galls." *The American Naturalist* 52: 155–76.

Farrow, E.S. 1918. *A Dictionary of Military Terms*. New York: Thomas Y. Crowell Co.

Fedchenko, A.P. 1877. "Journey in Turkestan, 2 (5). Formicidae." *Izvestiya Imperatorskago Obshchestva Lyubitelei Estestvoznaniya, Antropologii I Etnografii* 26: 1–20.

Federico, G. 1997. *An Economic History of the Silk Industry, 1830–1930*. Cambridge: Cambridge University Press.

Federico, G. 2016. "The Holy Grail of Economic History: Long-Run Economic Growth and the Industrial Revolution in Britain." *Rivista di Storia Economica* 32(1): 119–33.

Ferguson, G. 1954. *Signs & Symbols in Christian Art*. New York: Oxford University Press.

Ferrier, P.M., Rucker, R.R., Thurman, W.N., and Burgett, M. 2018. "Economic Effects and Responses to Changes in Honey Bee Health." ERR-246, United State Department of Agriculture, Economic Research Service. Available online: https://ageconsearch.umn.edu/record/276245/ (accessed January 21, 2022).

Field, J., Senechal, M., and Shaw, Ma. 2007. *American Silk, 1830–1930: Entrepreneurs and Artifacts*. Lubbock: Texas Tech University Press.

Fink, H.T. [1887] 1902. *Romantic Love and Personal Beauty*. London: Macmillan.

Fisher, S.W. and Meuti, M.E. 2018. "Contributions of Entomology to Natural Theology from the Late 17th to the Early 19th Century." *American Entomologist* 64(4): 242–50.

Fogg, G.E. 1992. *A History of Antarctic Science*. Cambridge: Cambridge University Press.

Forbes, P. 2009. *Dazzled and Deceived: Mimicry and Camouflage*. New Haven: Yale University Press.

Ford, E.B. 1945. *Butterflies*. London: Collins.

Forel, A. 1907. "Novo speco kaj novo gentonomo de formikoj." *Internacia Scienca Revu* 4: 144–5.

Forel, A. 1913. "Fourmis d'Argentine, du Brésil, du Guatémala & de Cuba." Reçues de MM. Bruch, Prof. v. Ihering, Mlle Baez, M. Peper et M. Rovereto. *Bulletin de la Société Vaudoise des Sciences Naturelles* 49: 203–50.

Formaggio marcetto. Available online: https://abruzzoturismo.it/en/pecorino-marcetto-or-cacio-marcetto-cheese (accessed November 13, 2021).
Formaggio saltarello. Available online: https://wn.com/formaggio_saltarello/wikipedia (accessed November 13, 2021).
Forsburg, L. 2015. "Nature's Invisibilia: The Victorian Microscope and the Miniature Fairy." *Victorian Studies* 57(4): 638–66.
Fowle, F. 2000a. "Richard Dadd, *The Fairy Feller's Master-Stroke*, 1855–64, Summary." TATE Britain, London. Available online: http://www.tate.org.uk/art/artworks/dadd-the-fairy-fellers-master-stroke-t00598 (accessed April 13, 2018).
Fowle, F. 2000b. "Our English Coasts, 1852 ('Strayed Sheep')." Available online: http://www.tate.org.uk/art/artworks/hunt-our-english-coasts-1852-strayed-sheep-n05665 (accessed April 10, 2018).
Frauenfelder, M. 2005. "Most Rotten Cheese." In *The World's Worst: A Guide to the Most Disgusting, Hideous, Inept, and Dangerous People, Places, and Things on Earth*, 21–23. San Francisco: Chronicle Books.
Frelinghuysen, A.C. 1999. *Louis Comfort Tiffany at the Metropolitan Museum of Art*. New York: The Metropolitan Museum of Art.
Freuchen, P. 1961. *Book of the Eskimos*. Canada: Ballantine Books.
Frith, J. 2012. "The History of Plague – Part 1. The Three Great Pandemics." *Journal of Military and Veterans Health* 20(2): 11–16.
Frost, S.W. 1937. "The Insect Motif in Art." *The Scientific Monthly* 44(1): 77–83.
Gabriel, J.H. 2000. *Micromosaics: The Gilbert Collection*. London: Philip Wilson Publishers.
Gage, K.L. and Kosoy, M.Y. 2005. "Natural History of Plague: Perspectives from More Than a Century of Research." *Annual Review of Entomology* 50: 505–28.
Gaskell, E. [1848] 1996. *Mary Barton*. London: Penguin Classics.
Gaskell, E. [1865] 1996. *Wives and Daughters*. London: Penguin Classics.
Gates, B. 2002. *In Nature's Name: An Anthology of Women's Writing and Illustration, 1780–1930*. Chicago: University of Chicago Press.
Geoffroy Saint-Hilaire, E. 1833. "Discours prononcés au nom de l'Institut, du Muséum d'Histoire Naturelle et de la Société Entomologique de France, su la tombe de M. Latreille, décédé de 6 f'vriere 1833." *Annales des Sciences Naturelles* 28: 433–5. Available online: https://www.biodiversitylibrary.org/page/6096369#page/439/mode/1up (accessed October 14, 2022).
Geoffroy-Schneiter, B. 2007. "René Lalique: Monsieur Butterfly." *Beaux Arts Magazine* 273: 82–7.
Gibbs, K.E. 2014. "Without Benefit of Insects: The Story of Edith M. Patch of the University of Maine." *Maine Agricultural and Forest Experiment Station Miscellaneous Publication* 763.
Gibbs, P. 1920. *Now it Can Be Told*. New York: Harper and Brothers.
Gilchrist, A. 1907. *The Life of William Blake*. London: John Lane.
Giordano, C. 2018. "Sainsbury Becomes First UK Supermarket to Sell Edible Bugs as Crickets Hit Shelves." Available online: https://www.independent.co.uk/news/uk/home-news/sainsburys-first-uk-supermarket-sell-edible-bugs-insects-crickets-a8638581.html (accessed October 14, 2022).
Gordon, E. 1914. *The Butterfly Babies' Book*. Chicago: Rand McNally & Company.
Gosse, P.H. 1861. *The Romance of Natural History*. Boston: Gould and Lincoln.
Gosse, P.H. 1884. *Evenings at the Microscope: or, Researches among the Minuter Organs and Forms of Animal Life*. London: Society for Promoting Christian Knowledge.

Grandville, J.J. [a.k.a. Jean Ignace Isidore Gerard]. 1844. *Un Autre Monde*. Paris: H. Fournier.
Grandville, J.J., Karr, A., Delord, T., and Fœlix, Le C^TE (a.k.a. Raban Louis-François). 1867. *Les Fleurs Animées*, vols. 1 and 2. Paris: Garnier Frères.
Gray, A. 1860. "[Review of] *On the Origin of Species*." *Atlantic Monthly* 6: 109–16.
Gray, J.H. 1878. *China: A History of the Laws, Manners, and Customs of the People*. London: Macmillan.
Grayson, D. 2006. "The Late Quaternary Biogeographic Histories of Some Great Basin Mammals (Western USA)." *Quaternary Science Reviews* 25: 2964–91.
Grinnell, G.B. [1923] 1972a. *The Cheyenne Indians*, vol. 1, *History and Society*. Lincoln: University of Nebraska Press.
Grinnell, G.B. [1923] 1972b. *The Cheyenne Indians*, vol. 2, *War, Ceremonies, and Religion*. Lincoln: University of Nebraska Press.
Grinnell, M.E. and Hawes, I.L. 1943. *Bibliography on Lice and Man: With Particular Reference to Wartime Conditions*. Bibliographical Bulletin Number 1. Washington, DC: US Government Printing Office.
Guelke, J.K. and Morin, K.M. 2001. "Gender, Nature, Empire: Women Naturalists in Nineteenth Century British Travel Literature." *Transactions of the Institute of British Geographers* 26: 306–26.
Hacker, J.D. 2011. "A Census-Based Count of the Civil War Dead." *Civil War History* LVII(4): 307–48.
Hamilton, I. 1920. *Gallipoli Diary*, vol. I, New York: George H. Doran Co.
Hardy, G.H. 1915. *The Book of the Fly*. London: William Heinemann.
Hare, J., ed. 1905. *A Photographic Record of the Russo-Japanese War*. New York: P.F. Collier & Son.
Harper's New Monthly Magazine. 1854. "The Mosquito War." Vol. IX, No. LI (August), 428–9. New York: Harper & Brothers.
Harper's Weekly. 1861. "Our New Major-General." (October 5) 5, no. 249, 640.
Harper's Weekly. 1862. "The Rebel Flea." (July 12) 6, no. 289, 435.
Harris, B.S. 1967. *The Collected Drawings of Aubrey Beardsley*. New York: Crescent Books.
Harris, J. 1940. "The White Knife Shoshoni of Nevada." In R. Linton (ed.), *Acculturation in Seven American Indian Tribes*. New York: Appleton-Century Co.
Harris, M. 1766. *The Aurelian: A Natural History of English Moths and Butterflies Together with the Plants on Which They Feed*. London: sold by J. Edwards.
Harris, T. [1841]. 1862. Report on the Insects Injurious to Vegetation. Cambridge: Cambridge University Press. Available online: https://www.biodiversitylibrary.org/item/78853#page/17/mode/1up (accessed July 5, 2023).
Hart, P. 2011. *Gallipoli*. London: Profile Books Ltd.
Hauptman, J., ed. 2005. *Beyond the Visible: The Art of Odilon Redon*. New York: The Museum of Modern Art.
Hegarty, S. 2006. "Maggots, Songbirds and Other Acquired Tastes." *The Irish Times*, April 1.
Heizer, R.F. 1945. "Honey-Dew 'Sugar' in Western North America." *The Masterkey* 19(5): 140–4.
Hemming, A.F. 1958. "Official lists of family-group, generic and specific names in zoology and of works approved as available for zoological nomenclature. Official indexes of rejected and invalid family-group, generic and specific names in zoology and of rejected and invalid works in zoological nomenclature." London: International Trust for Zoological Nomenclature.

Hendricks, G. 1974. *Albert Bierstadt: Painter of the American West*. New York: H.N. Abrams.
Henneberry, T.J. 2008. *Federal Entomology: Beginnings and Organizational Entities in the United States Department of Agriculture, 1854–2006, with Selected Research Highlights*. USDA, ARS, Washington, DC: Agricultural Information Bulletin 802. Available online: https://econpapers.repec.org/paper/agsuersab/309846.htm (accessed October 14, 2022).
Herrick, G.W. 1913. "Household Insects and Methods of Control." *The Cornell Reading-Courses* 3(49): 1–47. Ithaca: New York State College of Agriculture at Cornell University. Available online: https://play.google.com/books/reader?id=NpsgAQAAIAAJ&pg=GBS.PP8&hl=en (accessed October 14, 2022).
Heusser, A.H., ed. 1927. *The History of the Silk Dyeing Industry in the United States*. Paterson: The Silk Dyers' Association of America.
Heuzé, M. 2003. "Les Froment-Meurice: Le Génie d'Une Dynastie d'Orfèvres." *L'Objet d'Art* 378: 38–47.
Hicks, E. 1961. *Shellac: Its Origin and Applications*. New York: Chemical Publishing Company.
Hinman, W.F. 1887. *Corporal Si Klegg and His "Pard."* Cleveland: N.G. Hamilton Publishing Co.
Ho, S. 2019. "The Japanese Tradition of Raising and Eating Wasps." Available online: https://www.splendidtable.org/story/2019/02/08/the-japanese-tradition-of-raising-and-eating-wasps (accessed May 22, 2023).
Hobhouse, C. 1950. *1851 and the Crystal Palace*. London: John Murray.
Hobsbawm, E. 1962. *The Age of Revolution, 1789–1848*. London: Weidenfeld & Nicholson. Available online: https://files.libcom.org/files/Eric%20Hobsbawm%20-%20Age%20Of%20Revolution%201789%20-1848.pdf (accessed October 14, 2022).
Hobsbawm, E. 1975. *The Age of Capital, 1848–1875*. New York: Pantheon Books.
Hoffman, L.-F. 1964. *La Peste à Barcelone*. Paris: Presses Universitaires de France.
Hogue, C. 1975. "The Insect in Human Symbolism." *Terra* 13(3): 3–9.
Hogue, C. 1979. "The Bugfolk." *Terra* 17(4): 36–8.
Hogue, C. 1987. "Cultural Entomology." *Annual Review of Entomology* 32: 181–99.
Holland, W.J. 1898. *The Butterfly Book*. Norwood: Norwood Press.
Holt, V.M. 1885. *Why Not Eat Insects?* London: Field & Tuer. Available online: https://play.google.com/books/reader?id=awNbAAAAQAAJ&pg=GBS.PA3&hl=en (accessed 14 October, 2022).
Honour, H. and Fleming, J. 1995. *The Visual Arts: A History*, 4th edn. New York: Harry N. Abrams.
Hopkins, W.P. 1903. *The Seventh Regiment Rhode Island Volunteers in the Civil War, 1862–1865*. Providence: Snow & Farnham, Printers.
Howard, L.O. 1911. *The House Fly, Disease Carrier: An Account of Its Dangerous Activities and of the Means of Destroying it*. New York: Frederick A. Stokes Company.
Howard, L.O. 1917. "Entomology as a National Defense." *Entomological News* 28(5): 229.
Howard, L.O. 1919. "Entomology and the War." *The Scientific Monthly* 8: 109–17.
Howard, L.O. 1921. "A Fifty-Year Sketch History of Medical Entomology and its Relation to Public Health." In M.P. Ravendel (ed.), *A Half Century of Public Health*, 412–38. New York: American Public Health Association.
Howard, L.O. 1930. *A History of Applied Entomology (Somewhat Anecdotal)*. Washington, DC: Smithsonian Institution.

Howe, D.H. 1975. "American Victorianism as Culture." *American Quarterly* 27(5): 507–32.
Hsu, S.Y. 1969. "The Cultural Ecology of the Locust Cult in Traditional China." *Annals of the Association of American Geographers* 59: 731–52.
Hull, D.L. 1975. *Darwin and His Critics: The Reception of Darwin's Theory of Evolution by the Scientific Community*. Chicago: University of Chicago Press.
Humphreys, J. 1887. *Insect Ways on Summer Days*. London: Blackie and Son.
Hunt, S.B. 1867. "Camp Diarrhea and Dysentery." In A. Flint (ed.), *Contributions Relating to the Causation and Prevention of Disease, and to Camp Diseases*, 291–318. New York: U.S. Sanitary Commission.
Hunter, J. 1993. "Japanese Women at Work, 1880–1920." *History Today* 43(5): 49–55.
Hurt, R.D. 2002. *American Agriculture: A Brief History*. West Lafayette: Purdue University Press.
Huxley, L. 1900. *Life and Letters of Thomas Henry Huxley by his son Leonard Huxley*, vol. II. London: Macmillan and Co. Available online: https://archive.org/details/lifelettersofthо0000huxl/page/n10/mode/1up?ref=ol&view=theater (accessed October 14, 2022).
Huxley, L. 1901. *Life and Letters of Thomas Henry Huxley by his son Leonard Huxley*, vol. I. New York: D. Appleton and Company. Available online: https://archive.org/details/lifeandletterst13huxlgoog/page/n10/mode/2up (accessed October 14, 2022).
ICZN (International Commission on Zoological Nomenclature). 1905. *Règles internationales de la nomenclature zoologique* (International rules of zoological nomenclature). *Internationale Regeln der zoologischen Nomenklatur*. Paris: Rudeval.
Imms, A.D. 1907. "Report on a Disease of Bees in the Isle of Wight." *Journal of the Board of Agriculture* 14(3): 129–40.
Impelluso, L. 2004. *Nature and Its Symbols*. Los Angeles: Getty Publications.
Ingpen, A. 1839. *Instructions for Collecting, Rearing, and Preserving British & Foreign Insects: Also for Collecting and Preserving Crustacea and Shells*, 2nd edn. London: William Smith.
Isaacs, J. 2002. *Bush Food: Aboriginal Food and Herbal Medicine*. Frenchs Forest, NSW, Australia: New Holland Publishers.
Iseminger, G.L. 1994. "Silk Trains on the Great Northern Railway." *Minnesota History* 54(1): 18–31.
Jaffé, A. 1964. "Symbolism in the Visual Arts." In C. Jung (ed.), *Man and His Symbols*. London: Aldus. Available online: https://antilogicalism.com/wp-content/uploads/2017/07/man-and-his-symbols.pdf (accessed October 14, 2022).
James, R.R. and Li, Z. 2012. "From Silkworms to Bees." In F.E. Vega and H.K. Kaya (eds.), *Insect Pathology*, 2nd edn., 425–459. Amsterdam: Elsevier.
Jansen, L., Luijten, H., and Bakker, N., eds. 2009. "Vincent van Gogh: The Letters." Van Gogh Museum, Amsterdam. Available online: http://vangoghletters.org/vg/ (accessed May 22, 2023).
Jintasataporn, O. 2012. "Production Performance of Broiler Chickens Fed with Silkworm Pupa (*Bombyx mori*)." *Journal of Agricultural Science and Technology A* 2(4): 505–10.
Johnson, P. 2012. *Darwin: Portrait of a Genius*. London: Penguin.

Johnston, N. 2013. "The History of the Parliamentary Franchise." House of Commons Library. House of Commons Library. Available online: https://researchbriefings.files.parliament.uk/documents/RP13-14/RP13-14.pdf (accessed December 6, 2021).

Jones, W. 1898. *Crowns and Coronations: A History of Regalia*. London: Chatto & Windus.

Jordan, A., Patch, H.M., Grozinger, C.M., and Khanna, V. 2021. "Economic Dependence and Vulnerability of United States Agricultural Sector on Insect-Mediated Pollination Service." *Environmental Science and Technology* 55(4): 2243–53.

Junker, S. 2016. *Tribe: On Homecoming and Belonging*. New York: Hachette Book Group, Inc.

Kačāne, I. 2012. "Insects in the 19th Century British Culture: Pre-Raphaelitism and Aestheticism." *Comparative Studies* 4(1): 26–41.

Kakkilaya, B.S. 2015. "Saga of Malaria Treatment." Available online: https://www.malariasite.com/history-treatment/ (accessed December 4, 2017).

Keen, W.W. 1917. *The Treatment of War Wounds*. Philadelphia: W.B. Saunders and Company.

Keene, M. 2015. *Science in Wonderland: The Scientific Fairy Tales of Victorian Britain*. Oxford: Oxford University Press.

Kieffer, J.-J. 1913. "Nouveaux microhyménoptères de l'Afrique équatoriale." *Bollettino del Laboratorio di Zoologia Generale e Agraria della R. Scuola Superiore d'Agricultura* 7: 105–12.

Kiesenwetter, E. von. 1858. "Gesetze der entomologischen Nomenclatur." *Berliner Entomologische Zeitschrift* 2: xi–xxii.

Kingsley, C. 1909. *The Water-Babies: A Fairy Tale for a Land-Baby*. London: Macmillian & Co.

Kirby, J. and White, R. 1996. "The Identification of Red Lake Pigment Dyestuffs and a Discussion of Their Use." *National Gallery Technical Bulletin* 17: 56–80.

Kirby, W. 1802. *Monographia apum Angliæ: Or, An Attempt to Divide Into Their Natural Genera and Families, Such Species of the Linnean Genus Apis as Have Been Discovered in England; with Descriptions and Observations. To which are Prefixed Some Introductory Remarks Upon the Class* Hymenoptera, *and a Synoptical Table of the Nomenclature of the External Parts of These Insects*, vol. 1. Ipswich: J. Raw. Available online: https://www.biodiversitylibrary.org/bibliography/10346 (accessed October 14, 2022).

Kirby, W. 1836. *On the Power, Wisdom and Goodness of God as Manifested in the Creation of Animals and in their History, Habits and Instincts*. Philadelphia: Carey, Lea & Blanchard. Available online: https://www.biodiversitylibrary.org/bibliography/52198 (accessed November 29, 2022).

Kirby, W. and Spence, W. 1815. *An Introduction to Entomology: or, Elements of the Natural History of Insects*, vol. 1. London: Longman, Hurst, Rees, Orme, and Brown. Available online: https://www.google.com/books/edition/An_Introduction_to_Entomology/BWE5AQAAMAAJ?hl=en&gbpv=1&dq=Kirby%20Spence%201815%20An%20Introduction%20to%20Entomology&pg=PR1&printsec=frontcover (accessed October 26, 2022).

Kirby, W. and Spence, W. 1817. *An Introduction to Entomology: or, Elements of the Natural History of Insects*, vol. 2. London: Longman, Rees, Orme, Brown, and Green.

Kirby, W. and Spence, W. 1826. *An Introduction to Entomology: or, Elements of the Natural History of Insects*, vol. 3. London: Longman, Hurst, Rees, Orme, Brown, and Green.

Kirby, W. and Spence, W. 1815–1826. *An Introduction to Entomology: or, Elements of the Natural History of Insects*, 4 vols, London: Longman, Rees, Orme, Brown, and Green.

Kirby, W. and Spence, W. 1846. *An Introduction to Entomology; or, Elements of the Natural History of Insects*, 6th edn. Philadelphia: Lea and Blanchard.

Klein, H.S. 2012. *A Population History of the United States*, 2nd edn. Cambridge: Cambridge University Press.

Klose, V. 1963. "Sericulture in the United States." *Agricultural History* 37(4): 225–34.

Kotenok, I. and Chicherin, I. 1977. "Houseflies (*M. domestica* L.) as Transmitters of the Agent of Cholera." [in Russian] *Zh Mikrobiol Epidemiol Immunobiol*, Dec. (12): 23–7. Available online: https://www.ncbi.nlm.nih.gov/pubmed/602541 (accessed December 4, 2017).

Krasner, J. 1992. *The Entangled Eye: Visual Perception and the Representation of Nature in Post-Darwinian Narrative*. New York: Oxford University Press.

Kritsky, G. 2008. "Entomological Reactions to Darwin's Theory in the Nineteenth Century." *Annual Review of Entomology* 53: 345–60.

Kritsky, G. and Cherry, R. 2000. *Insect Mythology*. San Jose: Writer's Club Press.

Kuehn, F. 2003. "Class Insecta: Landlords of the Earth." *Ornament* 26(3): 54–7.

La Bree, B. 1898. *Camp Fires of the Confederacy*. Louisville, KY: Courier-Journal Job Printing Company.

Lamarck, J.-B. 1809. *Zoological Philosophy: An Exposition with Regard to the Natural History of Animals*, vol. 1. Paris: Dentu.

Lamarck, J.-B. 1815. *Flore française ou descriptions succinctes de toutes les plantes qui croissent naturellement en France disposées selon une nouvelle méthode d'analyse; et précédées par un exposé des principes élémentaires de la botanique*. Paris: Desray.

Lang, A., ed. 1889. *The Blue Fairy Book*. London: Longmans, Green & Co.

Langon, É.L. [Baron de Lamothe-Langon]. 1837. *Evenings with Prince Cambacérès*, vol. 1. London: Henry Colburn.

Langstroth, Rev. L.L. 1853. *Langstroth on the Hive and the Honey-Bee*. Northampton: Hopkins, Bridgman, & Company. Available online: http://livinghistoryofillinois.com/pdf_files/the_hive_and_the_honey-bee.pdf (accessed October 14, 2022).

Laplace, P.S. 1799–1825. *Traité de mécanique celeste*, vols. 1–5. Paris: Crapelet.

Larson, E. J. 1997. *Summer for the Gods: The Scopes Trial and America's Continuing Debate Over Science and Religion*. New York: Basic Books.

Lawrence, E.A. 1993. "The Sacred Bee, the Filthy Pig, and the Bat out of Hell: Animal Symbolism a Cognitive Biophilia." In S.R. Kellert and E.O. Wilson (eds.), *The Biophilia Hypothesis*, 301–41. Washington DC: Island Press.

Leach, W. 2014. *Butterfly People: An American Encounter with the Beauty of the World*. New York: Pantheon Books.

Leasor, J. 2001. *The Plague and the Fire*. Cornwall: The House of Stratus.

Le Bourhis, K., ed. 1989. *Costume from Revolution to Empire 1789–1815*. New York: The Metropolitan Museum of Art/Harry N. Abrams, Inc.

Lee, A.B. 1913. "Carmine and Cochineal Stains." Chapter XII In *The Microtomist's Vade-Mecum; A Handbook of the Methods of Microscopic Anatomy*. Philadelphia: P. Blakiston's Son & Co. Available online: https://www.biodiversitylibrary.org/bibliography/1481 (accessed December 2, 2021).

Le-May Sheffield, S. 2001. *Revealing New Worlds: Three Victorian Women Naturalists*. London: Routledge.
Leneveux, L. 1850. *Les Petits Habitants des Fleurs*. Paris: M[ME] V[VE] Louis Janet.
Leslie, F. [Henry Carter]. 1878a. *Frank Leslie's Illustrated Newspaper* (September 21), 1199 (XLVII): 33, 48.
Leslie, F. [Henry Carter]. 1878b. *Frank Leslie's Illustrated Newspaper* (September 28), 1200(XLVII): 49, 56–7.
Lesnik, J. 2018. *Edible Insects and Human Evolution*. Gainesville, FL: University Press of Florida.
Lesser, F.C. 1740. *Insecto-Theology: Or a Demonstration of the Being and Perfections of God, from a Consideration of the Structure and Economy of Insects*. Translated from the German by Pierre Lyonnet. Berlin: Humboldt-Universität.
Lewes, G.H. [1858] 1860. *Sea-Side Studies at Ilfracombe, Tenby, the Scilly Isles, and Jersey*. Edinburgh: William Blackwood and Sons. Available online: https://www.google.com/books/edition/Sea_side_Studies_at_Ilfracombe_Tenby_the/7JUKAAAAMAAJ?hl=en&gbpv=1&printsec=frontcover (accessed May 22, 2023).
Lewis, A. 1871. "On the Application of the Maxim *'communis error facit jus'* to Scientific Nomenclature." *The Entomologist's Monthly Magazine* 8: 1–5.
Lewis, A. 1875. "On Entomological Nomenclature and the Rule of Priority." *Transactions of the Entomological Society of London* 1875: i–xlii.
Lewis, L. and Thompson, A. 1868. "The Painted Lady." *The Mask, a Humorous and Fantastic Review*, September, 247–56. Available online: https://catalog.hathitrust.org/Record/100197501 (accessed October 26, 2022).
Leydon, L. 2013. "Nature Overflows in these Musty Rooms." *The New York Times* (October 14), Section D, p. 2.
Li, L.M. 1981. *China's Silk Trade: Traditional Industry in the Modern World 1842–1937*. Cambridge: Harvard University Council on East Asian Studies.
Library of Congress. 2023. "Merovingian Bees." Available online: https://hdl.loc.gov/loc.wdl/wdl.14235 (accessed May 26, 2023).
Liedtke, W. 2003. "Still-Life Painting in Northern Europe, 1600–1800." In *Heilbrunn Timeline of Art History*. New York: The Metropolitan Museum of Art. Available online: http://www.metmuseum.org/toah/hd/nstl/hd_nstl.htm (accessed October 14, 2022).
Lightman, B. 2000. "The Visual Theology of Victorian Popularizers of Science: From Reverent Eye to Chemical Retina." *Isis* 91(4): 651–80.
Ligon, B.L. 2006. "Plague: A Review of its History and Potential as a Biological Weapon." *Seminars in Pediatric Infectious Diseases* 17: 161–70.
Lindroth, C.H. 1973. "Systematics Specializes Between Fabricius and Darwin: 1800–1859." In R.F. Smith, T.E. Mittler, and C.N. Smith (eds.), *History of Entomology*, 119–54. Palo Alto: Annual Reviews Inc.
Linnaeus, C. 1758. *Systema Naturae*, 10th edn. Stockholm: Salvius.
Lippincott, J.B. 1876. *Lippincott's Magazine of Popular Literature and Science*, vol. 18. Available online: https://play.google.com/books/reader?id=3e8zAQAAMAAJ&pg=GBS.PA4&hl=en (accessed October 14, 2022).
Lippit, A.M. 2000. *Electric Animal: Toward a Rhetoric of Wildlife*. Minneapolis: University of Minnesota Press.
Liu, R.K. 1998. "The Symbolic Importance of Insects in Jewelry." *Ornament* 22(1): 38–43.

Lockwood, Jeffrey A. 2004. *Locust: The Devastating Rise and Mysterious Disappearance of the Insect that Shaped the American Frontier*. New York: Perseus Books Group.

Lockwood, J.A. 2009. *Six-legged Soldiers*. New York: Oxford University Press.

Lockwood, J.A. 2010. "The Fate of the Rocky Mountain Locus, *Melanoplus spretus* Walsh: Implications for Conservation Biology." *Terrestrial Arthropod Reviews* 3: 129–60.

Long, J. 2009. "Portable Pets: Live and Apparently Live Animals in Fashion, 1880–1925." *Costume* 43: 109–26.

Longvah, T., Mangthya, K., and Ramulu, P. 2011. "Nutrient Composition and Protein Quality Evaluation of Eri Silkworm (*Samia ricinii*) Prepupae and Pupae." *Food Chemistry* 128(2): 400–3. DOI: 10.1016/j.foodchem.2011.03.041.

Lowenberg, B.J. 1941. "Darwinism Comes to America, 1859–1900." *The Mississippi Valley Historical Review* 28(3): 339–68.

Lowie, R.H. 1939. "Ethnographic notes on the Washo." *American Archaeology and Ethnology* 36(5): 301–52.

Lubbock, A. 1886. *Diary, 1884–1888*, Supplementary Avebury Papers, Add. MSS 62691, fol. 13.

Lubbock, J. 1882. *Ants, Bees, and Wasps. A Record of Observations on the Habits of the Social Hymenoptera*. New York: D. Appleton & Company.

Luo, Z.-Y. 1997. "Insects as Food in China." *Ecology of Food and Nutrition* 36(2–4): 201–7.

Lyell, C. 1830. *Principles of Geology*. London: Murray.

Ma, D. 1996. "Modern Silk Road: Global Raw-Silk Market, 1850–1930." *The Journal of Economic History* 56(2): 330–55.

Ma, D. 1999. "The Great Silk Exchange: How the World Was Connected and Developed." In D.O. Flynn, L. Frost, and A.J.H. Latham (eds.), *Pacific Centuries: Pacific and Pacific Rim History Since the Sixteenth Century*, 38–69. London: Routledge.

Maas, J., Trimpe, P.W., and Gere, C. 1997. *Victorian Fairy Painting*. London: Royal Academy of Arts.

Mabberley, D.J. 2000. *Arthur Harry Church: The Anatomy of Flowers*. London: Merrell Publishers Ltd.

MacDonald, M.F., de Montfort, P., and Thorp, N., eds. 2003–2010. "Butterflies: A Chronology of Whistler's Butterfly Signatures." In *The Correspondence of James McNeill Whistler, 1855–1903*. Glasgow: University of Glasgow. Available online: http://www.whistler.arts.gla.ac.uk/miscellany/butterflies/ (accessed March 4, 2018).

MacGregor, A. 2014. "Introduction." In A. Gáldy and S. Heudecker (eds.), *Collecting Nature*. Newcastle upon Tyne: Cambridge Scholars Publishing.

Mackenzie, W. [writing as Margery Meanwell]. 1822. *The New Tom Thumb: With an Account of his Wonderful Exploits*. London: J. Harris & Son.

Macmillan, J.D. 1972. "Holman Hunt's Hireling Shepard: Some Reflections on a Victorian Pastoral." *The Art Bulletin* 54(2): 187–97.

Madsen, D.B. 1989. "A Grasshopper in Every Pot." *Natural History*, July: 22–5.

Madsen, D.B. and Kirkman, J.E. 1988. "Hunting Hoppers." *American Antiquity* 53(3): 593–604. Available online: https://doi.org/10.2307/281220 (accessed May 22, 2023).

Madsen, D.B. and Schmitt, D.N. 1998. "Mass Collecting and the Diet Breadth Model: A Great Basin Example." *Journal of Archaeological Science* 25(5): 445–55. Available online: https://doi.org/10.1006/jasc.1997.0245 (accessed October 14, 2022).

Mallis, A. 1971. *American Entomologists*. New Brunswick: Rutgers University Press.
Malthus, T.R. 1798. *An essay on the Principle of Population as it Affects the Future Improvement of Society, with Remarks on the Speculations of Mr. Godwin, M. Condorcet, and Other Writers*. London: J. Johnson. Available online: https://archive.org/details/essayonprincipl00malt (accessed October 14, 2022).
Manicardi, N. 2018. "il Saltarello e I suoi Fratelli in attesa della Dop." *Premiata Salumeria Italiana. 1. Edizioni Pubblicità Italia*. Available online: https://www.pubblicitaitalia.com/food/prodotti/premiata-salumeria-italiana/2018/1/16459
Manunza, L. 2018. "Casu Marzu: A Gastronomic Genealogy." In A. Halloran, R. Flore, P. Vantomme, and N. Roos (eds.), *Edible Insects in Sustainable Food Systems*. Berlin: Springer Verlag. Available online: https://link.springer.com/book/10.1007/978-3-319-74011-9 (accessed October 14, 2022).
Mapp, A.J., Jr. 2003. *The Faiths of Our Fathers: What America's Fathers Really Believed*. Lanham: Rowman and Littlefield Publishers, Inc.
Margrave, R.D. 1975. "The Role of the Paterson, NJ Silk Industry in the 19th-Century Atlantic Economy." *Northeast Historical Archaeology* 44: 53–8.
Marks, J.A. 1907. *The Cheerful Cricket and Others*. Boston: Small, Maynard & Co.
Marsden, G.M. 1980. *Fundamentalism and American Culture: The Shaping of Twentieth-Century Evangelicalism 1870–1925*. New York: Oxford University Press.
Marsden, G.M. 2006. *Fundamentalism and American Culture*. New York: Oxford University Press.
Martin, C. 2012. "Scientists to the End." *Nature* 481(7381): 264.
Martin, D. 2014. *Edible: An Adventure into the World of Eating Insects and the Last Great Hope to Save the Planet*. Boston: New Harvest, Houghton Mifflin Harcourt.
Masci, D. 2019. "For Darwin Day, 6 Facts about the Evolution Debate." *Pew Research Center*, February 11. Available online: https://www.pewresearch.org/fact-tank/2019/02/11/darwin-day/ (accessed May 18, 2020).
Matsumura, S. 1912. [*Thousand Insects of Japan Illustrated*. Supplement IV.] Tokyo: Keiseisha Shoten, Meiji.
Mazzette, R., Colleo, M. M., Riu, G., Piras, G., Piras, F., Addis, M. et al. 2010. "Production Under Controlled Conditions of 'Casu Marzu' Cheese: Effect of the *Piophila casei* Colonization on Microbial and Chemical Composition of the Cheeses." *Italian Journal of Food Safety* 1(7): 45–54.
McTier, R.S. 2013. *An Insect View of Its Plain: Insects, Nature and God in Thoreau, Dickinson and Muir*. Jefferson: McFarland & Company.
Meagher, J. 2004. "The Pre-Raphaelites." In *Heilbrunn Timeline of Art History*. New York: The Metropolitan Museum of Art. Available online: http://www.metmuseum.org/toah/hd/praf/hd_praf.htm (accessed May 22, 2023).
Meagher, J. 2008. "Still-Life Painting in Southern Europe, 1600–1800." In *Heilbrunn Timeline of Art History*. New York: The Metropolitan Museum of Art. Available online: http://www.metmuseum.org/toah/hd/sstl/hd_sstl.htm (accessed May 22, 2023).
Melville, R.V. 1995. *Towards Stability in the Names of Animals. A History of the International Commission on Zoological Nomenclature 1895–1995*. London: ICZN.
Melville, R.V. and Smith, J.D.D. 1987. *Official Lists and Indexes of Names and Works in Zoology*. London: International Trust for Zoological Nomenclature. Available online: https://www.biodiversitylibrary.org/page/2481944 (accessed October 14, 2022).

Menely, T. 2004. "Traveling in Place: Gilbert White's Cosmopolitan Parochialism." *Eighteenth-Century Life* 23(3): 46–65.

Meyer, C.A. 2013. "The Farm Debut of the Gasoline Engine." *Agricultural History Society* 87(3): 287–313.

Michelet, J. 1875. *The Insect*. Trans. W.H. Davenport Adams. London: T. Nelson and Sons. Available online: https://archive.org/details/insect__00mich (accessed October 14, 2022).

Mikl, A.Sue. 2004. "Fairy Painting in Nineteenth Century Art and Late Twentieth Century Art: A Comparative Study." MS thesis, University of Wisconsin-Stout, Menomonie, WI.

Millais, J.G. 1899. *The Life and Letters of Sir John Everett Millais*, vol. 1. London: Methuen & Co.

Millar, F. 1981. "The World of The Golden Ass." *The Journal of Roman Studies* 71: 63–75.

Miller, G.L. 1997. "Historical Natural History: Insects and the Civil War." *American Entomologist* 43(4): 227–45.

Miller, G.L. and Adler, P.A. 2009. "'... drenched in dirt and drowned in abominations ...": Insects and the Civil War." In *Evolution of Military Medical Entomology. Proceedings of the DOD Symposium on Evolution of Military Medical Entomology, ADA506261*, 7–14. Available online: https://apps.dtic.mil/sti/pdfs/ADA506261.pdf (accessed October 14, 2022).

Mills, F.J. 1868. *The Life of John Carter*. New York: Hurd and Houghton.

Mimolette Cheese. n.d. Available online: https://saucydressings.com/blog/mimolette-cheese/ (accessed November 23, 2021).

Mingay, G.E. 1986. *The Transformation of Britain, 1830–1939*. London: Routledge & Kegan Paul.

Mishraa, N., Hazarikaa, N.C., Naraina, K., and Mahanta, J. 2003. "Nutritive Value of Non-Mulberry and Mulberry Silkworm Pupae and Consumption Pattern in Assam, India." *Nutrition Research* 23(10): 1303–11. Available online: https://doi.org/10.1016/S0271-5317(03)00132-5 (accessed October 14, 2022).

Mite Cheese. n.d. Available online: https://www.milbenkaese.de/ (accessed November 23, 2021).

Mitsuhashi, J. 1997. "Insects as Traditional Foods in Japan." *Ecology of Food and Nutrition* 36(2–4): 187–99. Available online: http://dx.doi.org/10.1080/03670244.1997.9991514 (accessed May 22, 2023).

Mokyr, J. 2009. *The Enlightened Economy: An Economic History of Britain 1700–1850*. New Haven: Yale University Press.

Moore, D.T. 2004. "William Kirby (1759–1850), Entomologist and Naturalist." *Oxford Dictionary of National Biography* 31: 768–9.

Moore, J.A. 1993. *Science as a Way of Knowing: The Foundations of Modern Biology*. Cambridge, MA: Harvard University Press.

Moore, J.R. 1986. "Geologists and Interpreters of Genesis in the Nineteenth Century." In D. Lindberg and R. Numbers (eds.), *God and Nature: Historical Essays on the Encounter between Christianity and Science*, 322–350. Berkeley: University of California Press.

Moore, J.R. 1989. "Of Love and Death: Why Darwin 'Gave Up Christianity.'" In J.R. Moore (ed.), *History, Humanity, and Evolution: Essays for John C. Greene*, 195–229. Cambridge: Cambridge University Press.

Moote, A.L. and Moote, D.C. 2004. *The Great Plague: The Story of London's Most Deadly Year*. Baltimore: John Hopkins University Press.
Morgan, T.H. 1910. "Sex Limited Inheritance in *Drosophila*." *Science* 32(812): 120–2.
Morrell, J.B. 1976. "London Institutions and Lyell's Career: 1820–41." *The British Journal for the History of Science* 9(2): 132–46.
Mount, H. 2013. *How England Made the English: From Why We Drive on the Left to Why We Don't Talk to Our Neighbors*. London: Penguin Books.
Munn, G.C. 1987. "The Jewelry of René Lalique." *The Magazine Antiques* 131(6): 1288–91.
Murphy, K. 2017. "A Slice of Life." *Apollo The International Art Magazine*, March: 125–31.
Murray, J. 1881. *Life, Letters and Journals of Sir Charles Lyell* 1: 14–17.
Myers, W. 1918. Letters. The National WWI Museum and Memorial. https://collections.theworldwar.org/argus/final/Portal/Default.aspx?lang=en-US.
Newton, W.W. 1878. *Little and Wise: Lessons from the Ants, the Conies, the Locusts and the Spider*. London: Wesleyan Conference Office.
New York Times, The. 1890. "A Young Lady Artist's Mania for Capturing Musial Insects" (May 29).
New York Times, The. 1920. "Found Dead After 'Loss of $2,000,000,' Isaac Mendelson, Silk Merchant, Believed to Have Killed Himself in Office" (October 10), 16.
Ngapo, T., Champagne, C., Chilian, C., Dugan, M., Gariépy, S., Vahmani, P., and Bilodeau, P. 2021. "Pemmican, an Endurance Food: Past and Present." *Meat Science* 178: 108526. Available online: https://www.sciencedirect.com/science/article/abs/pii/S0309174021001029?via%3Dihub (accessed October 14, 2022).
Niiler, E. 2012. "He Lost the Race to the South Pole but Made Discoveries for Science." *Washington Post* (January 12).
Nodier, Charles. 1867. "*Un Renaud: Pris au Piège*." In J.P. Stahl (ed.), *Vie Privée et Publique des Animaux*, 247–66. Paris: J. Hetzel.
North, M. and Symonds, J.C. 1894. *Recollections of a Happy Life*, vol. 1. New York: Macmillan and Co.
Nouvel-Kammerer, O., ed. 2007. *Symbols of Power: Napoleon and the Art of the Empire Style 1800–1815*. New York: Abrams.
O'Brian, P. 1987. *Joseph Banks: A Life*. London: Collins Harvill.
O'Donnell, P.C. 1985. "'Fantasy and Realism Combined': The Manufacture of Silk at Pelgram and Meyer, 1872–1928." In Philip B. Scranton (ed.), *Silk City: Studies on the Paterson Silk Industry, 1860–1940*, 99–105. Newark: New Jersey Historical Society.
Oertel, E. 1976. "Early Records of Honey Bees in the Eastern United States." *American Bee Journal* 116: 70–1.
Oliphant, L. 1860. *Narrative of the Earl of Elgin's Mission to China and Japan in the Years 1857, '58, '59*. New York: Harper and Brothers.
Ormerod, E.A. 1904. *Eleanor Ormerod, LL.D. Economic Entomologist. Autobiography and Correspondence*. Ed. Robert Wallace. New York: Dutton and Company.
Ormerod, E.L. 1868. *British Social Wasps: An Introduction to Their Anatomy and Physiology, Architecture, and General Natural History*. London: Longmans, Green, Reader and Dyer.
Osborn, H. 1937. *Fragments of Entomological History*, vol. 1. Columbus, OH: The Spahr & Glenn Company.

Ospovat, D. 1981. *The Development of Darwin's Theory: Natural History, Natural Theology and Natural Selection, 1838–1859*. Cambridge: Cambridge University Press.

Oswandel, J.J. 1885. *Notes of the Mexican War, 1846–47–48*. Philadelphia. Available online: https://lccn.loc.gov/02015450 (accessed October 14, 2022).

Ouchley, K. 2010. *Flora and Fauna of the Civil War*. Baton Rouge: Louisiana State University Press.

Overstreet, R.M. 2003. "Presidential Address: Flavor Buds and Other Delights." *Journal of Parasitology* 89(6): 1093–107.

Owen, R. [1894] 2017. *Life of Richard Owen*, vol. II. London: Murray.

Oyster, J.H. 1912. *The Spatula Ink Formulary: Recipes and Directions for Making All Kinds of Inks for All Kinds of Purposes*. Boston: The Spatula Publishing Company. Available online: https://hdl.handle.net/2027/iau.31858047933233 (accessed December 2, 2021).

Packard, A.S. 1881. *Half Hour Recreations in Natural History: Half Hours with Insects*. Boston: Estes and Lauriat.

Palestine Department of Health. 1941. *A Review of the Control of Malaria in Palestine (1918–1941)*. Jerusalem: Government Printing Press.

Paley, W. 1802. *Natural Theology: Or, Evidence and Attributes of the Deity, Collected from the Appearances of Nature*. London: Wilks & Taylor.

Paneth, N., Vinten-Johansen, P., Brody, H., and Rip, M. 1998. "A Rivalry of Foulness: Official and Unofficial Investigations of the London Cholera Epidemic of 1854." *American Journal of Public Health* 88(10): 1545–53.

Pariset, É. and Mazet, A. 1820. *Observations sur la Fièvre Jaune, Faites à Cadix, en 1819*. Paris: Audot.

Parks, J.A. 2015. "Flowers, Fruits, Beasts and Birds: The British Passion for the Natural World." *Drawing* Winter: 56–65.

Parry, E.J. 1935. *Shellac: Its Production, Manufacture, Chemistry, Analysis, Commerce and Uses*. London: Sir Isaac Pitman & Sons.

Patil, S., Amena, S., Vikas, A., and Rahul, R. 2013. "Utilization of Silkworm Litter and Pupal Waste – an Eco-Friendly Approach for Mass Production of *Bacillus thuringiensis*." *Bioresource Technology*, 131: 545–7. DOI: 10.1016/j.biortech.2012.12.153.

Patterson, J.H. 1916. *With the Zionists in Gallipoli*. New York: George H. Doran Company.

Payne, M. 2016. *Marianne North: A Very Intrepid Painter*, 2nd edn. Richmond: Kew Publishing.

Pemberton, R. 1994. "The Revival of Rice-Field Grasshoppers as Human Food in South Korea." *Pan-Pacific Entomologist* 70(4): 323–7.

Perty, M. 1830–1834. *Delectus animalium articulatorum, quae in itinere per Brasiliam annis MDCCCXVII–MDCCCXX jussu et auspiciis Maximiliani Josephi I. Bavariae regis augustissimi peracto collegerunt Dr. J.B. de Spix et Dr. C.F.Ph. de Martius*. Munich: Impensis Editoris.

Pettid, M. 2008. *Korean Cuisine: An Illustrated History*. London: Reaktion Books.

Phillips, E.F. 1928. *Beekeeping*. New York: The Macmillan Company.

Philonoë [pseudonym of Anne Fadiman]. 1998. "At Large and at Small: Collecting Nature." *The American Scholar* 67(2): 7–13.

Pierce, W.D. 1921. "Chapter IV: A General Survey of the Seriousness of Insect Borne Diseases to Armies." In W.D. Pierce (ed.), *Sanitary Entomology: The Entomology of Disease, Hygiene and Sanitation*, 43–50. Boston: Richard G. Badger.

Pierce, W.D. and Hutchinson, R.H. 1921. "Chapter XXII: The Control of Human Lice." In W.D. Pierce (ed.), *Sanitary Entomology: The Entomology of Disease, Hygiene and Sanitation*, 312–29. Boston: Richard G. Badger.

Plancy, A. [Baron de Plancy]. 1904. *Souvenirs du Comte de Plancy (1798–1816)*. Paris: Paul Ollendorff.

Polaszek, A. et al. 2005. "A Universal Register for Animal Names." *Nature* 437: 477. Available online: https://www.nature.com/articles/437477a#citeas (accessed May 22, 2023).

Polaszek, A. and Cranbrook, G. 2006. "Insect Species Described from Alfred Russel Wallace's Sarawak Collections." *Malayan Nature Journal* 57: 433–62.

Polaszek, A. and Michel, E. 2010. "Linnaeus – Sherborn – ZooBank." In A. Polaszek (ed.), *Systema Naturae 250, The Linnaean Ark*, 163–72. London: CRC Press.

Porter, R., ed. 2003. *The Cambridge History of Science*, vol. 4. Cambridge: Cambridge University Press.

Potter, B. 2012. *Peter Rabbit and Eleven Other Favourite Tales*. New York: Courier Corporation.

Powers, S. 1877. "Centennial Mission to the Indians of Western Nevada and California." *Annual Report of the Smithsonian Institution for 1876*. Washington DC: Government Printing Office.

Prettejohn, E. 2012. "The Painting of Dante Gabriel Rossetti." In E. Prettejohn (ed.), *The Cambridge Companion to the Pre-Raphaelites*, 103–55. New York: Cambridge University Press.

Prinzing, F. 1916. *Epidemics Resulting from Wars*. London: Clarendon Press.

Punch. 1870. "Suffrage for Both Sexes." (April 2). London: Punch Publications Ltd.

Punch. 1871. "The Butterfly was a 'Lady'." (June 17), vol. LX: 245. London: Punch Publications Ltd.

Pyle, R.L. and Michel, E. 2010. "ZooBank: Reviewing the First Year and Preparing for the Next 250." In A. Polaszek (ed.), *Systema Naturae 250, The Linnaean Ark*, 173–84. London: CRC Press.

Quammen, D. 2006. *The Reluctant Mr. Darwin: An Intimate Portrait of Charles Darwin and the Making of His Theory of Evolution (Great Discoveries)*. New York: W.W. Norton & Company.

Quinby, M. 1865. *Mysteries of Bee-Keeping Explained*. New York: Orange Judd and Co.

Ramos-Elorduy, J. 1993. "Insects in the Diets of the Tropical Forest People in Mexico." *Food and Nutrition in the Tropical Forest Biocultural Interactions and Applications to Development* UNESCO 205–12.

Ramos-Elorduy, J. and Pino, J.M. 1993. "Biogeographical Aspects of Some Edible Insects from Mexico." *Abstracts of the III International Congress of Ethnobiology* 141.

Ray, J. 1691. *The Wisdom of God Manifested in the Works of the Creation*. London: S. Smith.

Redouté, P.J. [Pierre Joseph]. 1827–1833. *Choix des Plus Belles Fleurs et des Plus Beaux Fruits*. Paris: Ernest Panckoucke.

Reed, W., Vaughan, V.C., and Shakespeare, E.O. 1904. *Report on the origin and spread of typhoid fever in U.S. military camps during the Spanish War of 1898*, vol. 1. House of Representative Document No. 757. Washington, DC: Government Printing Office.

Religious Tract Society. 1839. *The History of Insects*. London: The Religious Tract Society.

Rennie, J. 1921. "Isle of Wight Disease in Hive Bees – Acarine Disease: The Organism Associated with the Disease *Tarsonemus woodi*, n. sp." *Transactions of the Royal Society of Edinburgh* 52: 768–79.

Robinson, F. 1856. *Diary of the Crimean War*. London: Richard Bently.

Root, A.I. 1920. *The ABC and XYZ of Bee Culture*. Medina, OH: A.I. and E.R. Root Co.

Rossiter, M.W. 1982. *Women Scientists in America*, vol. I, *Struggles and Strategies to 1940*. Baltimore: The Johns Hopkins University Press.

Roy, E.A. 2021. "Another Side of Shellac." In K. Devine and A. Boudreault-Fornier, *Audible Infrastructures*, 209–228. New York: Oxford University Press.

Rush, R. 1828. *20th Congress, 1st Session, House of Representatives. Letter from the Secretary of the Treasury, Transmitting the Information Required by a Resolution of the House of Representatives, of May 11, 1826, in Relation to the Growth and Manufacture of Silk, Adapted to the Different Parts of the Union*. Washington: Gales & Seaton. Available online: https://www.google.com/books/edition/Growth_and_Manufacture_of_Silk/uZMaAAAAYAAJ?hl=en&gbpv=1&dq=growth+and+manufacture+of+silk&printsec=frontcover (accessed December 13, 2021).

Ruttner, F. 1987. *Biogeography and Taxonomy of Honey Bees*. Berlin: Springer-Verlag.

Ryland, E.H. 1939. "America's 'Multicaulis Mania.'" *The William and Mary Quarterly* 19(1): 25–33.

Salinas, C.M. 2018. "Mexican Cochineal, Local Technologies and the Rise of Global Trade from the Sixteenth to the Nineteenth Centuries." In M. Perez Garcia and L. De Sousa (eds.), *Global History and New Polycentric Approaches: Europe, Asia and the Americas in a World Network System*, 255–73. Singapore: Palgrave Macmillan.

Salmon, M.A. 2000. *The Aurelian Legacy: British Butterflies and their Collectors*. Berkeley: University of California Press.

Sambourne, E.L. 1881. "Crinoletta Disfigurans. An Old Parasite in a New Form." *Punch* (December 6).

Saxby, F.W. 1898. "How to Photograph Through a Fly's Eye." *Knowledge: An Illustrated Magazine of Science* 21(NS13): 189.

Schachat, S.R. 2014. "Drawn Before Wonderland: Bizarre Illustrated Insects of the Nineteenth Century." *American Entomologist* 60(3): 162–5.

Schindler, R.A. 1988. "Art to Enchant: A Critical Study of Early Victorian Fairy Painting and Illustration." PhD dissertation, Brown University, Providence, RI.

Schmidt, A., Call L., Macheiner L., and Mayer H. 2019. "Determination of Vitamin B12 in Four Edible Insect Species by Immuno-affinity and Ultra-high Performance Liquid Chromatography." *Food Chemistry* 281: 124–9.

Scott, D.B. 1868. *The Mask*, vol.vol. 1. Princeton University. Available online: https://catalog.hathitrust.org/Record/000061081 (accessed May 22, 2023).

Scranton, P.B., ed. 1985. *Silk City: Studies on the Paterson Silk Industry, 1860–1940*. Newark: New Jersey Historical Society.

Scudder, S.H. 1890. *Volume XIII: The Tertiary Insects of North America*. Report of the U.S. Geological and Geographical Survey of the Territories. Available online: https://pubs.usgs.gov/unnumbered/70038967/report.pdf (accessed October 14, 2022).

Secord, A. 1994. "Corresponding Interests: Artisans and Gentlemen in Nineteenth-Century Natural History." *The British Journal for the History of Science* 27(4): 383–408.

Secord, A. 2015. "Elizabeth Gaskell's Social Vision: The Natural Histories of Mary Barton." In B. Marsden (ed.), *Uncommon Contexts: Encounters Between Science and Literature, 1800–1914*. Pennsylvania: University of Pittsburgh Press.

Secord, J.A. 2000. *Victorian Sensation: The Extraordinary Publication, Reception, and Secret Authorship of Vestiges of the Natural History of Creation*. Chicago: University of Chicago Press.

Sedgwick, A. 1845. "Vestiges." *Edinburgh Review* 82(3): 1–85.

Sedgwick, A. 1890. *The Life and Letters of Adam Sedgwick*, vol. 2. Cambridge: Cambridge University Press.

Seymour, R. 1831. *Cholera "Tramples the victors & the vanquished both."* National Institutes of Health, U.S. National Library of Health, Digital Collections. Available online: http://resource.nlm.nih.gov/101393375 (accessed March 6, 2018).

Sharp, D. 1873. *The Object and Method of Zoological Nomenclature*. London: Janson, Williams & Norgate.

Sharpe's London Magazine. 1846. "The Hearth Cricket" (April 11) 1, 24; British periodicals, p. 376.

Shepherd, J.A. 1991. *The Crimean Doctors: A History of the British Medical Services in the Crimean War*, vol. II. Liverpool: Liverpool University Press.

Sheppard, C.A. 2004. "Benjamin Dann Walsh: Pioneer Entomologist and Proponent of Darwinian Theory." *Annual Reviews of Entomology* 49(1): 1–25.

Sheppard, W.S. 1989. "A History of the Introduction of Honey Bee Races into the United States, Parts I and II." *American Bee Journal* 129: 617–19, 664–7.

Sherborn, C.D. 1902–1933. *Index Animalium; sive, Index nominum quae ab A.D. MDDCCLVIII generibus et speciebus animalium imposita sunt*. London: Trustees of the British Museum.

Shermer, M. 2002. *In Darwin's Shadow: The Life and Science of Alfred Russel Wallace*. Oxford: Oxford University Press.

Shindler, K. 2018. *A Museum at War: Snapshots of the Natural History Museum During World War One*. London: Natural history Museum.

Shipley, A.E. 1915. *The Minor Horrors of War*. London: Smith, Elder & Co.

Shultz, H.A. 1991. "100 Years of Entomology in the Department of Defense." In J. Adams (ed.), *Insect Potpourri: Adventures in Entomology*, 61–72. Gainesville: Sandhill Crane Press, Inc.

Silk Association of America Annual Report. 1913. Volume 41.

Silk Association of America Annual Report. 1921. Volume 49.

Silver, C.G. 1999. *Strange and Secret Peoples: Fairies and Victorian Consciousness*. New York: Oxford University Press.

Simkin, J. 2017. "Trench Rats." Spartacus Educational. Available online: http://spartacus-educational.com/FWWrats.htm (accessed January 2, 2018).

Simmonds, P. 1885. *The Animal Food Resources of Different Nations*. London and New York: E and F.N. Spon.

Simmons, J.E. 2016. *Museums: A History*. London: Rowman & Littlefield.

Simoons, F. 2014. *Food in China: A Cultural and Historical Inquiry*. Milton Park, Abingdon-on-Thames, Oxfordshire: Taylor & Francis Group.

Smallman-Raynor, M.R. 2004. "Impact of Infectious Diseases on War." *Infectious Disease Clinics of North America* 18(2): 341–68.

Smallwood, W.M. and Smallwood, M.S.C. 1941. *Natural History and the American Mind*. New York: Columbia University Press.

Smart, P. 1975. *The International Butterfly Book*. New York: T.Y. Crowell Company.

Smith, E.H. 1976. "The Comstocks and Cornell: In the People's Service." *Annual Review of Entomology* 21: 1–26. Available online: https://www.annualreviews.org/doi/pdf/10.1146/annurev.en.21.010176.000245 (accessed December 7, 2021).

Smith, J.D.D. 2001. *Official Lists and Indexes of Names and Works in Zoology Supplement 1986–2000*. London: International Trust for Zoological Nomenclature.

Smith, L. 2015. *Lewis Carroll: Photography on the Move*. London: Reaktion Books.

Sneden, R.K. 2002. *Eye of the Storm: A Civil War Odyssey*. New York: Simon and Schuster.

Snowden, F. 2008. *The Conquest of Malaria: Italy, 1900–1962*. New Haven: Yale University Press.

Sorensen, W.C. 1995. *Brethren of the Net: American Entomology, 1840–1880*. Tuscaloosa: The University of Alabama Press.

Sorensen, W.C., Smith, E.H., Smith, J.R., with Weber, D.C. 2018. *Charles Valentine Riley: Founder of Modern Entomology*. Tuscaloosa: The University of Alabama Press.

Soros, S.W. and Walker, S., eds. 2004. *Castellani and Italian Archaeological Jewelry*. New York: Yale University Press.

Souter, A. and McCain, S. 2018. "The Aesthetic Movement: Movement Overview and Analysis." TheArtStory.org. Available online: http://www.theartstory.org/movement-aesthetic-art.htm (accessed March 17, 2018).

Sowerby, A. 1925. *A Naturalist's Note-book in China*. Shanghai: North-China Daily News & Herald.

Spielman, A. and D'Antonio, M. 2001. *Mosquito: The Story of Man's Deadliest Foe*. London: Faber and Faber.

Stahl, P.J., ed. 1867. *Vie Privée et Publique des Animaux*. Paris: J. Hetzel. Available online: https://archive.org/details/viepriveetpubl00stahuoft/page/n7/mode/2up?ref=ol&view=theater (accessed October 14, 2022).

Stainton, H.T. 1856–1859. *Manual of British Butterflies and Moths*. London: John Van Voorst.

Stauffer, R.C., ed. 1975. *Charles Darwin's Natural Selection: Being the Second Part of His Big Species Book Written from 1856–1858*. Cambridge: Cambridge University Press.

Stebbins, T.E., Comey, J.L., and Quinn, K.E. 2000. *The Life and Work of Martin Johnson Heade: A Critical Analysis and Catalogue Raisonné*, 2nd edn. New Haven: Yale University Press.

Steiner, P.E. 1968. *Disease in the Civil War*. Springfield: Charles C. Thomas.

Stephens, J.F. 1829. *Illustrations of British Entomology. Haustellata*, vol. II. London: Baldwin and Cradock.

Stephenson, J. 1831. *Medical Botany: Or Illustrations and Descriptions of the Medicinal Plants of the London, Edinburgh, and Dublin Pharmacopœias*. London: John Churchill.

Steward, J.H. 1938. "Basin-Plateau Aboriginal Sociopolitical Groups." *Bureau of American Ethnology*, Bulletin No. 120. Washington, DC.

Steward, J.H. 1941. "Culture element distributions: XIII. Nevada Shoshoni." *University of California Anthropological Records* 4(2): 209–59.

Stewart, A.M. 1865. *Camp, March and Battle-field; Three Years and a Half with the Army of the Potomac*. Philadelphia, PA: J.B. Rodgers.

Stewart, S. 1984. *On Longing: Narratives of the Miniature, the Gigantic, the Souvenir, the Collection*. North Carolina: Duke University Press.

Stokstad, M. 1999. *Art History*, Revised edn. New York: Harry N. Abrams.

Stork, N.E., McBroom, J., Gely, C., and Hamilton, A.J. 2015. "New Approaches Narrow Global Species Estimates for Beetles, Insects, and Terrestrial Arthropods." *Proceedings of the National Academy of Sciences USA* 112: 7519–23.

Strickland, H.E. 1843. Report of a committee appointed "to consider of the rules by which the nomenclature of zoology may be established on a uniform and permanent basis." Report of the 12th meeting of the British Association for the Advancement of Science, 105–21.

Stuckley, C.F., Scott, W.P., and Lindsay, S.G. 1987. *Berthe Morisot: Impressionist*. New York: Hudson Hills Press.

Surkes, S. 2019. "Grasshoppers Swarm into Israel from Syria." *Times of Israel*. June 5. Available online: https://www.timesofisrael.com/grasshoppers-swarm-into-israel-from-syria/ (accesed May 23, 2023).

Sutton, M. 1988. *Insects as Food: Aboriginal Entomophagy in the Great Basin*. Anthropological Paper No. 33. Menlo Park: Ballena Press.

Swainson, W. 1834. *A Preliminary Discourse on the Study of Natural History*. London: Longman, Rees, Orme, Brown, Green and Longman.

Swayne, M. [Maurice Nicoll]. 1917. *In Mesopotamia*. London: Hodder and Stoughton.

Takkanen, A. 2019. "Samuel Slater: American Industrialist." *Encyclopaedia Britannica*. Available online: https://www.britannica.com/biography/Samuel-Slater (accessed November 3, 2019)

Talairach-Vielmas, L. 2014. *Fairy Tales, Natural History and Victorian Culture*. New York: Palgrave Macmillan.

Tannahill, R. 1973. *Food in History*. London, UK: Stein and Day.

Tennyson, A.L. 1850. "In Memoraim A. H. H." Available online: https://ia804701.us.archive.org/7/items/inmemoriambyalfr00tennuoft/inmemoriambyalfr00tennuoft.pdf (acessed July 5, 2023).

The National Gallery (London). 2023. The Execution of Lady Jane Grey, by Paul Delaroche, painting referenced in Chapter 4, page 119. Available online: https://www.nationalgallery.org.uk/paintings/paul-delaroche-the-execution-of-lady-jane-grey (accessed July 5, 2023).

The State of Queensland, Department of Agriculture and Fisheries. 2020. "The prickly pear story." Available online: https://www.daf.qld.gov.au/__data/assets/pdf_file/0014/55301/prickly-pear-story.pdf (accessed December 2, 2021).

Thomas, D.H., Kautz, R.R., Melhorn, W.N., Thompson, R.S., and Trexler, D.T. 1983. "The Archeology of Monitor Valley. 1. Epistemology." *American Museum of Natural History Anthropological Papers* 58(1): 1–194.

Thompson, C. and Pape, T. 2016. "Sherborn's Influence on *Systema Dipterorum*." *ZooKeys* 550: 135–52.

Thompson, D.A. 1913. *On Aristotle as a Biologist, with a Prooemion on Herbert Spencer: Being the Herbert Spencer Lecture Delivered Before the University of Oxford, on February 14, 1913*. Oxford: Clarendon Press.

328th Field Artillery, eds., United States Army. 1919. *The Red Guidon*. Milwaukee, WI: Hammersmith-Kortmeyer Co. Available online: https://babel.hathitrust.org/cgi/pt?id=nyp.33433082131131&view=1up&seq=7 (accessed May 23, 2023).

Times, The. 1858. "What a Pity it is That the Thermometer Fell Ten." (June 18), 9.

Tocqueville, A. de. [1838] 1863. *Democracy in America*, vol. 1. Trans. Henry Reeve, ed. Francis Bowen. Cambridge: Sever and Francis.

Tolini, M. 2002. "'Beetle Abominations' and Birds on Bonnets: Zoological Fantasy in Late-Nineteenth-Century Dress." *Nineteenth-Century Art Worldwide* 1(1): 1–20.

Tomes, N. 1998. *The Gospel of Germs: Men, Women, and the Microbe in American Life*. Cambridge, MA: Harvard University Press.
Topham, J. 1992. "Science and Popular Education in the 1830s: The Role of the *Bridgewater Treatises*." *The British Society for the History of Science* 25(4): 397–430.
Traveling Scriptorium. 2013. Iron Gall Ink. Published May 21. Available online: https://travelingscriptorium.com/2013/03/21/iron-gall-ink/ (accessed December 2, 2021).
Trivedy, K., Kumar, S. N., Mondal, M., and Kumar Bhat, C.A. 2008. "Protein Banding Pattern and Major Amino Acid Component in De-Oiled Pupal Powder of Silkworm, *Bombyx mori* Linn." *Journal of Entomology* 5(1): 10–16.
Tuxen, S.L. 1973. "Entomology Systematizes and Describes: 1700–1815." In R.F. Smith, T.E. Mittler, and C.N. Smith (eds.), *History of Entomology*, 95–118. Palo Alto, CA: Annual Reviews Inc.
Twain, M. 1869. *The Innocents Abroad*. Chapter XLIX (49).
Twidle, A. 1920. *Beautiful Butterflies of the Tropics: and How to Collect Them*. London: R.T.S.
UK Parliament. 2022. The Reform Act 1832. Available online: https://www.parliament.uk/about/living-heritage/evolutionofparliament/houseofcommons/reformacts/overview/reformact1832/ (accessed January 26, 2022).
United States Entomological Commission. 1878. Report of the United States Entomological Commission for the year 1877 relating to the Rocky Mountain Locust. Washington: G.P.O. p. 393. (USEC 178).
Usub, T., Lertsatitthanakorn C., Poomsa-ad, N., Wiset, L., Yang, L., and Siriamornpun, S. 2008. "Experimental Performance of a Solar Tunnel Dryer for Drying Silkworm Pupae." *Biosystems Engineering* 101: 209–21. Available online: http://dx.doi.org/10.1016/j.biosystemseng.2008.06.011 (accessed October 14, 2022).
Van Bruyssel, E. 1888. *The Population of an Old Pear-Tree, or Stories of Insect Life*. London: Macmillan.
Van Druten, T., ed. 2013. *Pierre-Joseph Redouté: Botanical Artist to the Court of France*. Rotterdam: Haarlem Teylers Museum.
Van Gogh Museum. 2018. "Roses." Amsterdam (Vincent van Gogh Foundation). Available online: https://www.vangoghmuseum.com/en/collection/s0187V1962 (accessed April 14, 2018).
van Huis, A., Van Itterbeeck, J., Klunder, H., Mertens, E., Halloran, A., Muir, G., and Vantomme, P. 2013. *Edible Insects: Future Prospects for Food and Feed Security. Food and Agricultural Organization of the United Nations*. FAO Forestry Paper #171.
Vanterpool, A. 2010. *Silk Trains of North America*. Edmonton, Alberta: Alberta Pioneer Railway Association.
Vega, C., curator. 2015. *Alice: 150 Years of Wonderland New York:* The Morgan Library & Museum. June 26–October 12. Available online: https://www.themorgan.org/exhibitions/online/alice (accessed October 14, 2022).
von Sydow, M. 2005. "Charles Darwin: A Christian Undermining Christianity?" In D.M. Knight and M.D. Eddy (eds.), *Science and Beliefs: From Natural Philosophy to Natural Science, 1700–1900*. Burlington: Ashgate.
Wakely, T. 1841. "The Poor-Law Amendment Act." *The Lancet* May 1: 193–6.
Waldbauer, G. 2009. *Fireflies, Honey, and Silk*. Berkeley: University of California Press.
Wallace, A.R. 1869. *The Malay Archipelago*. London: MacMillian.

Wallbank, T.W., Taylor, A.M., and Bailkey, N.M. 1971. *Civilization Past & Present*, 4th edn. Glenview, IL: Scott, Foresman and Company.
Walsh, B.D. 1864a. "On Certain Entomological Speculations of the New England School of Naturalists." *Proceedings of the Entomological Society of Philadelphia* 3: 207–49.
Walsh, B.D. 1864b. "On the Origin of Races and Species by Phytophagic Isolation." *Proceedings of the Entomological Society of Philadelphia* 3: 403–30.
Walsh, K. 2002. *The Representation of the Past: Museums and Heritage in the Post-Modern World*. London: Routledge.
Walton, O.C. and Stevens, M. 2018. "Avian Vision Models and Field Experiments Determine the Survival Value of Peppered Moth Camouflage." *Communications Biology*, 1: 118.
Wandersee, J.H. and Schussler, E.E. 1999. "Preventing Plant Blindness." *The American Biology Teacher* 61: 82, 84, 86.
Ward, A. 2013. "Cheese Confessionals: I Ate Casu Marzu, aka 'Maggot Cheese'." Serious eats, January 28. Available online: https://www.seriouseats.com/2013/01/casu-marzu-cheese-maggot-maggots-italian-what-it-tastes-like-illegal.html (last updated August 10, 2018).
Ward, M. 1859. *World of Wonders Revealed by the Microscope*. London: Groombridge and Son.
Warren Photographic. Peppered moth. WP11321/www.warrenphotographic.co.uk/11321-peppered-moths.
Watson, L.R. 1928. "Controlled Mating in Honeybees." *Quarterly Review of Biology* 3: 377–90.
Watts, I. [1715] 1904. "How Doth the Little Busy Bee." In Carmen et al. (eds.), *Worlds Best Poetry*, vol. 1, 108. Philadelphia: Morris & Co.
Way, A. *Caricature of Charles Darwin*. CUL DAR 204, f.29. Available online: https://www.cam.ac.uk/research/news/mr-darwin%E2%80%99s-postbag (accessed 31 January 2018).
Weintraub, S. 1997. *Uncrowned King: The Life of Prince Albert*. New York: The Free Press.
Weiss, H.B. 1930. *The Pioneer Century of American Entomology*. New Brunswick: Published by the author.
Werness, H.B. 2006. *The Continuum Encyclopedia of Animal Symbolism in World Art*. New York: The Continuum International Publishing Group Inc.
WHMA (Whistler House Museum of Art). 2016. "Whistler and his Butterfly Signature." January. Available online: http://www.whistlerhouse.org/index.php/in-the-news-2/111-whistler-butterfly-signature (accessed March 19, 2018).
Wijayasinghe, M.S. and Rajaguru, A.S. 1977. "Use of Silkworm (*Bombyx mori* L.) Pupae as a Protein Supplement in Poultry Rations." *Journal of the National Science Foundation of Sri Lanka* 5(2): 95–104.
Wilde, O. [1890] 2012. *The Picture of Dorian Gray and Other Writings*. London: Random House.
Wilde, O. 1904. *Salomé: A Tragedy in One Act*. London: Melmoth & Co.
Wiley, B.I. 2008a. *The Life of Billy Yank: The Common Soldier of the Union*. Baton Rouge: Louisiana State University Press.
Wiley, B.I. 2008b. *The Life of Johnny Reb: The Common Soldier of the Confederacy*. Baton Rouge: Louisiana State University Press.
Williams, W.S. 2016. *George Eliot, Poetess*. London: Routledge.

Willingham, E. 2011. "Of Lice and Men: An Itchy History." *Scientific American*. Available online: https://blogs.scientificamerican.com/guest-blog/of-lice-and-men-an-itchy-history/ (accessed February 7, 2018).

Wilson, J. 1835. *A Treatise on Insects*. Edinburgh: Adam and Charles Black. Available online: https://www.biodiversitylibrary.org/item/264264 (accessed October 14, 2022).

Winter, A. 1997. "The Construction of Orthodoxies and Heterodoxies in the Early Victorian Life Sciences." In B. Lightman (ed.), *Victorian Science in Context*, 24–50. Chicago: University of Chicago Press.

Withers, B.G. and Craig, S.C. 2003. "Chapter 2, The Historical Impact of Preventative Medicine in War." In P.W. Kelley (ed.), *Textbooks of Military Medicine: Military Preventive Medicine, Mobilization and Deployment*, vol. l., 21–57. Washington, DC: Office of the Surgeon General.

Wood, C. 2000. *Fairies in Victorian Art*. Woodbridge: Antique Collectors' Club.

Wood, I. 1994. *The Merovingian Kingdoms 450–751*. Harlow: Pearson Education Limited.

Wood, Rev. J.G. 1862. *Common Objects of the Country*. London: Routledge, Warne & Routledge.

Worth, J. 2002. *Call the Midwife. A Memoir of Birth, Joy, and Hard Times*. New York: Penguin Books.

Wright, A.E. 1900. "On the Results Obtained by the Anti-Typhoid Inoculations in the Beleaguered Garrison in Ladysmith." *Lancet* 2: 95.

Wright, C.S. and Priestly, R.E. 1922. *Glaciology. British (*Terra Nova) *Antarctic Expedition 1910–1913*. London: Harris & Sons.

Xu, Y. 2012. "Industrialization and the Chinese Hand-Reeled Silk Industry (1880–1930)." *Penn History Review* 19(1): 27–46.

YaoWang, C. 1989. "Animal Raising and Plant Cultivation on an Integrated Fish Farm." *Integrated fish farming in China*, NACA *Technical Manual* 7, Bangkok, Thailand: Aquaculture Centers in Asia and the Pacific. Available online: https://agris.fao.org/agris-search/search.do?recordID=XF2016009750 (accessed October 14, 2022).

Yhoung-Aree, J., Puwastien, P., and Attig, G.A. 1997. "Edible Insects in Thailand: An Unconventional Protein Source?" *Ecology of Food and Nutrition* 36(2–4): 133–49. Available online: http://dx.doi.org/10.1080/03670244.1997.9991511 (accessed October 14, 2022).

ACKNOWLEDGMENTS

Introduction

The editors acknowledge with gratitude the insightful comments of Professors Chris Otter, Bruce McPheron (The Ohio State University), Dr. Gary Miller (USDA-ARS Systematic Entomology Laboratory), and Bill Styer (independent scholar) on the introduction to this volume.

INDEX

abolishment of the slave trade 2–3
Aboriginal Australian food 94
"About Looking" (Berger) 156–7
Acacia kempeana shrub 94
Academy of Natural Sciences of Philadelphia 19–20
Adam Bede (Eliot) 196
The Adventures of Pinocchio (Collodi) 154, 155
Aedys aegypti see mosquitoes
Aesthetic Movement (mid-1870–1880s) 210
Africa
  Africanized honey bee 106
  edible insects 94–5
  see also colonial powers/colonies; slavery
Agassiz, Louis 18–19, 42, 144–5, 147
agrarian society 15–17
  see also food
Albert, Prince 11–12, 13
Aleppo oak (*Quercus infectoria*) 121–3
Alice Through the Looking Glass (Carroll) 182–92
allantoin 56
Allegory of Love, Cupid and Psyche (Goya) 206
Alma-Tadema, Sir Lawrence 215
alpha-linolenic acid (ALA) 82
Amazon Basin 31–3
American Entomological Society 19–20
American Entomology (Say) 19
American Philosophical Society 19–20
amino acids 90
analogies *see* symbols/symbolism
Andrews, Hazel 38
Anglican Church 129
  see also Church
aniline dyes 111, 119–20
animal kingdom in art 223–4
Antarctica 12–13, 28
anti-fly campaigns 50, 51
anti-Semitism 96
ants 33, 34–5, 37, 38–9, 89–90, 130–1, 151, 164–5, 173–4
apiculture 101, 102–5
  see also bees/beekeeping
Appleman, P. 126–7
Arago, Jacques 202
Armstrong, Isobel 187–8
Arnold, Matthew 142–3
art 199–228
  acquiring an eye for nature 214
  advancement of entomology 217–19
  among the plants 219–21
  animal kingdom 223–4
  beyond bees and butterflies 221, 223
  caring for insects 221
  fleas 202–4
  insect-fairy interactions 207–9

kaleidoscopes/butterflies 209–12
Lepidoptera 208–13, 215, 221, 223, 223–4
mocking insect collecting 216–17
mythology and arthropods 206–7
ornamentation 212–14
plague 204–6
fleas 202–4
us and them, otherness 228
Vincent Van Gogh 225–8
yellow fever 200–2
Artigue, Albert-Émile 215
Ashmead, William H. 33, 40–1
Assam silk moth (*Antheraea assamensis*) 91
Association of American Ornithologists 22
"association plantings" 90
Astrolabe expedition 30, 31
At the Bee Yard (Samokish-Sudkovskaya) 221
The Atlantic Monthly 145
The Aurelians (Cooke) 218
Australia, edible insects 94

Bacon, Edward W. 47
"bad air", miasmas 13–14, 49, 69, 70–1, 75
Baer, William 57–8
Baly, Joseph Sugar 35–6
Bank Holidays Act 182
Banks, Joseph 28
Barber, L. 129, 130
Baroque art 199
bars, anti-mosquito 72, 73
Basire, James 165
Bates, Henry Walter 32–3, 35, 144, 177–8
Beach at Cape May (Miller) 200, 201
Beagle, HMS 11, 13, 28, 30–1, 35, 128, 139, 140, 142, 143
Beardsley, Aubrey 210
Beauregard, P.G.T. 60, 61
bees/beekeeping 102–9
imports and diversity 105–7
new era in 110–11
technological advances in 109
see also honey bee
beetles 21, 35–6, 192–3, 174, 175, 224, 225, 226–7
ornamentation 212–13, 214
see also individual types of beetle ...
beetle-wings 197
BEF (British Expeditionary Forces) 54, 71, 72, 74, 76
"Beondegi" (silk moth pupae) 93
see also silk moth/worm
Berger, J. 156–7
Berlin Acclimatization Society 105
Bernard, Emile 227–8
Bettcher, C.W. 62
Bevan, Edward 221, 222
Bible 95, 127, 128, 136, 140, 148
"Bierstadt Butterflies" (Albert Bierstadt) 211
Bigelow, H.J. 14
Billings, J.D. 66
Biologia Centrali-Americana 35, 36
biting flies 68–74, 226–7
Black Death 204–6
Blackburn, Rev. Thos. 33
Blake, William 203–4
Blane, Juan Manuel 202
blister beetles 192, 226–7
"blue ointment" for lice 67
Böcklin, Arnold 205–6
Boer War 50, 54
Boisduval, Jean Baptiste 21, 30, 31
Bôle, Jeanne 215
Bombay Plague 204
Bonaparte, Napoleon 16, 212, 219–20
Book of the Fly (Hardy) 170–1
Book of Nature 161–2
Boston Society of Natural History 19–20
Botswana 95
Bowler, Henry Alexander 210
Bown, N. 167
box hives 102–4
Boxer Uprising (1910) 90
Boy with Butterfly Net (Matisse) 215
Bread-and-Butterfly 173, 190–2
Bridgewater Treatises 11, 135–7, 138
Britain 2–4
Entomological Society of London 12–13, 132
First World War 23–4, 47–8
imported species debacles 123–4
industrial revolutions 4–11
Natural History Museum, London 12, 13, 32–3, 34, 35–6, 42, 75, 76

Natural Theology
end of Age of Industry 146–8
and history 129–32
science and medicine 11–15
silk trade 113–15, 116–17
treaty port systems, silk trade 113–14
see also First World War
British East India Company 4
British Expeditionary Forces (BEF) 54, 72, 74, 76
British Museum *see* Natural History Museum
Brontë, Anne 10–11, 196–7
Brontë, Charlotte 10–11, 196
brooches 212, 213
Brooke, John Hedley 127
Brown, Edith 208–9
Browne, Janet 7, 128
Brullé, Auguste 30–1
bubonic plague 204–6
Buckley, Arabella 159–60
Budgen, L.M. 189, 193, 218
Bureau of Entomology, USDA 76
Burmeister 21
butter muslin nets 74
Butterflies: Design for the Gentle Art of Making Enemies (Whistler) 210, 211
butterflies 30–1, 186, 192–3, 196
art 206–7, 208, 209–11, 215, 224
collecting 215
kaleidoscopic 209–12
ornamentation 212–13
Bread-and-Butterfly 173, 190–2
lac products 120, 121
Butterfly Hunting (Rowlandson) 216, 217
The Butterfly (Portrait of Mabel Hamilton) (Collier) 210
The Butterfly was a 'Lady' (Sambourne) 197, 198
butterfly wings 206–8

cacti 118, 119, 123–4
Cactoblastis cactorum moth 41
cadaverine 56
calcium oxide/hydroxide 53
California Gold Rush (1848–1855) 17
Cambridge University 129
Camp Chase, Columbus, Ohio 68
camp sanitation 50, 51
Canadian Department of Agriculture 22
Cannadine, David 7
capitalism 23–4
Carabidae 39, 194
Carbó, Mariano Fortuny y 202, 203
caring for insects 221, 222
Carlyle, Thomas 9–10
Carniolan honey bees 106–7
Carpenter, W.M. 223
Carroll, Lewis 182–92, 197–8
casu marzu 97–8
Catching Butterflies (Eisman-Semenowsky) 215
caterpillars 95, 182–4
Caucasian honey bees 106–7
Caventou, Joseph 69–70
"Chamber Hive" 103–4
"Changeable Hive" 103–4
chapalinas (small grasshoppers) 89–90
Chartist Movement (1838–1848) 9–10
The Chase (Hopley) 216
"Chasing Butterflies" (Berg) 120, 121
Chasse au Papillon (1906) 216
La Chasse aux Papillons (Artigue) 215
Chasse aux Papillons (Morisot) 215
chattel slavery 15–16
"chatting"/"skirmishing" 65, 66
Chaudet, Antoine-Denis 206
cheese fly (*Piophila casei*) 97–8
Cherry, R. 150
Child Chasing Beetle with Butterfly Net (Allen & Ginter Company) 223, 224
child labor laws 117
childbirth 14
Childeric I 220
Chilperic I 221
The Chimes (Dickens) 192
China
edible insects 90–2
Neolithic, Yangshao culture 79
silk industry 113–14, 118
Chinese Communist Party (CCP) 90
Chinese Revolution (1911) 90
Choix des Plus Belles Fleurs et des Plus Beaux Fruits (Redouté) 219
cholera 13–14, 47–9, 49, 54–6

Christianity 15, 90, 127, 129–31, 130–1
 see also Natural Theology
A Christmas Carol (Dickens) 192
Church of Ireland 127
Church, the 129, 145–6
 see also Natural Theology
cicadas 88, 226
Civil War, US (1861–1865) 15–18, 17, 18
 disease 48–50
  biting flies 69, 70–2
  filth flies 51–2, 54, 56
  fleas 58–9, 60, 61
  lice 62, 65, 67
  metamorphosis 75, 77
Clerck, Carl 34
Cleveland, Grover 18
Closson, William Baxter 211
cochineal 111, 118–20
cockroaches 152–3, 170, 226–7
cocoon reeling 91, 112–13
Cohen, Morton 185
Coleoptera 21, 35
 see also beetles
Coleridge, Samuel 146
collecting insects 12, 35, 215
 mocking of/satire 216–17
Collier, John Maler 210
Collodi, Carlo 154, 155
colonial powers/colonies 2–4
 edible insects 90–2
 imported species debacles 123–4
 Oaxaca and cochineal 119–20
 symbolism 165
 see also Britain; United States
Colophospernum sp. 95
coloring agents 120
 see also dyes
Comstock, Anna 21–2, 110
Cook, James 28
corn earworms 90
Corn Laws 10
Corsica 98
cossid moths 94
The Cricket on the Hearth: A Fairy Tale of Home (Dickens) 192
crickets 87, 90–1, 94–5, 154–5, 192–3, 226
 see also individual types of cricket ...
Crimean War (1854–1856) 14–15, 54–6, 59, 66–7, 75–6
"Crystal Palace", Great Exhibition of 1851, Hyde Park 12
cultural hegemony 36–8
Cupid and a Butterfly (Thorvaldsen) 206, 207
Cupid Playing with a Butterfly (Chaudet) 206
Cupid and Psyche 206–7, 212
cynipid wasp (*Cynips gallae-tinctoriae*) 121–3

Dadant, Charles 109
Dadd, Richard 208
Dalton, Henry 212
Daniel Deronda (Eliot) 195
Dans le Parc (Morisot) 215
Dark Bee of Northern Europe (*Apis mellifera mellifera*) 105–6
Darwin, Charles 24–5, 30, 35, 43, 147–8
 On the Origin of Species 11, 38, 125, 128, 137–8, 144–5, 164, 174
 caricature 186, 187
 The Expression of Emotions in Man and Animals 184
 insect-mania 174
 natural selection 24–5, 137–45, 148, 174
 and Natural Theology 137–43
 reactions to 143–6
 taxonomy 175–6, 191–2
 "*Transmutation*" notebooks 139–40
 see also *Beagle*
Darwin Correspondence Project 144–5
"Darwin and his Hobby" (Way) 186, 187
"day flies" *see* filth flies
De Geer, Charles 34
Dead Sea 95, 96
Deere, John 16
Deism 15, 126
 see also Natural Theology
Dejean, Pierre François Marie Auguste 21
delousing 65, 66, 67
Descent of Man (Darwin) 141–2
"Designs after Nature" series (Sambourne) 197, 198
diarrhea 50, 54
Dickens, Charles 10, 192, 195–6

"Dickensian" 13
Diptera 21, 33, 42
 see also true flies
disease 47–77
 biting flies 68–74
 filth flies 47, 48, 50–8
 fleas 58–61
 Germ Theory of Disease 14–15, 47–8, 114
 "germs with legs" 170–2
 lice 62–8
 metamorphosis 74–7
 plague 202–6, 223
diversity of honey bees 105–7
"Dividing Hive" 103–4
Dodd, Alan Parkhurst 41
Dodd, Frederick Parkhurst 41
domestic insect-mania *see* insect-mania
Donovan, E. 92
Doolittle, G.M. 107–9
Dorian Gray (Wilde) 196–7, 221
The Doubt: Can these Dry Bones Live? (Bowler) 210
Dover Beach (Arnold) 142–3
Doyle, Richard 208, 209
Doyle, Sir Arthur Conan 50, 169
dragonflies/damselflies 21, 189, 208
Drame de la Peste (Géricault) 205–6
Drosophila melanogaster 20–1
Dulac, Edmund 217
dyes 111, 118–20, 121–3
dysentery 50, 54
Dytiscidae 90–1

Earl of Bridgewater 135–7
 see also Bridgewater Treatises
Edible insects: Future prospects for food and feed security (FAO Forestry Report) 95
edible insects *see* food
Edinburgh Review 180–1
Education Act (Forster's Act) 8–9
Eggleston, M.A. 68
Egypt 173–4
 beekeeping 101, 102
 "Egyptomania" 212
Eisman-Semenowsky, Émile 215
Eliot, George 10, 161, 178–80, 194–8
Elliott, I.H. 68, 69
Emery, Carlo 38
emperor moth (*Saturnia pyri*) 226
Empey, A.G. 65, 66
"*en tremblant*" 212–13
Endeavour, journey of James Cook 28, 30
Endoxyla leucomochla moth 94
Entomological Society of London 12–13, 132
Entomological Society of Pennsylvania 19–20
The Entomologist (Spratt) 216
The Entomologist's Dream (Dulac) 217
Episode de la Fièvre Jaune à Valence (José) 202
Episodes of Insect Life (Budgen) 189, 193, 218
Episodio de la Fiebre Amarilla en Buenos Aires (Blane) 202
Eri silk moth (*Samia cynthia ricini*) 91
escamole (ants) 89–90
An Essay on the Principle of Population as it Affects the Future Improvement of Society (Malthus) 139–40
Europe
 Dark Bee 105–6
 edible insects 96–8
Evans, Edmund 205, 208, 209
Evenings at the Microscope (Gosse) 188
The Execution of Lady Jane Grey (Delaroche) 118–19
The Expression of Emotions in Man and Animals (Darwin) 184

Fabricius, Johann Christian 28, 30
fairies 166–7, 207–9
Fairy Feller's Master-Stroke (Dadd) 208
Fairy Frisket; or, Peeps at Insect Life (Tucker) 168–9
The Fairy Queen Takes an Airy Drive in a Light Carriage, a Twelve-in-Hand, Drawn by Thoroughbred Butterflies (Doyle) 208, 209
fantasies of immersion 167
FAO (Food and Agriculture Organization) 94–5
filth flies 47, 48, 50–8
Fink, Henry T. 197
fireflies 213, 227

First World War (1914–1918) 23–4
disease 47–8
biting flies 71, 72, 74
filth flies 54, 57–8
fleas 59
lice 62, 63, 64, 65, 67–8
metamorphosis 76–7
Fitch, Asa 20, 146–7
Flanders during the First World War 65
flavonoids 92
fleas 12, 47, 48, 58–61, 152–3, 189, 202–4
Fleming, Alexander 47–8
flies 21, 47, 48, 152–3, 170–1, 187
see also individual types of fly ...
fly-blown 56, 57
food 79–100
African countries 94–5
Australia and Pacific Island countries 94
casu marzu 97–8
China and other Asian countries 90–2, 93
edible grasshoppers in Asia 92–3
Europe 96–7
Mexico and other Latin American countries 89–90
Middle East 95–6
US Industrial Revolution 80–9
wasp cultivation in Japan 93
Food and Agriculture Organization (FAO) 94–5
Ford, Henry Justice 208–9
Forel, Auguste-Henri 37–40
Förster, Arnold 40–1
Fort Sedgwick, Petersburg, V.A. 47, 48
fossils 13, 127
Franco-Prussian War (1870–1871) 54
Frank Leslie's Illustrated Newspaper 200–1
Franklin, John 204
French polishing 120
Froment-Meurice, François-Désiré 212
Frost, Jack 69, 70
fruit fly (*Drosophila melanogaster*) 20–1, 123
Furniss, Harry 184–5

Gaimard, Joseph Paul 30
gall-based inks and dyes 121–3
Gallipoli, invasion of (1915) 52
Gardner, Edward 169
Gardner, Martin 185
gas engines 16–17
Gaskell, Elizabeth 174–7, 196
gender 38
art 218–19
cocoon reeling 91
entomology in other nations 21–2
Industrial Revolution in Britain 10–11
science and medicine in Britain 14–15
Genera Insectorum (Wytsman) 44
Genesis 127, 136
Geological Society of London 128
Géricault, Jean Louis André Théodore 205–6
Germ Theory of Disease 14–15, 47–8, 114
"germs with legs", insects as 170–2
Ghost of a Flea (Blake) 203
Giant Peacock Moth (Van Gogh) 226
giant water bugs 86, 90–1
Gibbs, P. 65
Global Biodiversity Information Facility (GBIF), Copenhagen 30
Glover, Towend 75
gnats 187, 189, 190
Goble, Warwick 208–9
God 11, 16, 126, 205
see also Natural Theology
Godman, Frederick DuCane 35
Gonimbrasia belina moth 95
Goober Peas 77
Gorham, Rev. Henry Stephen 35, 36
Gosse, Philip Henry 174–5, 180, 188
Goya, Francisco 204–5, 206
Grammar Schools Act (1840) 8
grana fina 118–19
Grandville, J.J. 209, 216
The Graphic newspaper 10
grasshoppers 84–5, 86, 87, 89–95, 96, 99, 199, 226–7
Gray, Asa 140–1, 145, 147
Great Basin, North America 83–4, 86–7, 94–5
Great Britain *see* Britain
Great Expectations (Dickens) 195–6
Great Flood 127, 136
Great Plains 17, 85
Great Reform Act (1832) 9
Great Salt Lake, US 82–3, 84–5, 95
green beetles 226
The Green Butterfly (Moore) 210

green-bottle fly larvae 90–1
Grinnell, George Bird 88
ground beetle (*Carabus nemoralis*) 39, 194
growth, in knowledge 42–5
growth and Major World Powers, *c.* 1820 2–4
grubs 94
see also maggots
Guatemala 35, 119–20
Günther, Albrecht Karl Ludwig Gotthilf 42
gypsy/spongy moth 123–4

Halal food 95
Half Hours with Insects (Packard) 228
Hamilton, I. 58
Handbook for Palestine and Syria (Thomas Cook and Son) 96
Hardy, G.H. 170–1
Hardy, Thomas 10
Harewood Hospital, Washington, District of Colombia 72, 73
Harmony in Grey and Green: Miss Cicely Alexander (Whistler) 210
Harper's Weekly 60, 61, 69, 70
Harris, Thaddeus W. 20
Hart, P. 52
Hatch Act (1887) 18
Heade, Martin Johnson 221, 223
hebo (*Vespula flaviceps* wasp spp.) 93
hebo wasp spp. 93
Herodotus 133–4
Herrick, Glenn W. 74, 75
Herschel, John 145–6
Heteroptera 21
Hinman, Wilbur 60
The Hireling Shepherd (Hunt) 210
Hirn, Jean Georges 219
"historically black" institutions 18
hive design, honey bees 102–5
Hobsbawn, Eric 1
Hogue, Charles 150, 151, 153, 154, 155
Holt, V.M. 96–7, 98
Homestead Act (1862) 17
Homo sapiens 79, 209–12
Homoptera 21, 33
see also cicadas; scale insects
Honey Bee Act (1922) 106
honey bee (*Apis mellifera*) 39, 102–5
importation, propagation, and diversity 105–7
industrial production of queens 107–9
honey production 95, 110–11
see also honey bee
The Honey-Bee (Bevan) 221, 222
honeydew (aphid excrement) 86, 95
Hong Kong 90
Hopley, Edward 208, 216
Hospital de Apestados (Goya) 204–5
house cricket (*Acheta domesticus*) 193, 218
house fly (*Musca domestica*) 50–1, 55, 170–1
The House-Fly at the Bar 171
Howard, L.O. 20–1, 54, 55, 76, 147
Hruschka, Francesco de 109
Huguenots 80, 81
Humphreys, Jennett 213
Hunt, S.B. 70
Hunt, William Holman 210
Huskisson, Robert 208
Huxley, Thomas Henry 8–9, 217–18, 145
Hydrophilidae 90–1
Hymenoptera 33, 185
see also ants; bees/beekeeping; wasps

ICZN (International Commission on Zoological Nomenclature) 43
Idea durvillei from New Guinea 30, 31
imagination 161
imagination of the loving gaze 167
imago 157
Imperialism 23–4
imported species debacles 123–4
imports of honey bees 105–7
"Inclined Bottom Board Hive" 103–4
Index Animalium (Sherborn) 42–3, 44
industrial
mechanization/technological advances in beekeeping 109
production of honey bee queens 107–9
industrial melanism 6
Industrial Revolution 4–11, 80–9
inks 111, 121–3
see also dyes
Insect Ways on Summer Days (Humphreys) 213
insect-derived commodities *see* products
"Insectes Utiles" (Aralette & Co.) 118, 119
insect-mania 174–5, 192–8
Insecto-Theology (Lesser) 126
Insects at Home (Wood) 169

Insects and their Habitations: A Book for Children (Committee of General Literature and Education) 164, 165
"Insensibility during surgical operations induced by inhalation" (Bigelow) 14
internal combustion gas engine 16–17
International Code of Zoological Nomenclature 39–40
International Commission on Zoological Nomenclature (ICZN) 43
An Introduction to Entomology (Kirby & Spence) 19, 128, 132–7, 154–5, 161–2, 180–2, 186, 215
Ireland
  Church 127
  potato famine 10
Isle of Wight 106
Israel 96
Italian honey bees 106–7
Italy, raw silk export market 113–14

Jabberwocky taxonomy 190–2
Jaffé, A. 153
Japan
  edible insects 91, 92–3
  Russo-Japanese War (1905) 56, 57
  silk industry 118
  wasp cultivation 93
Japan Black 120
Jarvis, Frank P. 62
Jefferson, Thomas 19
Jenyns, Leonard 180–1
jewelry *see* ornamentation
Jewish people
  anti-Semitism 96
  food laws 95
John the Baptist 95
José, Aparicio 202
Junge Dame Beim Schmetterlingfangen (Kaemmerer) 215
Justinian, Plague of 204

Kaemmerer, Frederik Hendrik 215
kaleidoscopes 209–12
katydids (*Anabrus simplex*) 82, 83, 84–5, 94–5, 130–1
Keats, John 161
Keen, William 56
Kieffer, Jean-Jacques 41
Kilbourne, Frederick 75
Kirby, William 132–7, 140, 149, 215
  Bridgewater Treatises 135–7, 138
  see also Introduction to Entomology
"knowing" gaze 167
knowledge 27–45
  basis of biodiversity discovery 29–36
  cultural hegemony 36–8
  organizing the growth in 42–5
  paradigm shifts 38–40
  taxonomic practice 40–1
Koch, Robert 14
Korea 92–3
kosher food 95
Kritsky, G. 150

La Bree, B. 77
labor laws 117–18
lac products 120–1
Lamarck, Jean-Baptiste 127, 135–6
Langon, É.L. 220
Langstroth, L.L. 104–5, 109
Laplace, Pierre 135–6
Larson, E.J. 148
Lasteyrie, Charles-Philibert de 200–1
Latreille, Pierre André 21, 28, 41
Läuse, German postcard from the Eastern front 62, 64
Lefroy, Harold 75
Lejeune, E. 208–9
Leopoldina, Maria 31
Lepidoptera 21, 33, 208–13, 215, 223, 223–4
  see also butterflies; moths
Lesser, Friedrich Christian 126
Lewes, George Henry 189
Lewis, George 33
lice 53, 62–8, 88
Life and Her Children: Glimpses of Animal Life from the Amoeba to the Insects (Buckley) 159–60
Lightman, B. 157–9
lime 53
Lincoln, Abraham 18
Lindroth, C.H. 27, 38, 39, 42
Linnaean Society 132, 144
Linné, Carl von (Carl Linnaeus) 27–9, 30, 40, 43, 44
Linnell, John 203

Lipcott's Magazine 197
Lippit, A.M. 156–7
Lister, Sir Joseph 14
literature and language 173–98
  Alice Through the Looking Glass 182–90
  entomological eloquence 182–5
  microcosms 192–8
  taxonomy 175–80
    Jabberwocky 190–2
"little people", insects as 165–9
Little and Wise (Newton) 130–1
Locust Catcher/"hopper dozer" 85–6
locusts 85–6, 90–1, 94–5, 130–1, 189, 223
Loew, Hermann 21
Long, Stephen H. 18, 19, 30
Longchamps, Sélys 21
On Longing (Stewart) 182
Longpré, Paul de 219, 220
"Looking-Glass Insects" 187
  see also Carroll; microscopes
Louisiana Purchase (1812) 16
"loving" gaze 167
Lubbock, Alice 193
Lubbock, Sir John 180–1, 182, 193
Lyceum Theater, London 197
Lyell, Charles 18–19, 128, 136, 139, 144, 180–1
Lyell, Katherine 174
lysine 90

McCormick reaper 16–17
Macedonia during the First World War 71, 72
madora worms 95
Madsen, D.B. 87
maggots 50–1, 56, 57, 130–1
Maine Agricultural Experiment Station 22
Major World Powers, *c.* 1820 2–4
malaria 69–71, 72, 73, 75, 96
"malaria defeatism" 71
Malay Archipelago 13, 31–3, 35
male pseudonyms in literature 10–11
Mali 95
Malthus, Thomas Robert 139–40
"manna from heaven" 95
Manson, Patrick 75, 170
Mary Barton (Gaskell) 176–7, 196
The Mask (1866) 197
Matisse, Henri 215
mauveine 119–20
Mazet, André 202
Méaulle, Fortuné Louis 208–9
Mécanique Céleste (Laplace) 135–6
Meckel, Heinrich 69
melanism in peppered moth 6
Mendel, Gregor 123
Meneaud-Lissenburg, Dudley 52
Merovingian kings 220–1
Mesopotamia during the First World War 52, 72, 74
metamorphosis 10, 74–7, 154–5, 157, 182–4, 227
"Metdugi" (rice-field grasshopper) 92–3
Mexican Expedition (1916–1917) 54, 55
Mexican-American War (1846–1848) 61
Mexico and other Latin American countries, food 89–90
miasmatic contagion 47–8, 49
  see also "bad air", miasmas
Michelet, Jules 212
microscopes 129–30, 152, 166, 182–90
Middle Eastern food 95–6
Middlemarch (Eliot) 178–80, 195
A Midsummer Night's Dream (Shakespeare) 208
The Midsummer Night's Fairies (Huskisson) 208
Midwives Act (1902) 14–15
migratory grasshoper (*Melanoplus sanguinipes*) 87
migratory locusts 130–1
Miller, Alfred Jacob 200, 201
Miller, G.L. 56
Milne, General 72, 74
Mississippi, bivouac 47–50
Missouri Compromise (1820) 15–16
Modern Political Zionism 96
modernity/modernization 150–5, 156–7
  see also industrial
Monographia Apum Angliae (Kirby) 132
Moore, Albert Joseph 210
mopane worms 95
Morgan, T.H. 20–1, 123, 147
Morisot, Berthe 215
Morrill Land Grant Act (1862) 18
Moshav Alonei Habashan, Golan Heights 96
Moslem food laws 95

The Mosquito War (*Harper's New Monthly Magazine* 1854) 200
mosquitoes 71–2, 170, 226–7
malaria 69–71, 72, 73, 75, 96
yellow fever 53–4, 71–3, 200–2
see also biting flies
moths 41, 90, 111–18, 211
melanism 6
see also individual types of moth ...; silk moth/worm
mulberry silk moth *see* silk moth/worm
Munch, Edvard 202
Museo di Storia Naturale, Florence 34
Museum of Comparative Zoology 42
Muséum National d'Histoire Naturelle, Paris 34
mythology and religion 125–48
art 206–7
Bridgewater Treatises 134–7, 138
founding entomologists, Kirby & Spence 132–5
natural history/theology, UK 129–32
Natural Theology and its antecedents 126
relationship between religion and science 126–9

Napoleonic Wars (1803–1815) 47–8
National Gallery, London 118–19
Native Americans 83–4, 85–7, 88–90
see also Ute people of the Great Salt Lake
The Natural History and Antiquities of Selborne (White) 177–8
Natural History Museum, London 12, 13, 32–3, 34, 35–6, 42, 75, 76
natural history museums 34, 37
natural history/theology, UK 129–32
natural selection 24–5, 137–45, 148, 174
see also On the Origin of Species by Means of Natural Selection
Natural Theology 24–5, 32, 125, 126–7
and Charles Darwin 137–43
and its antecedents 126
status at end of Age of Industry 146–8
see also Darwin; Wallace
Natural Theology or Evidence and Attributes of the Deity (Paley) 11, 19, 126, 135–9
Nature Study Movement 22
Naturhistorisches Museum, Vienna 31, 34
Naturhistoriska Riksmuseet, Stockholm 34
nets
anti-mosquito 72, 73–4
catching butterflies 215, 216–7, 218
New South Wales colony, Australia 123–4
Newton, Isaac 145–6
Newton, William Wilberforce 130–1
Nicoll, Maurice 74
Nicolle, Charles 75
night air *see* "bad air", miasmas
Nightingale, Florence 14–15, 75–6
Ninety-Four Degrees in the Shade (Alma-Tadema) 215
Noah's Ark *see* Great Flood
Nodier, Charles 216
North, Marianne 221, 223
Novara Expedition (1857–1859) 31, 32
The Nursery Alice 182–3
nylon 111, 118

Oaxaca 118–20
Odonata 21, 189, 208
oil of citronella 74
Olive Trees (Van Gogh) 226–7
omega-3 and -6 fatty acids 82
On the Power, Wisdom and Goodness of God, as Manifested in The Creation of Animals and in Their History (Kirby) 135–7, 138
Opium Wars 90, 113–14
orchards, beekeeping 110–11
On the Origin of Species by Means of Natural Selection (Darwin) 38, 125, 128, 137–8, 143–5, 164, 174
Ormerod, Edward Latham 184–5
Ormerod, Eleanor 22
ornamentation 212–14
Ornithoptera 175
orthopterans, *see also* crickets; grasshoppers; katydids
Ospov, D. 139–40
Oswandel, J.J. 58, 61
otherness 228
Our English Coasts (Strayed Sheep) (Hunt) 210
Our New Major-general, Jack Frost 69, 70
Owen, Sir Richard 43, 129–30

Pacific Island countries, edible insects 94
Packard, A.S. 184, 189, 228
"The Painted Lady" short story 197
Paiute people of California 84, 86, 88
Palestine 96
Paley, William 11, 19, 126, 135–9
Paris Peace Treaty 24
partitioning of Africa 23–4
Pascoe, Francis Polkinghorn 35–6
Pasteur, Louis 14, 47–8, 114
Patch, Edith Marion 22
Paterson, New Jersey, silk factory 23, 117–18
Pax Britannica 3–4
Paykull, Gustaf von 34
pébrine see silk moth/worm
Pedro I 31
Peel, Robert 129–30
Pelletier, Joseph 69–70
peppered moth 6
Perkin, William 119–20
Perty, Josef Anton Maximilian 31
Die Pest (Böcklin) 205–6
pests 170–2
  fleas in art 202–4
  see also disease
Petrel, HMS 33
Pew Research Center 148
Philippine Islands 94
Philippine-American War (1899–1902) 72, 73
Phytophora infestans, plant pathogen 10
Pierce, W.D. 77
plague 202–6, 223
  see also disease
plants 219–21
  plant blindness 29
  teosinte, wild grass 90
Pliny the Elder 133–4
poetry 133–4, 142–3, 182
Le Poilu 62, 63
Pope, Alexander 167
Poppies and Bees (Longpré) 219, 220
The Population of an Old Pear-Tree, or Stories of Insect Life (E. van Bruyssel) 162–3
population growth 7
portraits 181, 210, 218, 225
potato famine 10
Potter, Beatrix 193, 194
Un Pouilleux (Carbó) 202, 203
Pre-Raphaelite Brotherhood 210
prickly pear cacti 118, 123–4
Principles of Geology (Lyell) 128
Prinzing, F. 59–60
Proceedings of the Entomological Society of Washington 41
products 101–24
  apiculture 102–5
  cochineal 118–20
  economies of scale, honey bee queen production 107–9
  gall-based inks and dyes 121–3
  importation, propagation, and diversity of honey bees 105–7
  imported species debacles 123–4
  insects as model organisms 123
  lac products 120–1
  new era in honey production 110–11
  silk industry 111–18
Protestantism 15
Psyche 154, 208
  see also Cupid and Psyche
Puck and the Moth (Hopley) 208
Punch magazine 180–1, 182, 196, 197–8, 214
putrescine 56
"putro-adynamic" fever 75
pyramids 173–4

queens, honey bees 107–9
Queensland colony, Australia 123–4
Quinby, Moses 102–5
quinine 69–70, 71, 120
Quoy, Jean René Constant 30

Rackham, Arthur 208–9
railroads 4–6, 17, 81, 99, 115
The Rape of the Lock (Pope) 167
raw silk exports *see* silk trade
Ray, John 126
The Record of Zoological Literature 42
Redon, Odilon 211
Redouté, Pierre-Joseph 219
Reed, Walter 53–4, 72, 75–6, 200
Reform Act (1867) 9
"*Règles*" 42
  see also zoological nomenclature

religion *see* mythology and religion
Un Renaud: Pris au Piège (Grandville) 216
Rennie, J. 106
Report on the Insects Injurious to Vegetation (Harris) 20
Republic of China 90
Riley, Charles Valentine 20, 34, 85–6
Riley Locust Catcher 85–6
Robinson, F. 59
The Rocking-Horse Fly 190, 191
Rocky Mountain locust (*Melanoplus spretus*) 85, 86
Rococo art 199–200
The Romance of Natural History (Gosse) 174–5
Roosevelt, Franklin Delano 117
Roosevelt, Teddy 115–16
rose chafer beetle (*Cetonia aurata*) 226
Roses (Van Gogh) 226
Ross, M.T. 208–9
Ross, Ronald 69, 75
Rossetti, Dante Gabriel 210
Rossiter, Margaret W. 21
Rothney, G.R. James 33
Rothschild, Dame Miriam 12
Rothschild, Nathanial (Charles) 12
Rowlandson, Thomas 216, 217
Royal Academy of Science 132
Royal College of Chemistry, London 119–20
Royal Entomological Society of London 37, 218
Royal Society of London 135, 145–6
Russia
  Anti-Jewish pogroms (1881–1884) 96
  Crimean War (1854–1856) 14–15, 54–6, 59, 66–7, 75–6
  Zoological Museum, St. Petersburg 34
Russo-Japanese War (1905) 56, 57
Russo-Turkish War (1828–1829) 59–60
Ryle, John 117

Salmonella typhi, bacteria 13–14, 54
Salome (Wilde) 210
Salt Lake City *see* Great Salt Lake
Salvin, Osbert 35
Sambourne, Edward Linley 180–1, 182, 197, 198, 214
Samokish-Sudkovskaya, Elena 221
Sanambele, Mali 95
Santschi, Felix 38
Sardinia 98
Sargent, John Singer 197
satire 216–17
  see also Punch magazine
Saturniidae moth spp. 91, 95
Saxby, Fred 187
Say, Lucy 21–2
Say, Thomas 19, 30
scale insects 118, 119, 120
scarab beetles 192–3
The Scarab's Serenade (Kaemmerer) 215
Der Schmetterlingsfänger (Spitzweg) 217–18
Schussler, E.E. 29
Science Museum 12
Scientific Queen-Rearing as Practically Applied being a Method by which the best of queen-bees are reared in perfect accord with Nature's ways (Doolitle) 107–9
Scott, Robert Falcon 12–13
The Scream (Munch) 202
Scudder, Samuel H. 194–5
Seated Nude: The Flea (Munch) 202
"The Second Coming" (Yeats) 24
Second Great Awakening 15
Second World War (1939–1945) 117–18
Sedgwick, Adam 144
Self-Portrait with a Straw Hat (Van Gogh) 225
sericulture 101, 112–13
  see also silk moth/worm
Seymour, Robert 47–8, 49
Shakespeare, William 133–4, 208
Sharp, David 35–6
Sharpe's London Magazine 192
shellac 111, 120–1
  see also lac products
Sherborn, Charles Davies 42–3, 44
Shipley, A.E. 65
Shirley (Brontë) 196
Sibylla Palmifera (Rossetti) 210
Silas Marner (Eliot) 195
"Silk City", Paterson, New Jersey 23, 117–18
silk moth/worm (*Bombyx mori*) 79–81, 91–2, 93, 114

silk industry/trade 23, 111–18
Simkin, J. 59
Simond, Paul-Louis 60
Sino-Japanese War 90
Siphonaptera 21
skep hives 102–3
"skirmishing" *see* "chatting"/"skirmishing"
Slater, Samuel 80
slavery 2–3, 15–16, 164–5
slender-bodied digger wasp (*Crabro cribrarius*) 141–2
Sloane, Sir Hans 34
Small Pear Tree in Blossom (Van Gogh) 226
small tussore moth (*Antheraea paphia*) 91
Smeathman, Henry 165
"Il Smelatore" honey extractor 109
Smith, Frederick 32–3
Smith, Herbert H. 33
Smith, Mary Russell 224
Smith, Theobold 75
Smith/Lever Act (1914) 18
Smithsonian Museum 33
Sneden, R.K. 58
Snow, John 13–14
Société Entomologique de France 21, 37
Society of Aurelians 218
sodium hydroxide 91
Solander, Daniel Carl 28
Der Sonntagsspaziergang (Spitzweg) 215
South Africa 34, 37
Spanish-American War (1898) 23, 53, 54, 72
The Spatula Ink Formulary 123
spectral animal 156–7
Spence, William 132–5, 140, 149, 213
  see also Introduction to Entomology
spiders 48–9, 111, 130–1, 137, 138, 163, 173–4, 194–6, 199, 206, 214
Spitzweg, Carl 215, 217–18
spongy moth (*Lymantria dispar*) 123–4
Spottiswoode, William 145–6
Spratt, George 216
Stainton, H.T. 175
steam technology 5–6
Stewart, Susan 182, 189
still-life painting 211–12, 219
stink bugs 90–1
Stoliczka, Ferdinand 33
Stothard, Thomas 167
A Street During the Plague with a Death Cart and Mourners (Evans) 205
Stricklandian Code 42–3
stylized butterflies 210
sub-Saharan Africa 94–5
subterranean cricket (*Brachytrupes membranaceus*) 95
Suez Canal 23–4, 114–15
sugar/honeydew (aphid excrement) 86, 95
Summer Amusement, Bug Hunting (Rowlandson) 216
supernatural beings 166–7
  see also fairies
"Suspended Hive" 103–4
Sutton, M. 86–7
Swayne, M. 52, 74
Sylvie and Bruno (Furniss) 184
symbols/symbolism 149–72
  insects as "germs with legs" 170–2
  insects as "little people" 165–9
  and modernity 150–5
  in popular entomology books 157–65
  visual representation of animals 156–7
Symonds, J.C. 223
Syria 96, 112–13, 212
Systema Naturae (Linné) 27–8

Taiping "Christian" ideology 90
Taiping Rebellion 90, 114
The Tale of Mrs Tittlemouse (Potter) 193
taxonomies 40–1, 175–80, 190–2
  see also individual taxonomies ...
Tegetmeier, W.B. 22
telegraph cables 114–15
The Tenant of Wildfell Hall (Brontë) 196–7
Tenniel, John 182–4, 190, 191, 208–9
Tennyson, Alfred 142
teosinte, wild grass 90
termites 95
terpenoids 92
Terra Nova expedition (1910–1913) 12–13
Terry, Ellen 197
Texas cattle fever 75
textile industry 4–5, 8, 111–18
  see also silk moth/worm
Thailand 91, 92–3, 99
Thomas Cook and Son travel agent 96
Thoreau, Henry David 19

Thorvaldsen, Bertel 206, 207
Thysanura 21
Das Tierrech 44
The Times of Israel 96
Tocqueville, Alexis de 15
La Toilette: Madame Fabre (Toulouse-Lautrec) 202
tombs, Egyptian beekeeping 101, 102
tortoise-shell butterfly 210
"totos" (lice) 62, 65
Toulouse-Lautrec, Henri de 202
Traditional Ecological Knowledge (TEK) 81–2, 87, 89, 94, 99
"*Transmutation*" notebooks (Darwin) 139–40
trapdoor spiders 137, 138
Treat, Mary 22
Treaty of Nanjing 90
treaty port systems, silk trade 113–14
true flies 32–3, 42
  see also flies
tryptophan 90
Tucker, Charlotte Maria 168–9
tussore/tussah moth (*Antheraea mylitta*) 91
Tuxen, S.L. 27
Twain, Mark 96
typhoid fever 50, 54, 55
"typho-malarial" fever 75

Under the Hill (Beardsley) 210
Union Cases, lac products 120–1
United Nations (UN) 94–5
United States 2–4
  abolishment of the slave trade 2–3
  agrarian society 15–17
  Army Yellow Fever Commission 53–4, 72, 200
  Department of Agriculture, USDA 18, 20–2, 67–8, 75–6, 106
  edible insects, Industrial Revolution 80–9
  Entomological Society of Pennsylvania 19–20
  First World War, anti-fly campaigns 50, 51
  immigrants 17, 23, 80–82, 86–89, 117
  lac products 120–1
  Mexican Expedition, typhoid vaccination 54, 55
  National Museum 40–1
  Natural Theology, end of Age of Industry 146–8
  response to Charles Darwin 144–5
  science in agriculture/entomology 18–21
  silk trade 23, 115–16, 117–18
  War Department 67–8
  yellow fever in art 200–2
  Yellow Fever Commission 200
  see also Civil War
urbanization *see* modernity/modernization
d'Urville, Dumont 30
Ussher, Archbishop 127
Utah's Lakeside Cave 87
Ute people of the Great Salt Lake 82–3, 84–5

vaccination 54, 55–6
van Bruyssel, E. 162–3
Van Gogh, Vincent 10, 225–8
Van Rensselaer, Martha 75
Varley, John 203
Vega, Carolyn 188
Venus Verticordia (Rossetti) 210
Vespula flaviceps wasp spp. 93
Vestiges of the Natural History of Creation (Anon.) 143–4
Vibrio cholera 54–5
  see also cholera
Victoria and Albert Museum 12
Victoria, Queen 11–12, 14, 143, 145–6, 187
vinyl 111
"Visionary Heads" pieces by Blake 203
"visual literalism" 153
visual representation of animals, symbols/symbolism 156–7
vitamin B12 82, 90

Wagoner 109
Wakefield, C.M. 33
Walden (Thoreau) 19
Walker, Francis 32–3, 75
Walker, J.E. 22
Wallace, Alfred Russel 13, 24–5, 31–3, 35–6, 144, 175
  see also *Beagle*
Walsh, Benjamin D. 144–5, 147

Wandersee, J.H. 29
Ward, Mary 187–8
"The Wasp in the Wig" 184, 185
wasps 93, 140–1, 184–5
see also individual types of wasp ...
"Wasp-Waist Mania" 197
water beetle (Dytiscidae) 90–1
water scavenger beetles (Hydrophilidae) 90–1
Watts, Issac 182
Way, Albert 186, 187
Wells, H.G. 186
Westminster Abbey 145–6
Westwood, John Obadiah 12, 42
Wheeler, William Morton 38
Whistler, Beatrice 212
Whistler, James McNeill 210, 211
White, Gilbert 177–8
Why Not Eat Insects? (Holt) 96–7
Wilde, Oscar 10, 196–7, 210, 221
Wiley, B.I. 51–2, 62
Willingham, Emily 62
Wilson, Woodrow 115–16
Winter, Alison 137–8
The wisdom of God manifested in the works of the creation, Collected from the Appearances of Nature (Ray) 126
witchetty/witjuti grubs 94
Wives and Daughters (Gaskell) 174–6
women *see* gender
Wood, John 169
A World of Wonders Revealed by the Microscope (Ward) 187–8
worms 90, 95, 103–4, 189, 193
see also caterpillars; maggots; silk moth/worm
Wright, Almroth 54
Wytsman, Philogène 44

Yeats, W.B. 24
yellow fever 53–4, 71–3, 200–2
Yersin, Alandre 60
Yersinia pestis 202
see also fleas
A Young Girl with a Butterfly Net (Bôle) 215
"A Young Lady Artist's Mania for Capturing Musical Insects" 192–3

Zacharias, John 56
Zionism 96
Zoological Museum, St. Petersburg 34
zoological nomenclature 39–40, 42–3
Zoological Record 35–6, 42, 44
Zoological Society of London 42
Zoologisk Museum, Copenhagen 34